STRATHCLYDE UNIVERSITY LIBRARY
30125 00079186 2

KU-345-677

Geochemical Exploration
Methods For Mineral Deposits

А. А. Беус, С. В. Григорян

ГЕОХИМИЧЕСКИЕ МЕТОДЫ ПОИСКОВ И РАЗВЕДКИ МЕСТОРОЖДЕНИЙ ТВЕРДЫХ ПОЛЕЗНЫХ ИСКОПАЕМЫХ

The most important principles of geochemical prospecting, based on both Soviet and non-Soviet experiences of the last decade, are discussed.

The regularities in the distribution of the chemical elements in crustal rocks, and the main factors responsible for the migration and concentration of elements by natural processes, are used as a theoretical basis for applying the techniques of geochemical prospecting. Special emphasis in the discussion of the above theoretical principles is placed on the logical integration of geochemical techniques with other methods in solving specific problems at various stages of exploration. A special section of the book is concerned with the application of statistics to geochemical prospecting.

This book is intended for geologists engaged in mineral prospecting, as well as for geology students.

МОСКВА · «НЕДРА» · 1975

Geochemical Exploration Methods For Mineral Deposits

A. A. Beus
AND
S. V. Grigorian
MOSCOW

Technical Editor: A. A. LEVINSON
Translator: RITA TETERUK-SCHNEIDER

APPLIED PUBLISHING LTD.
Wilmette, Illinois, U.S.A.

For information address:
Applied Publishing Ltd.,
P.O. Box 261
Wilmette, Illinois 60091, U.S.A.

Composed in Canada by Alcraft Printing Co. Limited,
Calgary, Alberta.

Printed in the United States of America by
The University of Chicago Printing Department,
Chicago, Illinois.

Library of Congress Catalog Card Number: 77-075045

ISBN 0-915834-03-0

Editor's Preface

Exploration geochemistry, as we know it today, originated in the Soviet Union in the 1930's. The early attempts in the late 1940's and early 1950's at using exploration geochemistry in the West were based on the techniques originally developed in the U.S.S.R., but modified for local environmental conditions such as those encountered in tropical areas. From these early beginnings, geochemical methods have become even more important in the exploration for mineral deposits in the Soviet Union, as well as in the rest of the world, and there is much we can learn from our Soviet colleagues.

It is regrettable that only two books in Russian on exploration geochemistry have ever been translated into English. Specifically, I refer to the books by I. I. Ginzburg and D. P. Malyuga which appeared in 1960 and 1964, respectively. Although individual articles by Soviet geochemists have appeared in translation, particularly in *Geokhimiya* and *Geochemistry International,* and perhaps a dozen or so articles have appeared in English in the Proceedings of various congresses and conferences, we in the West have been largely in the dark with respect to the details of the concepts, techniques, and advances of Soviet geochemists in the last 20 years. This up-to-date volume, which was first published in December 1975, will go a long way to correcting this deficiency. At the same time, it will point out to most exploration geochemists facets of this field never before considered in the West.

Few, if any, are more qualified than Professor A. A. Beus and Dr. S. V. Grigorian to attempt a compilation of Soviet experiences and techniques into a textbook. Professor Beus is the Professor of Geochemistry at The Moscow Polytechnical Institute and, in addition, he has spent seven years with the United Nations supervising programs concerned with the development of mineral resources in developing countries. Professor Beus is regarded as one of the leading exploration geochemists in the Soviet Union, with interests in both the theoretical and practical aspects of the subject. Dr. S. V. Grigorian is Deputy Director of the Institute of Mineralogy, Geochemistry and Crystallochemistry of Rare Elements in Moscow, and is also considered one of the leading geochemists in the Soviet Union today with particular interest in what we in the West would call "rock geochemistry." The expertise and clairvoyance of these two scientists will become evident to those who read this book and, therefore, further statements on the credentials of these two men are superfluous.

About one-third of this book is concerned with rock geochemistry, whereas such topics as exploration in glaciated terrain, soil classifications, and certain analytical techniques (e.g., atomic absorption) are barely mentioned, if at all. On the other hand, in addition to the above-mentioned emphasis on rock geochemistry, emphasis has also been placed on geochemical barriers, the distinction between supra-ore and below-ore primary halos, and other topics not covered in English textbooks on exploration geochemistry. The reason for this "imbalance" is easily explained. The authors, as stated in this Preface, have chosen to concentrate their efforts on topics not covered in existing textbooks published in the Soviet

Union (the two main textbooks in English on exploration geochemistry are available in translation to Soviet scientists). As a result, this book, which can be considered a supplement to existing textbooks, stresses material new in textbook form to Soviet geochemists, and new in any form to English-speaking geochemists.

Although the translator and I have tried to obtain the sense and detail of the Russian for presentation in an exact and idiomatic translation, we may not have succeeded in every instance. For example, I found certain parts of Chapter 5 somewhat difficult due, in part, to the fact that the concepts presented were new to me. In view of this difficulty, I was delighted when Professor Beus agreed to read the translation of the first 10 chapters in their entirety, and to make any corrections or modifications where necessary. Similarly, Dr. Grigorian graciously answered some questions by correspondence. The editing of Chapter 11 was done by Dr. R. B. McCammon, U.S. Geological Survey, Reston, Virginia, who is an expert in statistics applied to geology.

Most of this translation was completed during a sabbatical leave from The University of Calgary. The cooperation of Professor Beus, Dr. Grigorian and Dr. McCammon has already been mentioned and is hereby formally acknowledged. Special thanks are due to Miss Beverly A. Ross for her prompt typing and careful proof reading. And, last but not least, I appreciate the dedication of Mrs. Rita Teteruk-Schneider, the translator, to this project.

A. A. Levinson
Technical Editor

February, 1977

Editorial Note

Throughout this book, in accordance with usage in the Soviet Union, and because several tables and equations have been reproduced photographically from a copy of the Russian text:

1. a comma is used as the decimal sign instead of the mid-line or base-line point:
2. "lg" is used in place of "log".

All photographs have been reproduced from a copy of the published book. Accordingly, some of the figures may not be as clear as would have been the case if the original drawings had been used.

Preface

Geochemical methods of prospecting for mineral deposits, which were first used in the Soviet Union in the 1930's, have now become indispensable in geological exploration both in the Soviet Union and in other countries. These methods have now become a very effective tool in the hands of exploration geologists in practically all stages of the search for mineral deposits. It is now possible to look for deposits not only in outcrops, but also to conduct direct exploration for hidden deposits and ore bodies still unexposed by erosion, or those overlain by unconsolidated materials of various types.

Geochemical methods of exploration for mineral deposits have been developing at a rapid rate during the last decade. The scientific principles on which these methods are based are continually being refined. The range of problems whose solutions require the utilization of various modifications of geochemical techniques has been expanding steadily.

The large-scale implementation of geochemical methods in the field of geological exploration contributes to scientific and technological progress in geology, and leads to greater success in the search for new mineral deposits. This is fully in line with the directives of the 24th CPSU Congress which called on all geologists to speed up the practical implementation of the achievements in science and technology, to intensify geological exploration, and to reduce its costs.

A great number of Soviet geologists dealing with certain aspects of the search and exploration for metalliferous and nonmetalliferous deposits have found it necessary to resort to geochemical methods, applying them to the solution of various practical problems. However, the absence of a practical handbook in this field is often an obstacle to the successful utilization of these methods. It is felt that there is a need for a book which will summarize the rather extensive experiences which have accumulated on the application of various geochemical methods, both in the Soviet Union and abroad, and which will give specific practical recommendations. This must be done at a high theoretical level in conformity with the present level of fundamental geochemistry.

The authors, who have been instructing nonresident students in a course entitled "Geochemistry and Geochemical Prospecting" for more than five years, have realized that the absence of such a handbook has resulted in the lack of competence on the part of exploration geologists in this field. This book is intended mainly for them.

Accurate and effective use of geochemical methods is impossible without a knowledge of the main distribution patterns of the chemical elements in the surface regions of the Earth's crust, or without a knowledge of the laws of element migration under endogenic and exogenic conditions. This is why the authors found it necessary, before discussing applied geochemistry specifically, to begin with two chapters which discuss up-to-date material on the chemical composition of the Earth's crust and the major laws of the distribution of the chemical elements in rocks, as well as the main concepts which characterize the geochemical migration of the elements. The subject matter which illustrates the major principles of geochemical exploration is presented in conformity with the universally accepted classification of geochemical prospecting procedures, that is, on the basis of the materials being used for the geochemical surveys (e.g., soils, rocks, water). In view of the inadequate coverage in the literature on the various exploration methods for mineral deposits based on the use of primary geochemical halos, the authors have decided to place particular emphasis on this subject in the book. Primary geochemical halos (aureoles) are considered as the major criteria which should be used in all stages of the search and ultimate exploration for mineral deposits.

Successful application of geochemical methods depends on the combined use of various prospecting methods, as well as on their accurate implementation. Therefore, a special chapter which concludes with a discussion of the materials used in the individual methods, deals with these problems. At the end of the book, elementary methods for the statistical processing of geochemical data are given. Every geologist engaged in geochemical prospecting for mineral deposits should know these procedures.

Chapters 1 to 4, as well as 8 and 9, were written by A. A. Beus. Chapters 5 and 7 were written by S. V. Grigorian. Chapters 6, 10, and 11 were written by both authors.

The authors fully realize that the book is not without its shortcomings. They will, therefore, appreciate all constructive criticisms intended for further improvement of the book. Comments should be sent to the attention of the authors, in care of the Chair of Mineral Deposits, Faculty of Mining, All-Union Correspondence Polytechnical Institute, Pavel Korchagin Street No. 22, Moscow, USSR.

Contents

1

Chemical Composition of the Earth and the Main Characteristics of the Distribution of the Chemical Elements in the Earth's Crust

General Information

The composition of rocks exposed at the surface is the only source of direct information on the chemical composition of the Earth's outer shell. The composition of the Earth's deep interior is deduced on the basis of geophysical, chiefly seismic, studies. Based on these indirect data, the Earth has a concentric zonal structure (consisting of shells) and is divided into the crust, mantle (upper and lower), and core (outer and inner). These are shown in Fig. 1.

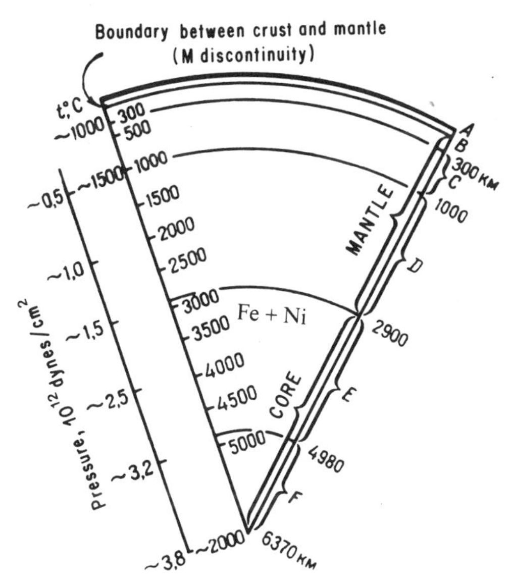

Fig. 1. The shells of the Earth (after A. P. Vinogradov).
A. Earth's crust. B. Upper mantle (olivine, pyroxene). C. Transitional layer. D. Lower mantle: MgO, FeO, SiO_2, and other oxides (the most dense packing of oxygen atoms). E. Outer core. F. Inner core.

That the Earth is divisible into shells is primarily revealed by the regular distribution of the densities of the material making up the different shells, and by the variations in seismic wave velocities with depth (Fig. 2). The average density of the Earth, which is 5.517± 0.004 g/cm^3, is much higher than the crustal density, which indicates that there is material in the Earth's interior which has much greater density than that of the Earth's crust (Fig. 3). However, no unambiguous explanation has been given so far for the fact that density increases with depth. It may be due to the compaction of the atomic structure of material under the effect of the enormous pressures within the Earth, or it may be due to changes in the chemical composition within individual shells. Both these factors are, in all probability, responsible to some extent for the increase in density of the Earth's material at depth; however, their quantitative role remains obscure. This is the reason for the main contradictions in attempts to construct hypothetical models of the chemical composition of the Earth's deep-seated shells.

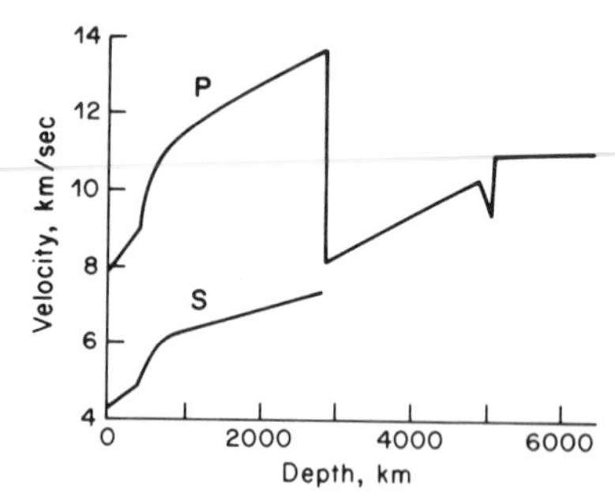

Fig. 2. Velocities of compressional (P) and shear (S) seismic waves within the Earth (after K. E. Bullen).

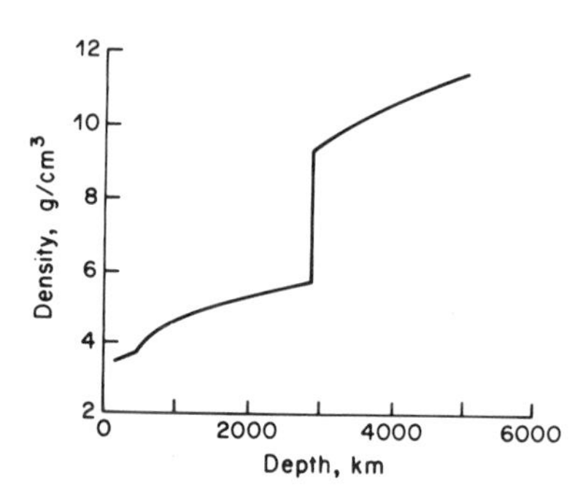

Fig. 3. Density distribution within the Earth (after K. E. Bullen).

For example, taking into account the great density of the Earth's core, many researchers still believe that it is composed largely of an iron-nickel alloy, similar to the material found in iron meteorites.

However, the most recent experimental data, which have shown that the density of iron depends on pressure, rule out the hypothesis for a purely iron core. A density which is incompatible with geophysical data was obtained for iron at pressures of millions of atmospheres (Demenitskaya, 1967). Some critics of the hypothesis of the Earth's iron-nickel core maintain that the chemical composition of the core is essentially the same as that of the lower mantle, and its high density is related to the phenomena of metallization of the material under the effect of superhigh pressures. Experimental data have recently been obtained on the physical properties of iron-silicon and iron-sulfide alloys at pressures corresponding to those within the Earth's core. It was found that under these conditions an iron-silicon alloy containing 14 to 20 weight percent of silicon is similar to the material making up the Earth's core, in terms of density and compressional wave velocity. Similar data were obtained for iron sulfides. These estimates should be considered with reservations, because various phase transformations, and melting (the outer core is in a liquid state, based on seismic evidence), were not taken into account in the development of this hypothesis.

The core accounts for 31.5% of the Earth's mass and, therefore, it is impossible to construct a reasonably satisfactory model characterizing the average chemical composition of the Earth, unless the problem of the composition of the core is solved. All presently available models of the Earth's average chemical composition are based on the hypothesis of an iron-nickel core.

Equally hypothetical is the composition of the lower mantle which, at the depth of 2900 km, is separated from the core by a surface with an abrupt change in the compressional seismic wave velocity (Fig. 2). It is assumed that the chemical composition of the lower mantle is approximately similar to that of ultrabasic or basic crustal rocks. Another hypothesis proceeds from the similarity of the average composition of Earth's mantle to that of the chondrites. Some investigators have suggested that there might be significant amounts of heavy metal sulfides in the lower mantle, however, no reliable confirmation of this hypothesis has been provided. At very high pressures, the oxides constituting the material in the lower mantle must have very dense oxygen packing.

An experimental study of the physical properties of minerals at high pressures enabled Anderson et al. (1972) to propose a model for the chemical composition of the lower mantle based on an interpretation of seismic wave velocities in the mantle and the average density of mantle material. According to these authors, the lower mantle is composed of 0.32 MgO + 0.18 FeO + 0.50 SiO_2 (molar quantities). This model would have been more valuable had it taken into account the probable presence of the oxides of other elements (aluminum, for example) in the lower mantle.

The boundary between the lower and upper mantle lies at a depth of 1000 km. The available data from seismic studies indicate that the upper mantle is heterogeneous in its structure. At a depth of about 400 km it is divided into two approximately equal layers by a surface with an abrupt change in density and in seismic wave velocity. This discontinuity in physical properties may be caused by phase transformations. The so-called *low-velocity layer* (for seismic waves), or *asthenosphere,* occurs within the upper mantle at a depth of 70 to 80 km beneath the oceans and at about 150 km beneath the continents. The physical properties of this layer, based on seismic evidence, suggest that it exists in a partly molten state. Great significance is presently attached to the low-velocity layer in the upper mantle, both from the point of view of the tectonic development of the Earth's surface shells, and its importance in the nature of the chemical composition of these shells. A close causal relationship probably exists between the evolution of the chemical composition of the upper mantle and the formation of the overlying crust. At present, the crust is considered to be the product of geochemical differentiation of the upper mantle under the effect of the Earth's gravitational and thermal fields.

The concepts concerned with the chemical composition of the upper mantle seem to be more definite by comparison with the hypotheses on the composition of the Earth's deep-seated shells, although upper mantle composition is still a widely discussed subject. Ultrabasic rocks, which occur in small amounts at the surface of the present day crust, are thought to be representative of upper mantle material which is accessible for direct geochemical and petrological studies.

However, if one recalls that the composition of the uppermost mantle must be considerably altered in the process of the differentiation and separation of the material constituting the Earth's crust, then the correspondence between the chemical composition of ultrabasic rocks which have penetrated into the crust, and the average composition of the upper mantle, is rather dubious. In recent years some scientists (Ringwood, Anderson, Harris, Beck, and others) have tried to develop models for the chemical composition of the upper mantle. All existing models of this type are based on the concept that ultrabasic rocks are residual products of the evolution of upper mantle material which has remained after the basalts were differentiated. At present, the model developed by Ringwood (1972), which is based on the basalt to peridotite ratio of 1:3, is widely accepted. This model of the composition of the upper mantle, called *pyrolite*, has been criticized from various points of view (O'Hara, 1973; Beus, 1972; Anderson et al., 1972) and can hardly be used as the basis for the characterization of the average composition of the upper mantle. For example, the mineral composition of the upper mantle based on the pyrolite model, has some physical properties which differ from those of upper mantle material, as inferred from seismic data.

Another hypothesis (first formulated by Goldschmidt) suggests the presence of rocks of an eclogite composition in the upper mantle. These rocks are known to be geochemical equivalents of gabbro-basaltic rocks, but they have higher densities (3.4 compared to 3.0) due to phase transformations under high pressure conditions. Although the hypothesis of a continuous eclogite layer in the upper mantle is now rejected by most scientists, the possibility of the presence of eclogite in this shell is recognized by practically everyone. Consequently, the upper mantle has, in all probability, a heterogeneous and quite complex composition closely related to the evolutionary features of specific sub-surface areas of the Earth. It is now recognized by all researchers that processes which are operative in the upper mantle are significant in controlling the formation and evolution of the crust.

Chemical Composition of the Earth's Crust

The discontinuity between the Earth's crust and the upper mantle is called the Mohorovičić surface, and it is characterized by an abrupt change in the velocities of the compressional (7.8 to 8.4 km/sec) and shear (up to 4.3 km/sec) seismic waves. The Mohorovičić surface is located about 30 to 35 km below the continents. In fact, however, its shape is complicated by downwarps and uplifts.*

---

*Some investigators include the part of the upper mantle down to the *low-velocity* layer as being in the lithosphere. This, however, reduces the definiteness of the concept of the lower crustal boundary thus creating additional difficulties in studies of the geochemical characteristics of both the lithosphere and the upper mantle.

The Earth's crust consists of three geospheres differing in composition and physical state; these are the lithosphere, hydrosphere, and atmosphere. The lithosphere, or the stony surface shell of the Earth, is the main object of study by geologists engaged in prospecting for mineral deposits.

The lithosphere, in turn, is divided into three layers (or shells) whose materials differ in composition and physical properties. The upper (sedimentary) layer is composed of various sedimentary rocks with a minor amount of Paleozoic or younger volcanic rocks. The next two crustal layers have been arbitrarily called *granitic* and *basaltic*. The boundary between them, as recognized from seismic data, is known as the *Conrad discontinuity*. The average thickness of the granitic layer below the continents is 10 to 11 km, as inferred from seismic data, whereas the ratio of the thicknesses of the granitic to the basaltic layers is estimated as 1:2. The granitic layer is absent below the oceanic crust.

The model used to determine the composition of the basaltic layer incorporates data on the properties of this layer obtained by geophysical methods. In addition, available experimental information on the densities of rocks which may exist under the conditions characteristic of the granulite and eclogite facies of metamorphism, and on the seismic wave velocities in such rocks, (Beus, 1972), have also been used. The resulting average composition of the basaltic shell is close to that of diorite. The average composition of the entire crust is also close to that of diorite.

Table 1. Average chemical composition of the lithosphere and its shells (weight percent).

Oxides	Entire lithosphere excluding sedimentary shell	Continental lithosphere, excluding sedimentary shell	Basaltic shell	Granitic shell	Sedimentary shell of continents (including volcanic rocks)*
SiO_2	57,4	69,6	53,3	66,1	49,82
Al_2O_3	15,4	15,3	15,4	15,2	12,97
Fe_2O_3	2,4	2,5	2,8	2,0	2,97
FeO	5,5	5,0	6,1	2,7	2,81
MgO	5,0	4,0	5,0	2,0	3,09
MnO	0,1	0,1	0,1	0,1	0,11
CaO	7,1	6,0	7,2	3,5	11,64
Na_2O	3,1	3,2	3,2	3,12	1,57
K_2O	1,9	2,3	1,8	3,24	2,03
TiO_2	1,1	1,0	1,2	0,60	0,65
P_2O_5	0,2	0,2	0,2	0,28	0,17
H_2O+	0,7	0,65	0,6	0,76	2,91
CO_2					8,73
SO_3	0,1	0,15	0,1	0,4	0,33
Cl					0,19

*After Ronov and Yaroshevskii (1967)

Table 1 lists data on the chemical composition of the granitic shell. Although these data are based on reasonably precise information from the average chemical composition of the rocks making up the granitic shell, they are, nevertheless, approximate. The main sources of error are: a) the absence of precise data on the quantitative ratio of granite and granodiorite even at the surface of the lithosphere; b) the absence of precise data on quantitative ratios among the metamorphic rocks, and relationships between them and intrusive rocks; and c) the total absence of data on possible variations in the ratios of the various rock types at depth. Therefore, inadequate knowledge of the chemical composition of the granitic shell is primarily the result of the lack of available geological information on the lithological composition of this shell. Even so, a comparison between the calculated average composition of the shell as a whole and the composition of its constituent rocks, shows a striking similarity between the average composition of the shell and that of granodiorite and paragneiss. This feature is a quantitative expression of the intimate relationship between the chemical composition of the arenaceous-argillaceous sedimentary rocks which become converted into crystalline schists and gneisses through the processes of progressive metamorphism, and the average composition of the granitic layer in the lithosphere. In conclusion, the presently available data suggest a dioritic average composition for the entire lithosphere and for its basaltic shell on the one hand, and a granodioritic composition for the granitic layer beneath the continental crust, on the other. Table 2 lists the lithological composition of the Earth's shell.

Table 2. Occurrence of rocks (percent of the mass of the Earth's shell)

Granitic shell				Sedimentary shell			
Igneous rocks	%	Metamorphic rocks	%	Sedimentary rocks	%	Volcanic rocks	%
Granite	63,7	Gneiss Crystalline schist	84,0	Shale and clay	53,0	Basalt	36,7
Granodiorite	21,5	Quartzite and sandstone	7,0	Sand and sandstone	25,4	Andesite	41,3
Quartz diorite	2,9	Amphibolite	8,2	Carbonate rocks	20,8	Dacite	0,8
Diorite	1,6	Carbonate rocks	0,8	Evaporite	0,8	Acid volcanic	21,0
Gabbro	9,0					Alkali volcanic	0,2
Peridotite Dunite	0,3						
Syenite	0,9						
Nepheline syenite	0,1						

Table 1 lists the oxide abundances of the 14 major chemical elements of the lithosphere and its shells, summed to 100%. In fact, however, the amount of these elements, including oxygen, is only about 99.5% of the mass of the lithosphere. The remaining 0.5% is accounted for by all the other chemical elements in the periodic table; their contents in the various types of rocks constituting the lithosphere, its granitic shell, and the continental lithosphere as a whole, are given in Table 3.* The average chemical composition of the major types of crustal rocks has been discussed by Beus (1972).

Vernadskii's decades, into which all elements are placed in accordance with their crustal abundances, are the basis of the most objective geochemical classification presently available (Table 4). Henceforth, those elements whose contents in natural formations is 1.0% or higher shall be called *major* (or rock-forming) *elements;* those elements present in amounts

*The average contents of trace elements in the granitic shell and in the continental lithosphere were calculated on the basis of the latest compilation by Green (1969), and from more recent publications.

Table 3. Average contents of chemical elements in the lithosphere and in its constituent rocks (weight percent).

Atomic number	Element	Continental lithosphere (excluding sedimentary cover)	Granitic shell	Granite	Granodiorite	Intermediate rocks	Basic rocks	Ultra-basic	Schist	Sedimentary rocks. Sandstone	Carbonate rocks
1	Hydrogen	0,10	0,10	0,06	0,09	0,11	0,12	—	0,40	0,25	0,09
2	Helium	$6\cdot10^{-5}$ cm³ per gram of rock weight									
3	Lithium	$2,0\cdot10^{-3}$	$3,0\cdot10^{-3}$	$3,8\cdot10^{-3}$	$3,0\cdot10^{-3}$	$2,5\cdot10^{-3}$	$1,5\cdot10^{-3}$	$0,2\cdot10^{-3}$	$6,6\cdot10^{-3}$	$1,5\cdot10^{-3}$	$0,5\cdot10^{-3}$
4	Beryllium	$1,5\cdot10^{-4}$	$2,5\cdot10^{-4}$	$3,5\cdot10^{-4}$	$2,5\cdot10^{-4}$	$1,8\cdot10^{-4}$	$0,4\cdot10^{-4}$	$0,2\cdot10^{-4}$	$3,0\cdot10^{-4}$	$0,n\cdot10^{-4}$	$0,n\cdot10^{-4}$
5	Boron	$0,7\cdot10^{-3}$	$1,0\cdot10^{-3}$	$1,5\cdot10^{-3}$	$1,2\cdot10^{-3}$	$0,9\cdot10^{-3}$	$0,5\cdot10^{-3}$	$0,3\cdot10^{-3}$	$10\cdot10^{-3}$	$3,5\cdot10^{-3}$	$2,0\cdot10^{-3}$
6	Carbon	$1,7\ 10^{-2}$	$3,0\cdot10^{-2}$	$3,0\cdot10^{-2}$	$3,0\cdot10^{-2}$	$3,0\cdot10^{-2}$	$2,0\cdot10^{-2}$	$1,0\cdot10^{-2}$	1,2	1,3	11,0
7	Nitrogen	$2,0\cdot10^{-3}$	$2,6\cdot10^{-3}$	$2,7\cdot10^{-3}$	$2,0\cdot10^{-3}$	$2,0\cdot10^{-3}$	$3,5\cdot10^{-3}$	$1,0\cdot10^{-3}$	$54,5\cdot10^{-3}$	$13,5\cdot10^{-3}$	$0,7\cdot10^{-3}$
8	Oxygen	46,6	48,1	48,7	48,0	47,0	44,5	43,7	49,0	51,5	49,2
9	Fluorine	$6,0\cdot10^{-2}$	$7,2\cdot10^{-2}$	$8,3\cdot10^{-2}$	$6,3\cdot10^{-2}$	$5,0\cdot10^{-2}$	$4,0\cdot10^{-2}$	$1,0\cdot10^{-2}$	$7,4\cdot10^{-2}$	$2,7\cdot10^{-2}$	$3,3\cdot10^{-2}$
10	Neon	$7,7\cdot10^{-8}$ cm³ per gram of rock weight									
11	Sodium	2,3	2,2	2,66	2,78	2,60	1,90	0,18	0,98	0,92	0,25
12	Magnesium	2,4	1,2	0,33	1,10	2,20	4,50	20,50	1,50	0,73	4,60
13	Aluminum	8,1	8,0	7,40	8,60	8,90	8,50	2,40	8,65	2,90	0,96
14	Silicon	27,7	30,9	34,0	30,5	27,5	23,0	20,0	27,5	34,7	3,4
15	Phosphorus	0,10	0,08	0,06	0,11	0,15	0,15	0,05	0,07	0,04	0,05
16	Sulfur	0,03	0,04	0,04	0,04	0,04	0,03	0,01	0,24	0,02	0,12
17	Chlorine	$1,0\cdot10^{-2}$	$1,7\cdot10^{-2}$	$2,0\cdot10^{-2}$	$1,3\cdot10^{-2}$	$1,0\cdot10^{-2}$	$0,6\cdot10^{-2}$	$0,5\cdot10^{-2}$	$1,80\cdot10^{-2}$	$0,1\cdot10^{-2}$	$1,5\cdot10^{-2}$
18	Argon	$2,2\cdot10^{-5}$ cm³ per gram of rock weight									
19	Potassium	1,8	2,70	3,50	2,52	1,50	0,70	0,05	2,70	1,32	0,28
20	Calcium	4,3	2,5	1,12	2,40	4,60	7,30	3,40	2,00	2,67	32,5
21	Scandium	$2,4\cdot10^{-3}$	$1,1\cdot10^{-3}$	$0,7\cdot10^{-3}$	$1,4\cdot10^{-3}$	$2,0\cdot10^{-3}$	$3,0\cdot10^{-3}$	$1,5\cdot10^{-3}$	$1,3\cdot10^{-3}$	$0,1\cdot10^{-3}$	$0,1\cdot10^{-3}$
22	Titanium	0,6	0,33	0,17	0,38	0,60	0,80	0,35	0,38	0,30	0,12

Table 3 (continued)

Atomic number	Element	Continental lithosphere (excluding sedimentary cover)	Granitic shell	Granite	Granodiorite	Intermediate rocks	Basic rocks	Ultra-basic	Schist	Sedimentary rocks. Sandstone	Carbonate rocks
23	Vanadium	$1,9\cdot10^{-2}$	$7,6\cdot10^{-3}$	$4,4\cdot10^{-3}$	$8,8\cdot10^{-3}$	$15\cdot10^{-3}$	$25\cdot10^{-3}$	$4,0\cdot10^{-3}$	$13\cdot10^{-3}$	$2,0\cdot10^{-3}$	$2,0\cdot10^{-3}$
24	Chromium	$1,2\cdot10^{-2}$	$0,34\cdot10^{-2}$	$0,1\cdot10^{-2}$	$0,22\cdot10^{-2}$	$0,55\cdot10^{-2}$	$1,7\cdot10^{-2}$	$16,0\cdot10^{-2}$	$0,9\cdot10^{-2}$	$0,35\cdot10^{-2}$	$0,11\cdot10^{-2}$
25	Manganese	0,09	0,07	0,04	0,07	0,12	0,12	0,10	0,08	0,04	0,04
26	Iron	5,7	3,6	1,83	3,30	5,50	8,40	8,70	4,80	2,80	0,83
27	Cobalt	$3,4\cdot10^{-3}$	$7,3\cdot10^{-4}$	$1,0\cdot10^{-4}$	$7,0\cdot10^{-4}$	$9,0\cdot10^{-4}$	$48\cdot10^{-4}$	$150\cdot10^{-4}$	$19\cdot10^{-4}$	$0,3\cdot10^{-4}$	$0,1\cdot10^{-4}$
28	Nickel	$9,5\cdot10^{-3}$	$2,6\cdot10^{-3}$	$0,45\cdot10^{-3}$	$1,5\cdot10^{-3}$	$5,0\cdot10^{-3}$	$13\cdot10^{-3}$	$200\cdot10^{-3}$	$6,8\cdot10^{-3}$	$0,2\cdot10^{-3}$	$0,2\cdot10^{-3}$
29	Copper	$6,5\cdot10^{-3}$	$2,2\cdot10^{-3}$	$1,0\cdot10^{-3}$	$2,6\cdot10^{-3}$	$4,0\cdot10^{-3}$	$8,7\cdot10^{-3}$	$1,0\cdot10^{-3}$	$4,5\cdot10^{-3}$	$0,1\cdot10^{-3}$	$0,4\cdot10^{-3}$
30	Zinc	$8,7\cdot10^{-3}$	$5,1\cdot10^{-3}$	$3,9\cdot10^{-3}$	$5,6\cdot10^{-3}$	$7,5\cdot10^{-3}$	$10,5\cdot10^{-3}$	$5,0\cdot10^{-3}$	$9,5\cdot10^{-3}$	$1,6\cdot10^{-3}$	$2,0\cdot10^{-3}$
31	Gallium	$1,7\cdot10^{-3}$	$1,9\cdot10^{-3}$	$2,0\cdot10^{-3}$	$2,0\cdot10^{-3}$	$1,7\cdot10^{-3}$	$1,7\cdot10^{-3}$	$0,15\cdot10^{-3}$	$1,9\cdot10^{-3}$	$1,2\cdot10^{-3}$	$0,4\cdot10^{-3}$
32	Germanium	$1,3\cdot10^{-4}$	$1,3\cdot10^{-4}$	$1,3\cdot10^{-4}$	$1,3\cdot10^{-4}$	$1,3\cdot10^{-4}$	$1,3\cdot10^{-4}$	$1,5\cdot10^{-4}$	$1,6\cdot10^{-4}$	$0,8\cdot10^{-4}$	$0,2\cdot10^{-4}$
33	Arsenic	$1,9\cdot10^{-4}$	$1,6\cdot10^{-4}$	$1,5\cdot10^{-4}$	$1,9\cdot10^{-4}$	$2,0\cdot10^{-4}$	$2,0\cdot10^{-4}$	$1,0\cdot10^{-4}$	$13,0\cdot10^{-4}$	$1,0\cdot10^{-4}$	$1,0\cdot10^{-4}$
34	Selenium	$1,0\cdot10^{-5}$	$1,4\cdot10^{-5}$	$1,4\cdot10^{-5}$	$1,4\cdot10^{-5}$	$1,4\cdot10^{-5}$	$1,3\cdot10^{-5}$	$0,5\cdot10^{-5}$	$5,0\cdot10^{-5}$	$0,5\cdot10^{-5}$	$0,8\cdot10^{-5}$
35	Bromine	$2,0\cdot10^{-4}$	$2,2\cdot10^{-4}$	$1,3\cdot10^{-4}$	$4,0\cdot10^{-4}$	$4,5\cdot10^{-5}$	$3,6\cdot10^{-4}$	$1,0\cdot10^{-4}$	$4,0\cdot10^{-4}$	$1,0\cdot10^{-4}$	$6,2\cdot10^{-4}$
36	Krypton	$4,2\cdot10^{-9}$ cm^3 per gram of rock weight									
37	Rubidium	$9,0\cdot10^{-3}$	$18\cdot10^{-3}$	$21\cdot10^{-3}$	$16\cdot10^{-3}$	$11\cdot10^{-3}$	$5,0\cdot10^{-3}$	$0,5\cdot10^{-3}$	$14\cdot10^{-3}$	$6,0\cdot10^{-3}$	$0,3\cdot10^{-3}$
38	Strontium	$3,8\cdot10^{-2}$	$2,3\cdot10^{-2}$	$1,1\cdot10^{-2}$	$4,4\cdot10^{-2}$	$4,5\cdot10^{-2}$	$4,7\cdot10^{-2}$	$0,1\cdot10^{-3}$	$3,0\cdot10^{-2}$	$0,2\cdot10^{-2}$	$6,1\cdot10^{-2}$
39	Yttrium	$2,6\cdot10^{-3}$	$3,6\cdot10^{-3}$	$4,0\cdot10^{-3}$	$3,4\cdot10^{-3}$	$2,9\cdot10^{-3}$	$2,1\cdot10^{-3}$	$n\cdot10^{-5}$	$2,6\cdot10^{-3}$	$4,0\cdot10^{-3}$	$3,0\cdot10^{-3}$
40	Zirconium	$1,3\cdot10^{-2}$	$1,7\cdot10^{-2}$	$1,8\cdot10^{-2}$	$1,6\cdot10^{-2}$	$1,4\cdot10^{-2}$	$1,1\cdot10^{-2}$	$0,45\cdot10^{-2}$	$1,6\cdot10^{-2}$	$2,2\cdot10^{-2}$	$0,2\cdot10^{-2}$
41	Niobium	$1,9\cdot10^{-3}$	$2,0\cdot10^{-3}$	$2,1\cdot10^{-3}$	$2,0\cdot10^{-3}$	$2,0\cdot10^{-3}$	$1,9\cdot10^{-3}$	$1,6\cdot10^{-3}$	$1,1\cdot10^{-3}$	$n\cdot10^{-5}$	$0,3\cdot10^{-4}$
42	Molybdenum	$1,3\cdot10^{-4}$	$1,3\cdot10^{-4}$	$1,3\cdot10^{-4}$	$1,2\cdot10^{-4}$	$1,1\cdot10^{-4}$	$1,5\cdot10^{-4}$	$0,3\cdot10^{-4}$	$2,6\cdot10^{-4}$	$0,2\cdot10^{-4}$	$0,4\cdot10^{-4}$
43	Technetium	—	—	—	—	—					
44	Ruthenium	Data not available									
45	Rhodium	Data not available									

Table 3 (continued)

Atomic number	Element	Continental lithosphere (excluding sedimentary cover)	Granitic shell	Granite	Granodiorite	Intermediate rocks	Basic rocks	Ultra-basic	Schist	Sedimentary rocks. Sandstone	Carbonate rocks
46	Palladium	$n \cdot 10^{-7}$	$n \cdot 10^{-8}$	$n \cdot 10^{-8}$	$n \cdot 10^{-8}$	$n \cdot 10^{-7}$	$2,0 \cdot 10^{-7}$	$5,0 \cdot 10^{-7}$	Data not available		
47	Silver	$9,0 \cdot 10^{-6}$	$4,8 \cdot 10^{-6}$	$3,7 \cdot 10^{-6}$	$5,1 \cdot 10^{-6}$	$7,0 \cdot 10^{-6}$	$11 \cdot 10^{-6}$	$6,0 \cdot 10^{-6}$	$7,0 \cdot 10^{-6}$	$n \cdot 10^{-6}$	$n \cdot 10^{-6}$
48	Cadmium	$1,9 \cdot 10^{-5}$	$1,5 \cdot 10^{-5}$	$1,3 \cdot 10^{-5}$	$1,6 \cdot 10^{-5}$	$1,8 \cdot 10^{-5}$	$2,2 \cdot 10^{-5}$	$0,1 \cdot 10^{-5}$	$3,0 \cdot 10^{-5}$	$0,n \cdot 10^{-5}$	$0,4 \cdot 10^{-5}$
49	Indium	$2,3 \cdot 10^{-5}$	$2,5 \cdot 10^{-5}$	$2,6 \cdot 10^{-5}$	$2,4 \cdot 10^{-5}$	$2,2 \cdot 10^{-5}$	$2,2 \cdot 10^{-5}$	$0,1 \cdot 10^{-5}$	$1,0 \cdot 10^{-5}$	$0,n \cdot 10^{-5}$	$0,n \cdot 10^{-5}$
50	Tin	$1,9 \cdot 10^{-4}$	$2,7 \cdot 10^{-4}$	$3,0 \cdot 10^{-4}$	$2,5 \cdot 10^{-4}$	$1,6 \cdot 10^{-4}$	$1,5 \cdot 10^{-4}$	$0,5 \cdot 10^{-4}$	$6,0 \cdot 10^{-4}$	$0,n \cdot 10^{-4}$	$0,n \cdot 10^{-4}$
51	Antimony	$2,0 \cdot 10^{-5}$	$2,0 \cdot 10^{-5}$	$2,0 \cdot 10^{-5}$	$2,0 \cdot 10^{-5}$	$2,0 \cdot 10^{-5}$	$2,0 \cdot 10^{-5}$	$1,0 \cdot 10^{-5}$	$15 \cdot 10^{-5}$	$0,n \cdot 10^{-5}$	$2,0 \cdot 10^{-5}$
52	Tellurium	$1,0 \cdot 10^{-7}$	$1,0 \cdot 10^{-7}$	$1,0 \cdot 10^{-7}$	$1,0 \cdot 10^{-7}$	$1,0 \cdot 10^{-7}$	$1,0 \cdot 10^{-7}$	$0,n \cdot 10^{-7}$	$10 \cdot 10^{-7}$	Data not available	
53	Iodine	$5 \cdot 10^{-5}$	$5 \cdot 10^{-5}$	$5 \cdot 10^{-7}$	$5 \cdot 10^{-7}$	$5 \cdot 10^{-7}$	$5 \cdot 10^{-5}$	$5 \cdot 10^{-5}$	$2,2 \cdot 10^{-4}$	$1,7 \cdot 10^{-4}$	$1,2 \cdot 10^{-4}$
54	Xenon	$3,4 \cdot 10^{-10}$ cm³ per gram of rock weight									
55	Cesium	$2,0 \cdot 10^{-4}$	$3,8 \cdot 10^{-4}$	$5,0 \cdot 10^{-4}$	$2,0 \cdot 10^{-4}$	$1,5 \cdot 10^{-4}$	$1,1 \cdot 10^{-4}$	$n \cdot 10^{-5}$	$5,0 \cdot 10^{-4}$	$n \cdot 10^{-5}$	$n \cdot 10^{-5}$
56	Barium	$4,5 \cdot 10^{-2}$	$6,8 \cdot 10^{-2}$	$8,4 \cdot 10^{-2}$	$4,5 \cdot 10^{-2}$	$3,8 \cdot 10^{-2}$	$3,3 \cdot 10^{-2}$	$0,4 \cdot 10^{-4}$	$5,8 \cdot 10^{-2}$	$n \cdot 10^{-3}$	$1,0 \cdot 10^{-3}$
57	Lanthanum	$2,5 \cdot 10^{-3}$	$4,6 \cdot 10^{-3}$	$5,5 \cdot 10^{-3}$	$4,0 \cdot 10^{-3}$	$3,0 \cdot 10^{-3}$	$1,5 \cdot 10^{-3}$	$n \cdot 10^{-5}$	$9,2 \cdot 10^{-3}$	$3,0 \cdot 10^{-3}$	$n \cdot 10^{-4}$
58	Cerium	$6,0 \cdot 10^{-3}$	$8,3 \cdot 10^{-3}$	$9,2 \cdot 10^{-3}$	$8,0 \cdot 10^{-3}$	$6,5 \cdot 10^{-3}$	$4,8 \cdot 10^{-3}$	$n \cdot 10^{-5}$	$5,9 \cdot 10^{-3}$	$9,2 \cdot 10^{-3}$	$1,2 \cdot 10^{-3}$
59	Praseodymium	$5,7 \cdot 10^{-4}$	$7,9 \cdot 10^{-4}$	$8,8 \cdot 10^{-4}$	$7,5 \cdot 10^{-4}$	$6,2 \cdot 10^{-4}$	$4,6 \cdot 10^{-4}$	$n \cdot 10^{-5}$	$5,6 \cdot 10^{-4}$	$8,8 \cdot 10^{-4}$	$1,1 \cdot 10^{-4}$
60	Neodymium	$2,4 \cdot 10^{-3}$	$3,3 \cdot 10^{-3}$	$3,7 \cdot 10^{-3}$	$3,2 \cdot 10^{-3}$	$2,7 \cdot 10^{-3}$	$2,0 \cdot 10^{-3}$	$n \cdot 10^{-5}$	$2,4 \cdot 10^{-3}$	$3,7 \cdot 10^{-3}$	$4,7 \cdot 10^{-3}$
61	Promethium	—	—	—	—	—	—	—	—	—	—
62	Samarium	$6,5 \cdot 10^{-4}$	$9,0 \cdot 10^{-4}$	$10,0 \cdot 10^{-4}$	$8,5 \cdot 10^{-4}$	$7,5 \cdot 10^{-4}$	$5,3 \cdot 10^{-4}$	$n \cdot 10^{-5}$	$6,4 \cdot 10^{-4}$	$1,0 \cdot 10^{-3}$	$1,3 \cdot 10^{-4}$
63	Europium	$1,0 \cdot 10^{-4}$	$1,4 \cdot 10^{-4}$	$1,6 \cdot 10^{-4}$	$1,4 \cdot 10^{-4}$	$1,2 \cdot 10^{-4}$	$0,8 \cdot 10^{-4}$	$n \cdot 10^{-5}$	$1,0 \cdot 10^{-4}$	$1,6 \cdot 10^{-4}$	$0,2 \cdot 10^{-4}$
64	Gadolinium	$6,5 \cdot 10^{-4}$	$9,0 \cdot 10^{-3}$	$10,0 \cdot 10^{-4}$	$8,5 \cdot 10^{-4}$	$7,5 \cdot 10^{-4}$	$5,3 \cdot 10^{-4}$	$n \cdot 10^{-5}$	$6,4 \cdot 10^{-4}$	$1,0 \cdot 10^{-3}$	$1,3 \cdot 10^{-4}$
65	Terbium	$1,0 \cdot 10^{-4}$	$1,4 \cdot 10^{-4}$	$1,6 \cdot 10^{-4}$	$1,4 \cdot 10^{-4}$	$1,2 \cdot 10^{-4}$	$0,8 \cdot 10^{-4}$	$n \cdot 10^{-5}$	$1,0 \cdot 10^{-4}$	$1,6 \cdot 10^{-4}$	$0,2 \cdot 10^{-4}$
66	Dysprosium	$4,6 \cdot 10^{-4}$	$6,5 \cdot 10^{-4}$	$7,2 \cdot 10^{-4}$	$6,1 \cdot 10^{-4}$	$5,2 \cdot 10^{-4}$	$3,8 \cdot 10^{-4}$	$n \cdot 10^{-5}$	$4,6 \cdot 10^{-4}$	$7,2 \cdot 10^{-4}$	$0,9 \cdot 10^{-4}$
67	Holmium	$1,3 \cdot 10^{-4}$	$1,8 \cdot 10^{-4}$	$2,0 \cdot 10^{-4}$	$1,8 \cdot 10^{-4}$	$1,5 \cdot 10^{-4}$	$1,1 \cdot 10^{-4}$	$n \cdot 10^{-5}$	$1,2 \cdot 10^{-4}$	$2,0 \cdot 10^{-4}$	$0,3 \cdot 10^{-4}$
68	Erbium	$2,6 \cdot 10^{-4}$	$3,6 \cdot 10^{-4}$	$4,0 \cdot 10^{-4}$	$3,2 \cdot 10^{-4}$	$2,8 \cdot 10^{-4}$	$2,1 \cdot 10^{-4}$	$n \cdot 10^{-5}$	$2,5 \cdot 10^{-4}$	$4,0 \cdot 10^{-4}$	$0,5 \cdot 10^{-4}$

Table 3 (continued)

Atomic number	Element	Continental lithosphere (excluding sedimentary cover)	Granitic shell	Granite	Granodiorite	Intermediate rocks	Basic rocks	Ultra-basic	Schist	Sedimentary rocks. Sandstone	Carbonate rocks
69	Thulium	$0,2\cdot10^{-4}$	$0,3\cdot10^{-4}$	$0,3\cdot10^{-4}$	$0,3\cdot10^{-4}$	$0,2\cdot10^{-4}$	$0,2\cdot10^{-4}$	$n\cdot10^{-6}$	$0,2\cdot10^{-4}$	$0,3\cdot10^{-4}$	$0,4\cdot10^{-5}$
70	Ytterbium	$2,6\cdot10^{-4}$	$3,6\cdot10^{-4}$	$4,0\cdot10^{-4}$	$3,2\cdot10^{-4}$	$2,8\cdot10^{-4}$	$2,1\cdot10^{-4}$	$n\cdot10^{-5}$	$2,6\cdot10^{-4}$	$4,0\cdot10^{-4}$	$0,5\cdot10^{-4}$
71	Lutecium	$0,8\cdot10^{-4}$	$1,1\cdot10^{-4}$	$1,2\cdot10^{-4}$	$1,0\cdot10^{-4}$	$0,8\cdot10^{-4}$	$0,6\cdot10^{-4}$	$n\cdot10^{-5}$	$0,7\cdot10^{-4}$	$1,2\cdot10^{-4}$	$0,2\cdot10^{-4}$
72	Hafnium	$2,6\cdot10^{-4}$	$3,5\cdot10^{-4}$	$3,9\cdot10^{-4}$	$3,2\cdot10^{-4}$	$2,8\cdot10^{-4}$	$2,2\cdot10^{-4}$	$0,5\cdot10^{-4}$	$2,8\cdot10^{-4}$	$3,9\cdot10^{-4}$	$0,3\cdot10^{-4}$
73	Tantalum	$1,0\cdot10^{-4}$	$2,1\cdot10^{-4}$	$2,5\cdot10^{-4}$	$1,8\cdot10^{-4}$	$1,2\cdot10^{-4}$	$0,5\cdot10^{-4}$	$0,2\cdot10^{-5}$	$0,8\cdot10^{-4}$	$n\cdot10^{-6}$	$n\cdot10^{-6}$
74	Tungsten	$1,1\cdot10^{-4}$	$1,9\cdot10^{-4}$	$2,2\cdot10^{-4}$	$1,7\cdot10^{-2}$	$1,2\cdot10^{-4}$	$0,7\cdot10^{-4}$	$0,1\cdot10^{-4}$	$1,8\cdot10^{-4}$	$1,6\cdot10^{-4}$	$0,6\cdot10^{-4}$
75	Rhenium	$7,0\cdot10^{-8}$	$7,0\cdot10^{-8}$	$6,7\cdot10^{-8}$	—	—	$7,1\cdot10^{-8}$	—	—	—	—
76	Osmium	—	Data not available		—	—	—	—	—	—	—
77	Iridium	$2,0\cdot10^{-8}$	$1,5\cdot10^{-8}$	$1,0\cdot10^{-8}$	—	—	$2,2\cdot10^{-8}$	—	—	—	—
78	Platinum	—	Data not available		—	—	$1,0\cdot10^{-5}$	$2,0\cdot10^{-5}$	—	—	—
79	Gold	$1,7\cdot10^{-7}$	$1,2\cdot10^{-7}$	$0,8\cdot10^{-7}$	$1,2\cdot10^{-7}$	$2,8\cdot10^{-7}$	$3,6\cdot10^{-7}$	$6,0\cdot10^{-7}$	$n\cdot10^{-7}$	$n\cdot10^{-7}$	$n\cdot10^{-7}$
80	Mercury	$4,6\cdot10^{-6}$	$6,6\cdot10^{-6}$	$6,7\cdot10^{-6}$	$6,7\cdot10^{-6}$	$7,5\cdot10^{-6}$	$6,5\cdot10^{-6}$	$6,4\cdot10^{-6}$	$6,6\cdot10^{-5}$	$7,4\cdot10^{-6}$	$4,5\cdot10^{-6}$
81	Thallium	$0,7\cdot10^{-4}$	$1,8\cdot10^{-4}$	$2,3\cdot10^{-4}$	$1,5\cdot10^{-4}$	$1,0\cdot10^{-4}$	$0,2\cdot10^{-4}$	$0,6\cdot10^{-5}$	$1,4\cdot10^{-4}$	$0,8\cdot10^{-4}$	$n\cdot10^{-6}$
82	Lead	$0,9\cdot10^{-3}$	$1,6\cdot10^{-3}$	$1,9\cdot10^{-3}$	$1,5\cdot10^{-3}$	$1,2\cdot10^{-3}$	$0,6\cdot10^{-3}$	$0,1\cdot10^{-3}$	$2,0\cdot10^{-3}$	$0,7\cdot10^{-3}$	$0,9\cdot10^{-3}$
83	Bismuth	$0,8\cdot10^{-6}$	$1,0\cdot10^{-6}$	$1,0\cdot10^{-6}$	$1,0\cdot10^{-6}$	$0,8\cdot10^{-6}$	$0,7\cdot10^{-6}$	$0,1\cdot10^{-6}$	Data not available		—
84	Polonium	—	—	—	—	—	—	—	—	—	—
85	Astatine	—	—	—	—	—	—	—	—	—	—
86	Radon		Data not available								
87	Francium	—	—	—	—	—	—	—	—	—	—
88	Radium		Data not available								
89	Actinium	—	—	—	—	—	—	—	—	—	—
90	Thorium	$7,3\cdot10^{-4}$	$1,4\cdot10^{-3}$	$1,7\cdot10^{-3}$	$1,2\cdot10^{-3}$	$8,5\cdot10^{-4}$	$4,0\cdot10^{-4}$	$4,0\cdot10^{-7}$	$1,2\cdot10^{-3}$	$1,7\cdot10^{-4}$	$1,7\cdot10^{-4}$
91	Protactinium	—	—	—	—	—	$1,0\cdot10^{-4}$	$1,0\cdot10^{-7}$	$3,7\cdot10^{-4}$	$4,5\cdot10^{-5}$	$2,2\cdot10^{-4}$
92	Uranium	$1,5\cdot10^{-4}$	$2,6\cdot10^{-4}$	$3,0\cdot10^{-4}$	$2,5\cdot10^{-4}$	$2,0\cdot10^{-4}$	—	—	—	—	—

of tenths of a percent shall be called *minor elements;* and all those elements whose content is less than 0.1% shall be called *trace elements.* Trace elements which form important economic concentrations of minerals in certain types of deposits will be referred to as *ore trace elements.*

Table 4. Distribution of the average contents of the chemical elements in the continental lithosphere based on Vernadskii's decades (see Glossary for definition).

Decade	Average content in decade (in mass concentration, %)	Logarithm of content in decade	number of elements	Element
I	>10(20—50)	1,...	2	O, Si
II	10^{0}—10^{1}	0,...	6	Al, Fe, Ca, Mg, Na, K
III	10^{-1}—10^{0}	$\bar{1}$,...	4	Ti, P, H, C
IV	10^{-2}—10^{-1}	$\bar{2}$,..	9	Mn, S, F, Ba, Sr, V, Cr, Zr, Cl
V	10^{-3}—10^{-2}	$\bar{3}$,...	14	Ni, Rb, Zn, Cu, Co, Ce, Y, La, Nd, Sc, N, Li, Ga, Nb
VI	10^{-4}—10^{-3}	$\bar{4}$,...	25	Pb, B, Th, Sm, Gd, Pr, Dy, Er, Yb, Hf, Br, Cs, Sn, As, Be, Ar, U, Ge, Mo, Ho, He, Eu, Tb, W, Ta
VII	10^{-5}—10^{-3}	$\bar{5}$,...	8.	Lu, Tl, I, In, Sb, Tm, Gd, Se
VIII	10^{-6}—10^{-5}	$\bar{6}$,...	5	Ag, Hg, Bi, Ne, Pt
IX	10^{-7}—10^{-6}	$\bar{7}$,...	4	Pd, Te, Au, Os
X	10^{-8}—10^{-7}	$\bar{8}$,...	3	Re, Ir, Kr
XI	10^{-9}—10^{-8}	$\bar{9}$,...	1	Xe
XII	10^{-10}—10^{-9}	$\overline{10}$,...	1	Ra

Table 5 presents a geochemical classification of the crustal elements.

It follows from Table 5 that the vast majority of chemical elements in the lithosphere have a marked affinity for oxygen and they form a large group of *oxyphile* elements. This group includes practically all the major and minor elements of the lithosphere, and thus they form the chemical basis of crustal material (the *lithophile* subgroup of elements). A smaller number of oxyphile elements tend to accumulate in natural formations in association with iron, and this group is known as the *siderophile* elements.

Table 5. Geochemical classification of crustal elements.

Major ($\bar{x}$>1%)		Minor ($\bar{x}$=0.n%)	Trace elements ($\bar{x}$<0.n%)	
I. Oxyphile a) lithophile	O, Si, Al, Fe Mg, Ca, Na, K	Mn, Ti, P, (C)	Mineral-forming elements: Li, Be, B, F, Sr, Ba, Y and elements of the rare earth group, Zr, Nb, Ta, Sn, Cs, W, Th, U	Dispersed: Ga, Ge, Rb, Hf, Sc, (Tl), Ra
b) siderophile	(Fe)	—	V, Cr, Co, Ni	—
II. Sulfophile (Chalcophile)		—	S, Cu, Zn, As, Se, Mo, Ag, Sb, Te, Hg, Pb, Bi	(Ga), (Ge), Cd, In, Re, Tl
III. Noble	—	—	Pd, Os, Ir, Pt, Au	Ra, Rh
IV. Hydrophile	(O)	H	Cl, Br, I, (S)	
V. Atmophile	(O)	C	He, N, Ne, Ar, Xe, Rn	

The *sulfophile* group of elements includes those chemical elements which have a marked affinity for sulfur. These elements are usually present in rocks in the form of sulfides and have a tendency to accumulate together with sulfur in sulfide deposits. The *noble* elements, which include the platinoids and gold, generally occur in nature as native metals. The trace elements of the first three groups are divided into *mineral-forming* elements and *dispersed* elements. The mineral-forming elements (for example, selenium, tellurium, tantalum, beryllium, etc.), even though their average contents within the crust is extremely low, form numerous minerals.*

Dispersed elements, in contrast to the mineral-forming elements, do not form minerals; rather they occur in the structures of other minerals as isomorphous admixtures, or they may form isolated minerals which are extremely rare in nature. The dispersed occurrence of some trace elements is due mainly to the similarity of size and properties of their atoms and ions with those of more abundant elements. As a result, ions of dispersed trace elements can freely replace those of the respective more abundant elements in the lithosphere, which leads to their dispersion in mineral structures of the major elements. The most typical pairs are gallium - aluminum, rubidium - potassium, germanium - silicon, and hafnium - zirconium. The amount by which a dispersed element is accumulated in the structure of a host mineral is determined by the physicochemical characteristics of the mineral-forming process. Therefore, the ratio of the major element to the substituting dispersed element may, in some cases, be used as an indicator of the conditions under which the mineral and the enclosing rock were formed. Ratios of this type (for instance, K/Rb or Al/Ga) which are used for petrological or geochemical purposes are called *indicator ratios*.

Some dispersed elements have dual characteristics owing to specific features of their atomic structure, and they may occur both in oxide and in sulfide minerals. These include gallium, germanium, and thallium.

The *hydrophile* group of elements includes the most typical elements of the hydrosphere and, in most cases, constitute the anionic part of chemical compounds which occur in aqueous solutions. Elements of the atmosphere and of the gaseous component of the solid lithosphere, including the inert gases, form the *atmophile* group of chemical elements.

In contrast to the granitic and basaltic shells of the lithosphere, the sedimentary shell is characterized by a greater diversity in chemical composition because of the variability in the composition of its constituent rocks. This is due to the phenomena of supergene differentiation of material which take place when bedrock is weathered on the Earth's surface, and also during the processes of sedimentation. A characteristic chemical feature of the sedimentary shell is an appreciable accumulation of carbon dioxide tied up within carbonate rocks, such rocks being characteristic formations of this sedimentary crustal shell.

Another feature of the sedimentary shell is the abundance of water filling intergranular spaces and capillary cracks in sedimentary rocks. This so-called pore fluid accounts for more than one-fourth of the lithospheric sedimentary shell. The total amount of the pore fluid in the sedimentary shell is equal to 23.5% of the entire mass of the World Oceans. The composition of the pore fluid must be close to that of seawater, as indicated by data obtained from many resistivity measurements, all of which were found to be quite similar for

*This does not rule out the possibility that the greatest proportion of the mineral-forming elements can occur in the Earth's crust as disseminated admixtures within the crystal structures of other minerals.

both types of water. The pore fluid forms a continuous linkage between the sedimentary shell of the lithosphere and the hydrosphere, the latter being the water shell on the Earth. The waters of the hydrosphere are divided into two contrasting groups on the basis of chemical composition: (1) salt water of the oceans and seas, and (2) fresh water of surface drainages and lakes. Salt water is much more abundant. There is a continual and close interaction between the hydrosphere and the atmosphere — the gaseous shell enveloping the Earth.

The biosphere — the concept includes all manifestations of life on the Earth — is of particular importance in the life and activity of man. The biosphere, which is within the realm of biogeochemistry, is concerned with the associations between animal and plant organisms (including microorganisms) which exist on the Earth's surface. The total mass of the living matter is extremely small and accounts for only about 0.01% of the entire crustal mass. It consists largely of phytoplankton and terrestrial vegetation.

Unlike other geospheres, the principal constituents of living matter are oxygen (about 70%), carbon (about 18%), and hydrogen (about 10%). These three major elements of life, together with calcium (0.5%), potassium (0.3%), nitrogen (0.3%), magnesium (0.07%), phosphorus (0.07%), sulfur (0.05%), chlorine (0.04%), sodium (0.02%), and iron (0.02%), account for 99.4% of living matter. The remaining 0.6% of living matter is made up of the so-called trace elements, and includes practically all the remaining elements in Mendeleev's periodic table. The abundance characteristics of the trace elements in various parts of the biosphere are closely associated with certain specific conditions, in particular, with the chemical composition of the environment. This is why living organisms, mainly terrestrial plants, are used for studying certain compositional features of the surface areas of the lithosphere and in exploration for mineral deposits.

Abundances of the Chemical Elements in Natural Formations. Geochemical Provinces

Geochemists, in their characterization of the distribution of a certain chemical element in rocks and other natural formations, usually have restricted themselves to an estimation of the average content of the object being studied. However, the average content, although an important geochemical characteristic of element distribution, provides only limited information. Unknowns, such as the variability of element distribution in rocks, ores, soils, etc., are important geochemical features of element distribution. It is impossible to understand the major geochemical properties of an element in a natural body, or even to assess the accuracy of a derived average percentage, unless the variability of the element distribution is known.

Geological sciences have now reached the stage of development where mathematically processed data is employed to characterize the statistical regularities in the distribution of each element, from which well-founded conclusions about the chemical composition of particular formations can be drawn. These data make it possible, with the help of statistical operations, to (1) assess the reliability of results; (2) compare element distributions being studied; (3) reveal natural relationships between elements; and (4) determine the probability of occurrence in a particular geochemical association of any contents of an element, or combination of elements, which might be of interest to researchers.

The results of the analytical studies of geochemical specimens, which are sampled systematically during field investigations, are the source of information on the chemical composition of natural formations.

The sets of geochemical data based on these results, in conjunction with the petrographic properties of the rocks (or using other characteristics if waters or plants were sampled), are processed by conventional statistical methods. This is done to determine the parameters of distribution of the contents of the chemical elements, or their logarithms (depending on the distribution law).

The main estimates of the *distribution parameters* are the arithmetic mean, the variance, and the standard deviation of the contents (in the case of the normal distribution law), or the arithmetic mean, the variance, and the standard deviation of the logarithms of the contents (in the case of lognormal distribution).

The above estimates of parameters make it possible to describe mathematically the distribution of the contents of a chemical element in natural formations of any size, from a thin bed to the entire Earth's crust. The accuracy of the determination will ultimately depend on the amount of the geochemical information used, i.e., the number of samples representing the object being studied. The parameters of distribution of the chemical elements are classified as *global, regional,* and *local,* depending on the scale of the natural object being studied. (Table 6).

Table 6. Statistical parameters of distribution of the chemical elements within the Earth's crust.

Global	Regional	Local
Characteristic of types and groups of rocks on the crustal scale The estimation of the average content for the types or groups of rocks, or for the Earth's crust as a whole, is called the *clarke* of the corresponding rocks, the clarke of the lithosphere, or of the Earth's crust. These parameters are used for the assessment of the deviation of the regional parameters from the norm.	Characteristic of types and groups of rocks which are representative of a specific region or geochemical province. These parameters are used to evaluate the deviation of the local parameters from the norm.	Characteristic of individual massifs, suits, etc. within a limited area. In the case of rocks uninfluenced by endogenic or exogenic mineralization processes or by any ore element concentrations, they describe the geochemical background.

In 1889, Clarke, an American geochemist, was the first to calculate the average composition of the most abundant elements in the lithosphere. To perpetuate his memory, Fersman proposed calling the average amount of a given chemical element in a certain cosmochemical or geochemical system (the Earth's crust, the Earth, the solar atmosphere, etc.), the *clarke of the chemical element.* It is expressed as a percentage of the weight of the given system, or as a percentage of the total amount (or volume) of the constituent atoms. This designation is commonly used in the Soviet geochemical literature, whereas in the

West it has not replaced the conventional mathematical term "the average percentage of . . .". The *global* parameters of the distribution of the chemical elements in the Earth's crust may be used in geochemical studies as a measure for evaluating the geochemical characteristics of the element distribution within individual large-scale structural areas of the Earth's crust called *geochemical provinces.* The concept of the geochemical province was first introduced by Fersman (1933 - 1934) who proposed this term to describe geochemically homogeneous regions characterized by certain associations of the chemical elements. Beus (1968) extended this concept to *large-scale crustal units characterized by common features of geological and geochemical evolution expressed in the chemical composition of the constituent geological complexes, as well as in the endogenic and exogenic metalliferous and nonmetalliferous concentrations of the chemical elements.* The distribution in time and space of various types of ore concentrations typical of a given geochemical province, which may have developed in different stages of its geological evolution, determine the *metallogenic specialization* of a geochemical province. Thus, metallogenic analysis of any region should be based on comprehensive data which takes into account both the geochemical features of the evolution of the geological complexes, and the geologic-tectonic environments of their formation.

The geochemical characteristics of rocks representative of entire geochemical provinces are described by regional estimates of the distribution parameters of their constituent elements. The *regional* parameters may, in some cases, be close to global, but in other cases the differences are significantly large (Tables 7 and 8). Consequently, within the Earth's crust geochemical provinces exist with anomalous distribution characteristics of one or several chemical elements, which differ appreciably from the average data characterizing given types of rocks on the crustal scale.

Within a geochemical province there may be considerable differences in the distribution of individual elements, or element associations, for rocks of the same type but of different age. In these cases, the values of the regional distribution parameters, determined for individual complexes of different age, will characterize the geochemical features of a specific tectono-igneous or depositional stage in the development of the given geochemical province. These differences are particularly well illustrated in the distribution of trace elements. This makes it possible, in some cases, to use such differences for subdividing igneous or sedimentary complexes of different age, which cannot be done by conventional petrographic or geological methods. The distribution characteristics of trace elements indicate clearly whether or not a given plutonic or volcanic rock may have undergone any post-igneous alterations, even though these alterations may not have caused any appreciable change in the appearance of the rock.

Usually only limited areas of the Earth's surface, within a certain geochemical province, are investigated by geochemical methods during field studies. *Local* distribution parameters of the chemical elements, characterizing the geochemical features of a given geological object, are estimated for specific geological units in the area being studied (igneous massifs, volcanic series, sedimentary and metamorphic suites or complexes). The estimates of the local distribution parameters of trace elements in rocks which have not been influenced by processes leading to the formation of ores or subsequent exogenic changes of ore concentrations, characterize the so-called *geochemical background* for a given area. The concept of

Table 7. Values of global, regional, and local distribution parameters of selected elements in granites.

Character of value	Region and composition of rocks	Arithmetic mean $\bar{x}$ of content	Arithmetic mean $\bar{x}$ of logarithms of content	Standard deviation S of content	Standard deviation S of logarithms of content	Limits of fluctuation of contents (with the probability of 0,01 for the boundary values)
		Titanium (percent)				
Global	Platformal regions of the Earth (data from 30 regions)	0,20	—	0,04	—	0,11 - 0,29 average for regions
Regional	Ukrainian Crystalline Massif (based on 60 samples from different complexes)	0,18	0,886	—	0,283	0,03 - 0,59
Local	Hornblende-biotite granite of the rapakivi formation from the eastern part of the Korsun-Novomirgorod pluton	0,31	—	0,07	—	0,15 - 0,47
		Lithium (ppm)				
Global	Various contents (based on 150 samples)	38	1,504	—	0,255	8 - 130
Regional	Ukrainian Crystalline Massif (based on 40 samples from different complexes)	40	1,528	—	0,294	4 - 174
	Eastern Transbaikalian Region	90	1,740	—	0,444	5 - 590
Local	Hornblende-biotite granite of the rapakivi formation from the eastern part of the Korsun-Novomirgorod pluton	33	—	13,6	—	1 - 65
	Granites of the Khangilai-Shilinski Massif in the eastern Transbaikalian Region	120	2,147	—	0,165	60 - 340

the geochemical background is based on statistically derived estimates of the distribution of trace elements in rocks of a certain region or area unaffected by any endogenic or exogenic ore element concentrations. The local distribution parameters of the chemical elements within various geological formations, combined in space, determine the regional distribution parameters of these elements, which are representative of the geochemical province in general. If one knows the local parameters of element distributions for certain geological complexes and the areal abundance of different rock types within a province, one can readily estimate the regional parameters characterizing the geochemical features of the province.

In addition to the usual estimates of the distribution parameters which are used for the comparison and description of geological formations, the so-called *concentration ratio* (CR) or *clarke of concentration* (CC) is used in some cases. It is the ratio of the average content of

Table 8. Global and regional distribution parameters of rock-forming elements in granites (in percent).

Element	Values of global parameters (average for 65 regions of the Earth)			
	Value of arithmetic mean* $\bar{x}$	Standard deviation s	Coefficient of variation v	Limits of fluctuation of regional average values (with the probability 0.01 for the boundary values)
O	48,7			
Si	34,0±0,2	0,70	0,02	32,4—35,6
Al	7,4±0,1	0,45	0,06	6,35—8,45
Fe	1,85±0,08	0,23	0,12	1,31—2,39
Mg	0,33(—0,507)**	(0,141)	0,22	0,15—0,66
Ca	1,12±0,06	0,24	0,22	0,56—1,68
Na	2,70±0,04	0,40	0,15	1,77—3,63
K	3,60±0,12	0,42	0,12	2,62—4,58
Ti	0,18±0,01	0,05	0,28	0,05—0,30
P	0,06(—1,256)	0,170	0,26	0,02—0,14
H	0,058	0,009	0,16	0,037—0,079

*Error in the determination of the arithmetic mean has been estimated with a 5% confidence level.
**Values in parentheses are the logarithmic estimates of the parameters for the distributions obeying the log-normal law.

Continuation of Table 8

Element	Values of regional parameters (Transbaikalian Region, average for 160 samples)			
	Value of arithmetic mean* $\bar{x}$	Standard deviation s	Coefficient of variation v	Limits of fluctuation of regional average values (with the probability 0.01 for the boundary values)
O				
Si	33,8±0,2	1,21	0,04	31,0—36,6
Al	7,8±0,15	0,95	0,12	5,6—10,0
Fe	1,45(0,104)	(0,234)	0,37	0,36—4,46
Mg	0,38(—0,609)	(0,411)	0,69	0,03—1,78
Ca	1,02(—0,056)	(0,303)	0,49	0,17—4,47
Na	2,70±0,11	0,70	0,26	0,10—4,30
K	3,83±0,14	0,92	0,24	1,69—5,97
Ti	0,14(—0,928)	(0,305)	0,49	0,02—0,61
P	—	—	—	—
H	—	—	—	—

the element in the material studied (rock type, massif, complex, etc.) to the regional or global average content of this element. It is advisable to determine the concentration ratio by comparing the local average contents with the corresponding regional parameter estimates, because use of the crustal clarkes may, in some cases, furnish only approximate information on the geochemical features of the rock complex in question. It should be remembered that the content of trace elements in similar types of rocks differs from province to province; therefore, the same CR value determined with respect to the global average value ($CR = x_{local}/x_{global}$) may have totally different geochemical meaning in different regions. Let us assume, for example, that within the Ukrainian Crystalline Massif and in the Eastern Transbaikalian Region, the contents of lithium in certain granite massifs were determined to be 100 ppm, which is much higher than the average content typical of this element in granites of the lithosphere (CR = 2.6). However, in the Eastern Transbaikalian Region this content of lithium is typical of Mesozoic granites found in this province. Consequently, the regional concentration ratio is, in this case, only slightly higher than unity: $CR_{regional} = x_{local}/x_{regional} = 100/90 = 1.1$

Based on the known estimates of the regional lithium distribution parameters in granites of the Ukrainian Crystalline Massif, the probability is less than 0.02 that lithium contents equal to or higher than 100 ppm will be found (i.e., less than two cases out of a hundred). This content of lithium in granites should be regarded as a strong local geochemical anomaly, the cause of which merits thorough investigation. This anomaly is clearly expressed in the regional concentration ratio: $CR_{regional} = 100/40 = 2.5$.

Individual concentration ratios of trace elements in rocks may, in some cases, be used for determining more complex multiplicative or additive geochemical ratios which characterize the behavior of certain groups of chemical elements in rocks:

$$CR_{MULT.} = CR_1 \cdot CR_2 \cdot CR_3 \cdot \ldots \ldots CR_n$$

$$CR_{ADD.} = CR_1 + CR_2 + CR_3 + \ldots \ldots CR_n$$

These ratios may, for example, be used for calculating the geochemical index which describes the ratio of the multiplicative or additive indexes of elements which accumulate in rocks, to their global or regional average abundances, or to elements which occur in these rocks in amounts smaller than prescribed by their global (or regional) averages. The resulting index may then be used as a criterion for assessing the geochemical similarity or differences between the rock variety being studied and other similar rock types.

Table 9 presents the results obtained by Ishevskaya (1973) who used the multiplicative and additive geochemical factors for subdividing similar-appearing granites in the eastern part of the Korsun-Novomirgorod Pluton (in the Ukraine). These granites had been divided by geological mapping into hornblende-biotite ovoid granites of the rapakivi formation and biotite nonovoid granites. Both varieties had been classified as being part of the Korosten complex. Geochemical study has shown, however, that the hornblende-biotite ovoid granites of the rapakivi formation contain a facies of metasomatically altered granites which differ only slightly from the original unaltered rocks in external appearance, but which have a specific distribution pattern of the trace elements. However, the non-ovoid biotite porphyroblastic granites have been found to be distinctly different from the

Korosten rapakivi granites in their trace element distribution patterns, but definitely similar to the older microporphyroblastic granitoid rocks of the Kirovograd-Zhitomir complex (see Table 9).

Table 9. Numerical values of geochemical indexes used for the subdivision of granites.

Rock	$\frac{CR\,(Zr \cdot Y \cdot Nb \cdot U \cdot Yb \cdot Be)}{CR\,(V^2 \cdot Ni^2 \cdot Ag^2)}$	$\frac{CR\,(Zr + Y + Nb + U + Yb + Be)}{CR\,2 \cdot (V + Ni + Ag)^*} \times 100$
Granites of the Korsun-Novomirgorod Pluton:		
a) hornblende-biotite granites with ovoids;	10^1	107
b) the same, metasomatically altered;	10^3	224
c) biotite porphyroid granites (non-ovoid)	10^{-3}	36
Microporphyroblastic granitoids of the Kirovograd-Zhitomir complex	10^{-3}	36

*The "2" is used in order to normalize the three elements in the denominator with the six elements in the numerator.

Ishevskaya (1973) also suggested a method for comparing geological complexes based on a successive series of trace elements. The method is as follows: diagrams are plotted for the rocks being compared in order to determine the position of their trace elements in a series which depends on the elemental content of the rock (concentration ratios of the elements relative to the global or regional average content may also be used). Table 10 gives the standard series of trace elements used for granites, as well as for intermediate and basic igneous rocks.

If one compares an empirical series of trace elements obtained by means of geochemical studies with the standard series in Table 10, it is easy to recognize the main geochemical characteristics of the rocks being studied which make them different from rocks with the theoretical average composition. The empirical series of trace elements for the granites from the Korsun-Novomirgorod Pluton (mentioned in Table 9) are as follows:

			←		←		←				
	Ba	Zr	Zn	La	Pb	Y	Cu	Nb	U	V	Ni
1.	533	**144**	86	37	35	21	19	10	5	3,9	**3,8**
				→		→		→		→	

	←					
Sn	Co	Mo	Sc	Yb	Be	Ag
2,8	2,7	2,7	1,8	1,8	1,2	**0,04**
			→	→	→	

			←	←	←				←		
	Ba	Zr	Zn	La	Pb	Y	Cu	Nb	U	Sn	Yb
2.	230	**180**	72	71	60	**42**	14	11	9	5,3	**5**
	→							→			

Ni	V	Mo	Be	Co	Sc	Ag
3,9	2,6	2,5	**2,3**	**1,6**	1,5	**0,03**
	→				→	

	Ba	Zr	←Pb	La	Zn	V	Cu	Ni	Y	Nb
3.	530	92→	56	40→	**34**	18→	10	**6,5**	4,5→	3,5→

←Co	U	Sn	Mo	Yb	Sc	Be	Ag
3,0	**3,0**	**2,8**	**1,3**	1,1→	0,8→	0,7→	**0,05**

Note: Arrows pointing to the left denote increased contents relative to the average value for granites; those to the right indicate lower contents; bold numbers indicate contents coinciding with, or close to, the average for granites.

In this case there is a definite similarity in the series for the unaltered and metasomatically altered granites of the rapakivi formation, despite appreciable differences in the absolute values of the average contents. The unaltered rapakivi rocks (1, above) differ from an average granite (Table 10) by having higher contents of zinc (86 against 39) and lead (35 against 19), and to a lesser extent cobalt (2.7 against 1), and by the lower contents of rare-earth elements (La 37 against 55; Yb 1.8 against 4) and beryllium (1.2 against 3.5). In the metasomatically altered rapakivi granites (2, above), in addition to zinc and lead which are present in appreciably higher quantities, uranium is also high; the contents of the rare-earth elements and beryllium are close to the clarke for granites (the estimated average content of beryllium in Precambrian granites of the lithosphere is 3.0 ppm). The order in the trace elements series changes considerably for the biotite porphyroid granites (3, above). This variety differs from the average granite, as well as from the rapakivi granites, by having much lower contents of zirconium, beryllium, yttrium, and the rare-earth elements. Only lead is present in appreciably higher quantities.

Modes of Occurrence of the Chemical Elements in Rocks

Four principal modes of occurrence of the chemical elements in igneous rocks may be distinguished:

1) as an independent structural component in the stoichometric formula of a mineral;
2) as an isomorphic (isomorphous) admixture in rock-forming or associated minerals;
3) within gaseous-liquid inclusions in minerals;
4) within capillary and pore solutions.

Of major importance in sedimentary rocks is the occurrence of trace elements in the adsorbed form which is related to the adsorption ability of finely dispersed particles.

Elements present in igneous rocks as independent structural components of minerals

This group primarily includes the main rock-forming elements (O, Si, Al, Fe, K, Na, Ca, Mg) of Vernadskii's first two decades (Table 4), which make up the bulk of rocks. The overwhelming majority of atoms of these elements in the Earth's crust are tied up in the structures of the rock-forming minerals which are oxygen-containing compounds (mainly silicates). Several of the minor elements of the third decade (titanium and phosphorus) are concentrated in rocks in the form of their own minerals (Table 11). Many trace elements occur mainly as isomorphous admixtures in various minerals. Under certain conditions

Table 10. Standard series of trace elements for the average compositions of igneous rocks, $n \cdot 10^{-4}\%$ (= ppm)

Rocks																	
Granites	Ba	Rb	Zr	Sr	Ce	La	V	Y	Zn	Li	Nb	Ga	Pb	Th	B	Cr	Cu
	840	210	180	110	92	55	44	40	39	38	21	20	19	17	15	10	10
	Sc	Cs	Ni	Yb	Be	Sn	U	Ta	Tl	W	As	Ge	Mo	Co	Ag		
	7	5	4,5	4	3,5	3	3	2,5	2,3	2,2	1,5	1,3	1,3	1	0,04		
Intermediate rocks	Sr	Ba	V	Zr	Rb	Zn	Ce	Cr	Ni	Cu	La	Y	Li	Se	Nb		
	450	380	150	140	110	75	65	55	50	40	30	29	25	20	20		
	Ga	Pb	B	Co	Th	Yb	As	U	Be	Sn	Cs	Ge					
	17	12	9	9	8,5	2,8	2	2	1,8	1,6	1,5	1,3					
	Ta	W	Mo	Tl	Ag												
	1,2	1,2	1,1	1,0	0,07												
Basic rocks	Sr	Ba	V	Cr	Ni	Zr	Zn	Cu	Rb	Co	Ce	Sc	Y	Nb			
	470	330	250	170	130	110	105	87	50	48	48	30	21	19			
	Ga	Li	La	Pb	B	Th	Yb	As	Mo	Sn	Ge	Cs	U				
	17	15	15	6	5	4	2,1	2	1,5	1,5	1,3	1,1	1				
	W	Ta	Be	Tl	Ag												
	0,7	0,5	0,4	0,2	0,1												
Ultrabasic rocks	Ni	Cr	Co	Zn	Zr	V	Nb	Sc	Cu	Sr	Li	Ga	Ge	Sr	As	Pb	
	2000	1600	150	50	45	40	16	15	10	10	2	1,5	1,3	1	1	1	
	Rb	Sn	Ba	Mo	Ta	Be	W	(Y, Cs, La, Ce, Yb)	Tl	Th	U						
	0,5	0,5	0,4	0,3	0,2	0,2	0,1	0,x	0,06	<0,01	<0,01						

Table 11. Principal minerals formed by minor elements and trace elements in rocks of the lithosphere.

Element	Granites and granodiorites	Intermediate rocks	Basic rocks	Alkaline rocks	Ultrabasic rocks
Titanium	ilmenite, sphene	ilmenite, sphene	ilmenite, titanomagnetite	ilmenorutile, sphene, complex titanium silicates	—
Phosphorus	apatite, monazite	apatite	apatite	apatite	—
Sulfur	sulfides of iron, copper, lead, zinc, etc.				
Boron	tourmaline	—	—	—	—
Zirconium	zircon	zircon	—	zircon, complex zirconium silicates	—
Lithium	lithium micas, amblygonite, spodumene (in metasomatically altered granites)	—	—	—	—
Beryllium	beryl, bertrandite, phenacite, chrysoberyl (in metasomatically altered granites)	—	—	—	—
Fluorine	fluorite, topaz	—	—	villiaumite, fluorite	—
Chromium	—	—	chromite	—	chromite
Manganese	secondary oxides	secondary oxides	secondary oxides	secondary oxides	secondary oxides
Copper, zinc, lead, nickel	sulfides	sulfides	sulfides	sulfides	sulfides
Arsenic	arsenopyrite	arsenopyrite	arsenopyrite	—	—
Molybdenum	molybdenite	—	—	molybdenite	—
Tin	cassiterite	—	—	—	—
Rare-earth elements	monazite, allanite, xenotime, rare-earth niobates	—	—	loparite, complex silicates	—
Niobium and tantalum	Columbite-tantalite, pyrochlore-microlite, rare-earth tantalo-niobates (in metasomatically altered granites)	—	—	loparite	—
Thorium	monazite, thorite	—	—	thorite	—
Uranium	uranium oxides, phosphates, etc.	—	—	—	—

they may also form their own minerals which, in some varieties of igneous rocks (especially those which have undergone metasomatic alteration), may concentrate the bulk of the elements which are present in the rock. Recent mineralogical studies carried out with the aid of the most advanced precision instruments have revealed a much greater distribution of ultra-fine mineral inclusions formed by several elements in both primary and secondary minerals (for instance, in micas), than was previously realized. The trace elements in these minerals had previously been attributed entirely to the phenomena of isomorphism.

Isomorphic form of element occurrence in igneous rocks

Numerous chemical elements are present in rocks generally as isomorphic admixtures in various minerals. (These elements include: Rb, Sc, Hf, Ra, Re, Ag, Cd, Ga, In, Tl, Ge, Sb,

Bi, Se, Te, many rare-earth elements, and some others). For other elements, the isomorphic form of occurrence is decidedly predominant (Li, Cs, Ba, Sr, Y, V, Nb, Ta, W, Mo, Mn, Zn, and some others). Consequently, studies of the isomorphic form in which an element occurs in various minerals is one of the major ways of understanding the distribution patterns of the chemical elements in rocks. Such studies make it possible to ascertain the proportions of each element concentrated in its own minerals, and that dispersed as an isomorphic impurity in the rock-forming and associated minerals constituting various phases and facies of the rock complex being studied. In view of the widespread phenomena of removal and redeposition of trace elements, which occur during the recrystallization reactions and postmagmatic alteration of primary igneous rocks, the data obtained from these studies are of great significance not only for characterization of the geochemical history of the elements, but also for finding criteria to be used for assessing the ore-producing potential of the complex being investigated.

Studies of the isomorphic form of element occurrence in rocks consists of isolating monomineralic fractions from the rocks and their quantitative analysis for the chemical elements of interest. The data obtained from quantitative petrographic and mineralogical studies, together with the results of analyses of the monomineralic fractions, are used to compile the so-called distribution balance for an element in rocks. These tables are an important constituent part of certain geochemical studies, and may, in some cases, also help in studies of the distribution characteristics of valuable components in ores having a complex composition (an example of the distribution balance of a trace element in rocks is given in Table 12). Generalizations from the data on the distribution patterns of trace element isomorphic admixtures in minerals from rocks of endogenic origin make it possible to identify concentrator-minerals for each trace element (minerals which concentrate maximum amounts of this trace element), and minerals which contain the bulk of the trace element in the rock (Table 13).

Table 12. Distribution balance of lead in granite (adamellite) from the Chatkai Region of Tien Shan. From Kozyrev et al.

Mineral	Content of mineral in rock, %	Content of lead in mineral, (ppm)	Quantity of lead within rock due to mineral, %
Quartz	28,0	Not determined	—
Plagioclase	29,0	11	18,8
Potassium feldspar*	37,9	34	75,8
Biotite	3,7	40	8,7
Magnetite	0,2	11	0,1
Zircon**	0,004	110	0,02
Sphene	0,003	Not detected	—

Abundance of lead in rock, ppm:
1) calculated value: 17.6
2) determined by analysis: 17

* Potassium feldspar — carrier of the majority of the lead in the rock.
** Zircon — concentrator of lead in the rock.

Table 13. Mineral-concentrators and mineral-carriers of the bulk of the trace elements in the rocks.

Element	Mineral
Lithium	Micas (in acid rocks), especially biotite; hornblende (in intermediate and basic rocks)
Rubidium	Biotite (concentrator of rubidium in intermediate and acid rocks); potassium feldspar (carrier in granitoid rocks); micas (in intermediate rocks)
Beryllium	Plagioclases (carrier of beryllium in intrusive rocks); muscovite (concentrator of this element)
Fluorine	Micas and hornblendes (in the absence of its own minerals)
Chromium	Pyroxenes and magnetite (in ultrabasic and basic rocks in the absence of chromite)
Nickel and cobalt*	Magnesium pyroxenes and olivine (in ultrabasic and basic rocks); biotite (in intermediate and acid rocks)
Copper*	Pyroxenes and amphiboles, as well as magnetite and biotite (in intrusive rocks)
Zinc*	Biotite and amphibole (in intermediate and acid rocks); magnetite (in basic rocks). The role of tourmaline is being studied.
Niobium and tantalum	Ilmenite, zircon, cassiterite, sphene (in granitoid rocks). Biotite is an interesting indicator. Other micas are of minor significance.
Molybdenum	Potassium feldspar and plagioclase (carriers in igneous rocks). The highest concentrations are found in magnetite, ilmenite, and sphene
Tin	Biotite (in granitoid rocks), muscovite (in muscovite granites), tourmaline
Cesium	Biotite and, to a lesser extent, muscovite (in granites)
Tungsten	Biotite (in granitoid rocks)
Lead	Orthoclases (in acid and intermediate rocks). Maximum concentrations are found in zircon (and in some other accessory minerals)

*Biotite and other minerals often contain extremely small inclusions of sulfides of these elements whose quantitative role is difficult to assess.

From the above list it follows that biotite is the best mineral indicator of the geochemical characteristics of siliceous and intermediate igneous rocks. It gives an indication of the presence of lithium, cesium, copper, zinc, niobium, tantalum, tin, and tungsten. Of lesser importance are: muscovite (beryllium, fluorine, tantalum, tin); plagioclase (beryllium, molybdenum); potassium feldspar (lead, rubidium, molybdenum); sphene, ilmenite, and zircon (niobium, tantalum, molybdenum, and tin); magnetite (molybdenum, lead, chromium); and pyroxenes and amphiboles (copper and zinc). The role of hornblende and tourmaline deserve further attention.

In basic and ultrabasic rocks the most interesting indicators of chromium, nickel, cobalt, and copper (in addition to sulfides) are pyroxenes, amphiboles, and to a lesser extent, olivine.

Of particular interest is the so-called mobile or soluble form of occurrence of trace elements in rocks. This form is determined by treating the rock with water or other solvents which do not destroy, or only slightly modify, silicates (solutions of sodium carbonate, ammonium carbonate, weak solutions of hydrochloric acid, etc.). This type of treatment results in the leaching out of those compounds or elements which are present in the rock in capillary and interstitial solutions, in gaseous-liquid inclusions, as well as within easily soluble minerals (in particular, micro-inclusions and extremely small inclusions of sulfides, native elements, some phosphates, etc.). For instance, if granitoid rocks are treated by a weak (1:50) solution of hydrochloric acid with an addition of sodium chloride (1 gram per liter), 30 to 50% of the lead and 70 to 90% of the zinc will go into solution. The less readily soluble molybdenum sulfide can be leached out only when treated with aqua regia. This treatment removes 30 to 80% of the molybdenum from a granite. For uranium, a treatment of the feldspars and quartz from granitoid rocks with a solution of ammonium carbonate makes it possible to leach out 70 to 100% of the total amount of this element from the rock.

The readily soluble form of occurrence of trace elements in sedimentary rocks and soils is of considerable practical interest. In this case, the greatest proportion of the elements are present as ions adsorbed on the surfaces of small, dispersed particles which are characteristic constituents of sedimentary formations and soils.

2

Geochemical Migration of the Elements

The distribution of the chemical elements in any geological formation in the Earth's crust is characterized by variations in element abundances relative to their local, regional, and global averages. If the average percentage (clarke) of an element in the lithosphere is taken as a certain average norm, then deviation from this norm toward lower values will be *dispersion,* and toward higher values *concentration,* of the given element. The highest concentration of a chemical element in a certain crustal area which could have an economic significance is called a *deposit* of the element. Both concentration and dispersion of the elements are due to the movement of their atoms in the Earth's crust, such movement being called *migration.*

Geochemical migration, as defined by Fersman, is the movement of atoms of the chemical elements within the Earth's crust, usually resulting in their dispersion or concentration.

A distinction is made between *internal factors* of migration, resulting from physical and chemical properties of atoms of the chemical elements, and *external factors,* which include the thermodynamic and chemical environments of migration.

Internal Factors of Migration

The internal factors of migration include thermal, gravitational, as well as chemical and radioactive properties of the atoms.

Thermal properties

Thermal properties of atoms (including properties of bonding) are responsible for fusibility and volatility of the chemical elements and their compounds in geologic and cosmic processes. These properties are especially important for the migration of the elements during the stellar stage of the evolution of matter, at temperatures higher than the melting and boiling points of the elements. The melting properties of chemical compounds in the lower crust and upper mantle determine the laws of formation of magmatic melts and the characteristics of migration of the elements in magmas. In the case of mercury, iodine, and all gaseous elements, the properties of bonding play an important role during their migration in the surface shells of the Earth's crust, including the supergene zone.

Gravitational properties

Gravitational properties of atoms and ions are responsible for their migrational features in the Earth's gravitational field. Redistribution of the chemical elements based on the specific volumes of the particles, plays a leading role in the migration of these elements within the Earth's crust and upper mantle, and is regarded as the major factor in the evolution of the Earth's outer shells and in the formation of the present shell structure (Beus, 1972). The role of gravitational properties of the elements becomes obvious during their migration in magmatic melts. In these cases, a local gravitational differentiation within a magma chamber results in the accumulation of those chemical elements (platinum and elements of the platinum group) or minerals (chromite and others) which have higher densities, in the lower levels of the chamber. The processes of gravitational separation of liquid melts into immiscible liquids with different densities, are known in petrology under the general term *liquation*.

Chemical properties

Chemical properties of the elements play a leading role in their migration within the crust, and such migration occurs mainly in the liquid phase (in melts, as well as in supercritical and hydrothermal aqueous solutions). Element migration in the gaseous (excluding supercritical aqueous solutions) and solid phases is definitely of minor significance. In all cases, the character and the manner of migration of the chemical elements depend on the stability of the migrating compounds in specific physicochemical environments. Most types of mineral deposits, other than oil and gas, owe their existence to the concentration of certain chemical elements from aqueous solutions in which migration of these elements took place. This is why studies on the behavior of the chemical elements in aqueous solutions, whose properties are close to those of natural solutions, are of major concern to geochemists interested in the processes of ore formation. It is impossible to understand the phenomena of deposition of ore material, unless this subject area is understood.

In characterizing the mobility of the chemical elements in aqueous solutions we shall stress again that it depends mainly on the stability of the specific chemical compounds (mode of transport) in which these elements migrate.

The concept of *stability* must be clarified. The main point is that the composition of natural solutions moving in fissures and in interstitial pores of rocks at various levels of the Earth's crust, is quite complex. This applies, in particular, to hydrothermal and supercritical solutions which migrate under conditions of high temperatures and pressures within the deep parts of the lithosphere. These solutions, which are in continuous interaction with the surrounding rocks, are relatively saturated with all the major elements of the rocks; in addition, they usually contain such components, as chlorine, fluorine, and various forms of the carbonate-ion, which readily migrate. The various chemical reactions occurring in complex solutions of this type, result in the formation of insoluble chemical compounds which precipitate from the solutions as minerals. In order to be transported over a considerable distance, a chemical compound must not react with other components, otherwise insoluble minerals may be formed. Compounds complying with such requirements are called *migration-stable*. These compounds, and the elements composing them, may be transported in natural processes over considerable distances from the place of their formation. The same happens with individual dissociated ions of some elements, provided they do not react with other components in the solution to produce insoluble compounds. Therefore, mobility of the elements in aqueous solutions is determined by the chemical properties of the dissociated ions of these elements and by their migration-stable compounds with other elements.

From classical chemistry it is well known that the properties of chemical compounds are determined largely by the properties of their constituent atoms and by the nature of the chemical bonding between them. We shall not go into the details of chemical bonding which is a subject considered in general chemistry, but rather we will review certain features of the ionic (electrostatic) and atomic (covalent) bonds, which are most common in natural compounds and which are essential in geochemical migration. For example, from the point of view of migration, the most characteristic feature of those compounds with predominant ionic bonding between the atoms is their ability to dissociate in aqueous solutions. Thus, it is essentially dissociated ions of such compounds which migrate in solutions:

$$\frac{\text{solid phase}}{KCl} \rightleftarrows \frac{\text{solution}}{K^{+} + Cl^{1-}} \ .$$

Compounds with predominantly covalent bonding generally do not dissociate in aqueous solutions and migrate in the combined form as so-called *complex ions* or *complex radicals*.

Typical examples of complex ions are $[CO_3]^{2-}$, $[SO_4]^{2-}$ and $[NO_3]^{-}$

$$\frac{\text{solid phase}}{Na_2[CO_3]} \rightleftarrows \frac{\text{solution}}{2Na^{+} + [CO_3]^{2-}}$$

The ability of chemical elements to form ionic or covalent bonds depends entirely on the properties of their atoms, more specifically, on the bonding energy of the atoms with their orbital electrons. Alkali metals are characterized by the weakest bonding energy with outer electrons. The strongest bonding energy with outer electrons is exhibited by the typical acid-forming elements which appear in the upper right hand corner of the periodic table, as well as by oxygen. Interacting with atoms of other elements, they display a pronounced tendency toward capturing valence electrons, and if they succeed in such capture, they become ionized turning into negatively charged anions. In order to quantitatively express the ability of the atoms to retain their own, and to capture extraneous, outer electrons, the value of their *electronegativity* (EN) is used, which is measured in kcal or kcal/g-atom (Table 14).

The specific type of chemical bonding is determined, to a large extent, by differences in the electronegativities of the interacting atoms. The greater the difference, the easier it is for the element with a high electronegativity to "win over" the valence electrons of another less electronegative element. Thus, the low electronegativity values of the alkali metals reflect their weak bonding with valence electrons. The metals of this group, when chemically reacting with halogens, oxygen, or sulfur, readily lose their outer electrons forming typical ionic electrostatic bonds. As the difference in the electronegativity of the

Table 14. Electronegativity (EN) of some chemical elements, kcal/g-atom (from Povarennykh).

EN	Element	EN	Element	EN	Element
90	Cs^+	180—190	Pb^{2+}, Ag^+, Cu^+, Fe^{2+}	250	Ti^{4+}, Cr^{3+}
97—100	Rb^+, K	190—200	Sb^{3+}, Bi^{3+}, Co^{2+}	260—270	Si^{4+}, Sn^{4+}
115—120	Ba^{2+}, Na^+	200—205	Zn^{2+}	270—280	S^{4+}
120—130	Li^+, Sr^{2+}, Ca^{2+}, La^{3+}	205—210	Be^{2+}, W^{4+}, Ta^{5+}, U^{6+}, Ni^{2+}	280—290	B^{3+}
140—150	Ce^{3+}	210—220	Hg^{2+}, Al^{3+}, As^{3+}	290—300	As^{5+}
160—170	Th^{4+}	220—230	V^{3+}, Nb^{5+}, Mo^{4+}	300—310	P^{5+}
170—180	Mg^{2+}, U^{4+}	230—240	Fe^{3+}, Cu^{2+}, Hg^+, Pb^{4+}	310—315	H^+
				370—380	C^{4+}, S^{6+}
				460	Cl^{7+}
				530	O^6
				605	F^{7+}

reacting atoms becomes smaller, an acid-forming element does not have enough energy to remove the valence electrons from the atoms with which they interact. In these cases, the tendency is for atomic or covalent bonding. The reacting atoms share their valence electrons having opposed spins, so that these paired electrons belong simultaneously to both atoms constituting the pair.

The molecules of the gases H_2, O_2, Cl_2, and F_2 are the simplest examples of covalent bonding. Two atoms with equal electronegativity may be linked by an atomic (in this particular case covalent) bond. Cases in which bonding of atoms with equal electronegativities occur are, however, quite rare in nature. Bonding of atoms with different electronegativities is much more common. In these cases, the so-called *mixed bonds* are produced due to the interaction of different atoms. This involves the transfer of some valence electrons of the less electronegative element to the element with a higher electronegativity; the remaining electrons of the interacting atoms become linked by pairs forming covalent bonding. Strictly speaking, all chemical bonds between elements are mixed. The degree of covalent bonding does not exceed a few percent in the alkali fluorides, whereas in oxides, such as quartz, the degree of covalent bonding between silicon (EN = 270 kcal/g-atom) and oxygen (EN = 530 kcal/g-atom) reaches 53%.*

From Table 14 it is seen that the most significant group of chemical elements, including the majority of the elements which are of economic interest, are characterized by average values of electronegativity ranging from 190 to 270 kcal/g-atom. This group comprises all the so-called amphoteric elements (zirconium, tungsten, beryllium, zinc, aluminum, molybdenum, etc.) which may exhibit both alkaline and acidic properties depending on the composition of the solutions. The amphoteric properties of the elements determine their

*Elements with electronegativities from 260 to 400 kcal/g-atom form oxygen compounds in which covalent bonding prevails. Especially characteristic of such bonding is the formation of complex oxygen anions such as $[BO_3]^{3-}$, $[PO_4]^{3-}$, $[CO_3]^{2-}$, $[SO_4]^{2-}$, etc. Therefore, the above-mentioned elements are often classified as the complex-forming group.

ability, which is important from the point of view of geochemical migration, to form complex compounds stable in aqueous solutions with certain compositions. These complex compounds, also called *acidocomplexes,* are characterized by a predominance of covalent bonding between the amphoteric element and the acid atom or radical (which, in this case, is called the addend). The compounds in aqueous solutions are characterized by strong covalent bonding and, therefore, the activity of the constituent atoms in the solution is sharply decreased.* This particular feature of atoms with covalent bonding in aqueous solutions plays a rather significant role in the migration of many chemical elements. The chemical element bound in a migration-stable complex ion behaves as if it were outside the influence of the components within the solution which are capable of chemical reaction with other components. This may be illustrated with the example of the complicated ammonium and chlorammonium complexes of platinum discussed by Lebedev (1957). The compound $[Pt(NH_3)_6]Cl_4$ dissociates in solution into $[Pt(NH_3)_6]^{4+}$ and $4Cl^-$, and all Cl may be precipitated by $AgNO_3$ as AgCl. However, the compound $[Pt(NH_3)_4Cl_2]\,Cl_2$ only releases $2Cl^-$, during dissociation, with the result that only half of the total amount of Cl is precipitated by $AgNO_3$. And, finally, the compound $[Pt(NH_3)_2\,Cl]$ does not dissociate at all in aqueous solution, and $AgNO_3$ does not precipitate any chlorine.

The formula of an acidocomplex compound is usually given schematically as $B_m[MA_n]$, where B is the alkaline element (K, Na, Li, Rb, Cs, Ca); M is the complex-forming amphoteric element (W, Zn, Be, U, Ta, Nb, Al, Hg, Mo, etc.); and A is the electronegative addend (Cl, F, $S[CO_n]^{m-}$, etc.). These acidocomplexes dissociate in aqueous solutions according to the scheme: $B_m[MA_n] \rightleftarrows {}_mB^+ + [MA_n]^{m-}$.

More complicated acidocomplexes may have a standard formula $[MA_n]A_m$, dissociating in solution into $[MA_n]^+$ and mA^-.

As long as the environment favors stability of a complex ion, all its constituent may migrate freely in aqueous solution, and in particular, these elements may be transported over considerable distances. The environments favorable for the stability of complex particles, as well as the factors responsible for their decomposition, will be discussed later in this book, where the external factors of migration will be considered.

Consequently, migration of the elements in natural solutions may, in conformity with the chemical properties of the elements, occur (1) in the ionic form (mainly strong bases) and also, (2) in the form of complex ions binding the amphoteric elements and the complex-forming agents with acidic elements possessing the maximum electronegativity. In this connection, two major properties of compounds capable of migration in natural solutions can be specified. These compounds must be soluble and migration-stable to a certain extent, i.e., they must not have any tendency to enter into chemical reactions in the specific physicochemical environment.

---

*Activity, or effective concentration of an element in a solution, takes into account the interaction of all the ions present, in particular, their influence on the ability of the given element to participate in chemical reactions typical of the element. Activity is expressed by the formula: $a = f \cdot c$, where a is the activity; c is the concentration of the element; and f is the factor of proportionality called the activity factor. In very dilute solutions in which the effect of extraneous ions is negligible, f is close to unity, and the activity of the element becomes equal to its concentration.

Radioactive properties

During the course of the Earth's evolution, the radioactive properties of the isotopes of some elements (uranium, thorium, potassium, rubidium, rhenium, etc.), result in a decrease in the number of their atoms, with an accompanying increase in the number of atoms of the radioactive decay products. This is illustrated with the following examples:

$$U^{238} \longrightarrow Pb^{206} + 8He^{4}$$
$$U^{235} \longrightarrow Pb^{207} + 7He^{4}$$
$$Th^{232} \longrightarrow Pb^{208} + 6He^{4}$$
$$K^{40} + e \longrightarrow Ar^{40}$$
$$Rb^{87} \longrightarrow Sr^{87} + \beta$$
$$Re^{187} \longrightarrow Os^{187} + \beta$$

External Factors of Migration

The external factors of migration are characterized by the thermodynamic conditions of the medium in which the chemical elements migrate and in which certain inherent properties of their atoms are expressed. The external factors of migration include the gravitational properties of the Earth, the chemical environments of migration, the temperature, and the pressure. The external factors of migration are based on the main laws of physical chemistry, a knowledge of which makes it possible to analyze the characteristics and the course of the natural processes of mineral formation, including the processes of the formation of ore deposits.

Gravitational properties of the Earth

Gravitational properties of the Earth are the factors determining migration of the chemical elements on a global scale. This factor is, in all probability, the major one responsible for the formation of the present zones within the Earth, although the mechanism by which this is accomplished, especially for the inner parts of the Earth, is still not adequately understood.

Chemical environment of migration

The chemical environment of migration is the most important external factor affecting migration of the chemical elements within the Earth's crust. With respect to the natural aqueous solutions which play the leading role in the formation of mineral deposits, one should bear in mind that the movement of these solutions within the Earth's crust occurs in the rock medium which interacts with the aqueous solutions to a certain extent. There is, consequently, every reason to regard natural solutions as complex systems which are more or less saturated with the components of the enclosing rocks, i.e., alkalis, silicon, aluminum, iron, etc. These solutions also contain certain concentrations of such mineralizers as carbonic acid, chlorine, and fluorine which are constituents of many mineral species. The mineralizers also can be detected in volcanic exhalations. Any geochemical assumptions which characterize migration of the elements in natural solutions must take into account the complex nature of their chemical composition. According to the law of mass action for an ideal solution, the rate of any reaction taking place in this solution is directly proportional to the product of concentrations of the reacting substances.* However, in natural solutions of complex composition, the overall concentration of an element cannot be used as the criterion for evaluating the reaction rate because of the interaction between the components. In such cases, the concept of thermodynamic activity which characterizes the effective concentration of the given chemical element in a non-ideal solution is used. Consequently, *the rates of chemical reactions in natural solutions are directly proportional to the activities of the reacting substances*. In some cases, when there is a relatively high overall concentration in the solution, a certain chemical element may remain totally inactive and not enter into chemical reactions, because its activity may actually be too low for the particular physicochemical conditions.

The chemical reaction in natural solutions, which results in the formation of insoluble products (minerals), depends only on the activities of the reacting components, which in some cases differ sharply from their overall concentrations in the solutions. Therefore, the geochemist studying the parageneses of minerals is generally unable to determine the concentrations of the chemical elements in the mineral-forming solution. However, with the help of mineralogical data from paragenetic studies he may be able to objectively estimate the relative activities of the reacting components and their variations in the process of mineral formation.

An insoluble compound (mineral) may be isolated from a mineral-forming solution if the product of activities of its constituent ions in the solution exceeds the solubility product (SP) of the compound.** Since the overwhelming majority of natural, especially high-temperature, minerals have very low solubilities and extremely small SP values, it follows that the migration of their components in natural solutions may take place only under condition of very low activities of the components.

---

*The law of mass action consists of the following: the rate of a chemical reaction is directly proportional to the product of concentrations (activities) of the reacting substances. In a general case, when m molecules of substance A and n molecules of substance B interact, the equation of the reaction rate will be: $v = k\, a_A \cdot a_B$, where a_A and a_B are activities of the components A and B, respectively; k is the constant of the reaction rate equal to the rate of reaction, provided the activity of each reacting substance is equal to unity. The rate-of-reaction constant depends on the temperature. From the law of mass action, the conclusion can be made that in reversible reactions, in a state of chemical equilibrium, the mathematical product of the activities of the reaction products, divided by the [mathematical] product of the activities of the initial substances, at a constant temperature, is a constant value called the equilibrium constant K (in ideal solutions, concentrations of substances may be used in place of activities). In the reversible reaction $m\mathrm{A} + n\mathrm{B} \rightleftarrows {}_p\mathrm{C} + {}_q\mathrm{D}$ the equilibrium constant is:

$$\frac{a_{\mathrm{C}}^{p} \cdot a_{\mathrm{Д}}^{q}}{a_{\mathrm{A}}^{m} \cdot a_{\mathrm{B}}^{n}} = K.$$

**The solubility product SP, is the product of concentrations (activities) of ions in a saturated solution of an insoluble electrolyte. At a constant temperature, the SP of a given compound is a constant value. For instance, in a saturated solution of AgCl, at room temperature, the concentration of silver ions $[Ag]^+$ is equal to the concentration of chlorine ions $[Cl]^-$, and amounts to $1 \cdot 10^{-5}$ g·ion/liter. In this case SP $= [Ag]^+ \cdot [Cl]^- = 1 \cdot 10^{-10}$.

As mentioned in the discussion of the internal factors of migration, complexing is one of the effective ways of lowering the activity of chemical elements in solution. Results from a high-temperature experiment on synthesizing beryl $Al_2Be_3Si_6O_{18}$ (Beus and Dikov, 1967) can be used to illustrate the role of complex compounds in suppressing the activity of an element present in a solution in considerable amounts. In view of the possible role of fluorine complexes of beryllium in the transport of this element (beryllium) in high-temperature solutions, the synthesis was performed in a fluorine medium in the presence of sodium fluoroberyllate, $Na_2[BeF_4]$. The solution which was placed in an autoclave also contained silicon and aluminum in quantities corresponding to the stoichiometric ratio of these elements in the beryl formula. However, notwithstanding the presence of the elements necessary for beryl crystallization in the solution, it was quite difficult to synthesize the mineral. Because a stable fluoroaluminum complex formed in the solution within the autoclave, the activity of aluminum became sharply suppressed, with the result that only beryllium silicates — bertrandite and phenacite — could crystallize from the solution which acted as though aluminum was totally absent. Beryl could only be synthesized when conditions were found which favored the decomposition of the fluoroaluminum complex.

In order to obtain some idea of the behavior of mobile complex compounds of amphoteric metals in mineral-forming solutions whose composition is continually changing, it is worthwhile considering certain properties of complex compounds, inferred from the law of mass action. Stability of complex ions in a solution is determined by the value of the so-called *instability constant of the complex, $C_{in.}$, which is expressed by the equation:*

$$\frac{a_{M} \cdot a_{A}^{n}}{a_{MA_n}} = C_{in.} \text{ (or } K_{H})$$

where a_{MA_n} is the activity of a complex ion on the solution; a_M is the activity of the dissociated cations of the complexing element (Pb, Cu, Zn, Sn, Be, Ta, Mo, etc.); and a_A^n is the activity of the addend anions.

Assuming an invariable value for the instability constant $C_{in.}$ an increase of the activity of the addend anions relative to a_A^n in the solution, contributes to the stability of the complex, by suppressing the activity of the dissociated ions of the complexing metal. Consequently, a medium with an increased activity of the addend anions (F^-, Cl^-, CO_3^{2-}, HCO_3^-, S^{2-}, etc.) is the most favorable environment for the transport of metal acidocomplexes in hydrothermal solutions.

Accordingly, a decrease in the activity of the addend anions in the solution, relative to a_A^n, leads to the decomposition of the complex radical $[MA_n]$ because of the necessity of a simultaneous increase in the activity of the dissociated cations of the metal a_M. Decomposition of a complex may also result from a decrease in the activity of the dissociated cations of the complexing mineral a_M at a constant a_A^n.

Consequently, decomposition of acidocomplexes of the transition elements during processes involving changes in the composition of the mineral-forming solutions, which in turn determines the possibility of isolating the complexing metal in a solid phase, may generally be caused by two factors: (1) reaction of the dissociated ions of the acidocomplex with certain components of the solution, resulting in the formation of an insoluble compound in a solid phase and (2) hydrolysis of the complex as a result of an increase in the pH of the solution.*

Insoluble products of the reaction form as a result of interaction between the dissociated cations of the acidocomplex $[MA_n]$ and certain components of the solution, provided the activity of the dissociated ions a_M exceeds a critical value based on the solubility product, SP, which is required to form the insoluble compound (mineral). Thus, the possibility of isolating a certain ore mineral from solution is in this case determined largely by the activities of the dissociated ions of the complexing metal, i.e., by the instability constant of its complex.

Below is the simplest example of the formation of an insoluble phosphate of a trivalent cation M during the process of decomposition of the acidocomplex MA_n:

$$MA_n + PO_4^{3-} \rightleftarrows M^{n+} + nA^- + PO_4^{3-} \rightleftarrows MPO_4 + nA^-.$$

The rate of the direct reaction leading to the formation of the phosphate amounts to $v_{forward} = k_{forward}\, a_M \cdot a_A^n \cdot a_{PO_4}$, where $k_{forward}$ is the rate constant of the forward reaction. At a constant activity of the phosphate ion (a_{PO_4} = const) the reaction rate will depend entirely on the value of the product $a_M \cdot a_A^n$ or (according to the equation of the instability constant), on the value of $k_{in.} \cdot a_{MA_n}$.

The probability of equilibrium in this reaction is determined by the probability of equalizing the rates of the forward and reverse reactions: $v_{forward} = v_{reverse}$; in other words, $k_{forward} \cdot a_M \cdot a_A^n \cdot a_{PO_4} = k_{reverse} \cdot a_{MPO_4} \cdot a_A^n$.

In this equation $a_M \cdot a_{PO_4}$ is the solubility product of the difficulty soluble phosphate MPO_4; in other words, the possibility of equilibrium is totally dependent on the value of SP_{MPO_4}. An increase in a_M relative to this value will immediately result in a shift of the reaction to the right and in the precipitation of MPO_4.

It follows from the equation of the dependence of the forward reaction on the instability constant of the acidocomplex, $v_{forward} = k_{forward} \cdot a_M \cdot a_A^n \cdot a_{PO_4} = k_{forward} \cdot K_{in} \cdot a_{MA_n} \cdot a_{PO_4}$ that if a solution contains acidocomplexes of several elements producing insoluble compounds with one component of the solution, then the element whose acidocomplex has the maximum instability constant will precipitate easiest. On the other hand, the element forming a stable complex with a minimum value for the instability constant will remain in the solution for the longest period of time. In this case, the ratio of the elements M/Me, which separate in a solid phase at each stage of the hydrothermal process, is determined by the ratio of activities of their dissociated cations in the solution $\frac{a_M}{a_{Me}}$.

In turn, this ratio depends on the instability constants and on the activities of the acidocomplexes MA_n and MeA_n, i.e., it is the function of $\frac{K_H \cdot a_{MA_n}}{K_{H_2} \cdot a_{MeA_n}}$.

---

*The hydrogen ion index, pH, is the negative logarithm of the activity of hydrogen ions in the solution: pH = log a_{H^+}. The activity of hydrogen ions a_{H^+} in pure water at 25°C equals 10^{-7}, i.e., its pH equals 7. In acid solutions $a_{H^+} > 10^{-7}$ (for instance, 10^{-5}, 10^{-2}, etc.); in other words, the more acid the solution, the higher activity of the hydrogen ions in it, and the smaller the pH. In alkaline solutions $H^+ < 10^{-7}$ and, consequently, the pH of alkaline solutions is larger than 7 (8, 9, 11, and so forth).

Consequently, fractional crystallization may result in separation of elements having similar properties, but whose complexes have different stabilities (Beus, 1958).

A knowledge of the nature of the specific complex compounds in which particular metals are transported, would make it possible to predict theoretically the order of precipitation of the ore elements into the solid phase as the constituent of specific minerals.

However, present knowledge of the nature of complex compounds in which metals are transported in the endogenic postmagmatic processes is, in most cases, only known in very general terms. Most of this knowledge is based on studies of mineral paragenesis, and on the alteration characteristics of the enclosing rocks which have reacted with hydrothermal solutions. For instance, the widespread occurrence of minerals containing fluorine (topaz, fluorite) in some high-temperature deposits of apogranitic and greisen formations, suggests an active role of fluorine complexes in the transport of metals associated with these types of deposits (tin, tungsten, beryllium, tantalum, etc.), although the precise composition of the complexes which had been present in the solutions remains problematic. Yet, notwithstanding the general nature of possible conclusions, numerous practical and valuable inferences can be made concerning the behavior of the metals transported in such solutions. For example, on contact with carbonate rocks, because of the resultant isolation of fluorine in the solid phase (as fluorite), the acidocomplexes in the solutions must decompose, and this is accompanied by an abrupt increase in the activity of the complexing metals and, consequently, by the formation of ore minerals. Natural occurrences of this type (tin and rare metals) do exist.

In some cases, the characteristics of mineral paragenesis and the composition of the gaseous-liquid inclusions within the minerals, suggest the possibility that chloride, as well as the more complicated fluoro-carbonate-chlorine, and chloride complexes, might participate in the transport of metals. In particular, the abundance of NaCl in the composition of the gaseous-liquid inclusions within ore and gangue hydrothermal minerals suggests that chloride complexes are an important form of transport of ore elements, such as lead, copper, etc. The mobility and stability of some chloride complexes have been studied experimentally (Helgeson, 1967). Under conditions of high activity of sulfur in solutions, the important forms of transport are, in all probability, various types of sulfide complexes whose decomposition leads to the formation of various sulfide minerals in the solid phase. All these complex compounds with different addends possess different chemical properties, different stabilities under the conditions of a varying acidity-alkalinity regime of the solution, and are, in all probability, interchangeable in their role as transporters of certain metals.

In view of the general tendency toward an increase in alkalinity of mineral-forming solutions as a result of interaction with the enclosing rocks, the interrelationship between the instability constant and pH is of particular interest, because it may explain the behavior of complex compounds under the conditions of varying alkalinity of solutions.

It can be shown, from the law of mass action, that under the conditions of varying alkalinity of a solution, the stability of acidocomplexes of some amphoteric elements is directly proportional to the activity of ions of the addend in the solution, and inversely proportional to the activity of OH^- (i.e., to the increase in pH), as well as to the rate of the hydrolysis reaction of the complex (Beus, 1958). Therefore, variation in pH is an important factor always to be considered in the concentration of ore minerals. It regulates the hydrolysis of complex compounds of some elements and their separation in a solid phase when there is a regular increase in the alkalinity of the mineral-forming solutions.

This factor is very distinctly expressed in mineral deposits whose formation is associated with processes involving the interaction of initially high-temperature acid solutions with enclosing rocks more or less enriched with alkalis. An increase in the alkalinity of the solutions as they interact with the enclosing rocks is, in these cases, the main factor responsible for the decomposition of the complex compounds. The increased alkalinity is also responsible for the higher activity of the complexing metals in the solution and, consequently, for the reactions leading to mineral formation. These phenomena have been studied and described for tin, beryllium, niobium, tantalum, and rare earths in apogranite and pegmatite, for tungsten and beryllium in greisen deposits, etc. Experimental and calculated data also indicate decomposition of lead-chloride complexes and precipitation of galena from hydrothermal solutions in the process of interaction with the enclosing rocks, resulting in an increase of the ratio Na^+/H^+ (Helgeson, 1967).

In addition to the above types of complex compounds, acidocomplexes also exist which are stable in solutions with increased alkalinity, but which decompose as the acidity of the solution increases, thus illustrating the wide range of the conditions favoring the transport of metals and their concentration from solutions.

The acidity-alkalinity regime of post-magmatic solutions is of major significance in the migration of the chemical elements. It is, therefore, necessary to consider certain patterns of variation in the acid-alkali properties of solutions in the processes of their evolution depending on changes in external factors (temperature and pressure). It is obvious that as the initial acid solutions interact with the enclosing rocks, the alkalinity of the reacting solutions, as a rule, increases. However, what are the factors which determine the appearance of acid aqueous solutions during the processes of endogenic mineral formation? In some cases it is possible to establish, with a reasonable degree of confidence, that the increase in acidity is not due to an introduction of additional amounts of the solution from depth, but rather is the result of the evolution of the same mineral-forming solution which previously had no acid properties.

Let us now discuss a theoretical model of a high-temperature supercritical solution separating from a magmatic melt. In accordance with the elementary laws of physical chemistry, this solution must be relatively saturated with all components of the magmatic melt. They may be divided into three major groups, in terms of chemical properties: (1) strong bases (K, Na, Ca, etc.); (2) amphoteric and complexing elements (Si, Al, etc.); and (3) acidic elements (F^-, Cl^-, CO_3^{-2}, S^{-2}, etc.).

It should be remembered that the above groups of elements have sharply differing solubilities in the supercritical (gaseous) aqueous solution. The ions of the non-volatile strong bases are weakly soluble in the supercritical solution, so that the solution rapidly becomes saturated, and the ionic activity in the gaseous phase increases, in contrast to the volatile acids whose solubilities in this phase are higher. Korzhinskii (1964) used such differences between non-volatile bases and volatile acids to explain the relatively high activity of strong bases in supercritical aqueous solutions, which could be assumed on the nature of their interaction with the surrounding rocks. However, external factors also exist which must have the same effect.

According to the Le Chatelier principle*, under the high pressure conditions existing in the crustal interior, where supercritical aqueous solutions separate from magmas (possibly also from the upper mantle), the gaseous phase must have a pronounced tendency toward forming groups or complexes of ions, atoms, and molecules in the solution. In this way the supercritical solution compensates, to a certain extent, for the effect of the pressure. The reaction between the acidic anions present in the solution and the complexing elements M must, under these conditions, be shifted toward the formation of a corresponding acidocomplex $M + nA \rightarrow MA_n$.

The major rock-forming elements, such as silicon and aluminum, are active complexing elements. Iron is also capable of forming complexes with sulfur. Consequently, it may be expected that all acid atoms and molecules present in the supercritical solution will form complexes, and the activity of acids in these solutions must be sharply decreased.

It follows that the supercritical aqueous solution, in high pressures regions, is likely to contain complexing elements (Si, B, etc.) and amphoteric elements (Al, most ore trace elements, etc.) mainly bound within high-temperature complexes with acid elements and oxygen. The alkali metals (Alk) in such solutions are present in ionic form, as expected from the dissociation of high-temperature complexes following the usual scheme Alk. $[MA_n] \rightarrow$ Alk.$^+$ + MA_n^-, where Alk. is the alkali metal. A decrease in the activity of acids in the solution entails a relative increase in the activity of dissociated ions of the alkali metals**, which ultimately determines the character of the interaction of the supercritical solutions with the surrounding rocks.

Evidence of the activity of such solutions can be seen in regions where the earliest high-temperature metasomatic alteration processes of igneous or metamorphic rocks have been developed. These processes, which find expression in the early development of potassium feldspar followed by albitization of rocks, occur under conditions of high activity of strong bases, primarily potassium.

As supercritical solutions develop into hydrothermal solutions due to a decrease in temperature, the internal structure of the solution experiences major rearrangement, and its chemical properties change. As was shown by Korzhinskii, conversion from a supercritical to a hydrothermal solution is accompanied by an abrupt increase in the solubility of strong bases, which causes a decrease in the activity of the latter. The solubility of acids in a hydrothermal solution, on the other hand, decreases markedly and their activity increases. As a result, the opposing changes in the solubilities of bases and acids during the change from supercritical to hydrothermal solutions, determine the increase in the activity of acids in solutions. In addition, a drop in temperature and the appearance of the liquid H_2O phase lowers the stability of high-temperature complexes. Therefore, one may expect a complete

*Le Chatelier's principle, which is also known as the principle of shift in the equilibrium, states: if some condition of the chemical equilibrium changes, the equilibrium will shift in such a direction as to tend to restore the original conditions. In accordance with this principle, an increase of pressure in a gaseous system shifts the homogeneous equilibrium toward the reaction leading to a decrease in the number of molecules.

**In accordance with Korzhinskii's principle, an increase in the overall activity of acids in solutions is responsible for an increase in the relative activity of weaker bases relative to stronger ones. Conversely, a decrease in the overall activity of acids leads to an increase in the relative activity of stronger bases relative to weaker ones. This rule, when used in studies of mineral paragenis, makes it possible to draw important conclusions about the composition of natural solutions during their geochemical evolution.

decomposition of some acidocomplexes which are stable in supercritical high-pressure solutions, but which dissociate in the liquid phase releasing anions of strong acids. Consequently, as the temperature of the system decreases and the supercritical natural solutions change to hydrothermal, their acidity must increase, reaching a maximum at the beginning of the hydrothermal stage of the postmagmatic process.

This phenomenon, which is caused by a decrease in temperature, is particularly well illustrated by the change from post-magmatic high-temperature potassium metasomatism in siliceous intrusive rocks to sodium metasomatism. The phenomenon of high-temperature acid leaching (greisenization) takes place under conditions of greatly increased acidity of solutions (Korzhinskii, 1953; Beus, 1963).

The tendency of amphoteric elements to form complexes in supercritical solutions, as well as in acidic environments during the early stage of the hydrothermal process, favors migration of the ore trace elements. However, acid hydrothermal solutions mark the beginning of a new important stage in the hydrothermal process: active interaction of acid solutions with the enclosing rocks. This interaction results in the removal of large amounts of alkali and alkali-earth metals from the surrounding rocks, which leads to the neutralization of the solutions and to the progressive increase in the activity of strong bases in these solutions. The increase in the alkalinity of the solutions is especially well defined at their frontal contact with the enclosing rocks. This *reaction barrier,* characterized by an abrupt change in the acidity-alkalinity regime of the mineral-forming hydrothermal solutions, is the zone where decomposition of many complex ions occurs, followed by the formation of certain minerals, including ore minerals.

With respect to the chemical environment of migration, it is necessary to stress the role of the enclosing rocks as they are responsible for changes in the chemical composition of the mobile aqueous solutions and for the progress of the chemical reactions within the solutions.

The high-silicon rocks, such as quartzite or sandstone, generally do not cause major changes in the composition of the mineral-forming solutions and, therefore, do not favor rapid decomposition of the mobile acidocomplexes of the ore elements, or crystallization of the ore minerals. When these rocks interact with alkaline solutions, this sometimes results in feldspathization.

Siliceous and alkaline igneous and metamorphic rocks relatively enriched with strong bases (granitoids, syenites, gneisses, schists) which interact with the natural solutions, are the sources of bases for solutions. The extent of the removal of bases from rocks is determined by the relative activity of the acids in the mineral-forming solutions. If the acidity of the mineral-forming solutions is especially high, this may lead to silicification of the rocks, corresponding to the maximum activity of silicon in the solutions (Korzhinskii's principle). Interaction of hydrothermal solutions with the groups of rocks in question, leads to a gradual increase in the alkalinity of the solutions, which has a negative effect on the stability of many mobile complexes of the ore trace elements contributing, in a general way, to precipitation of ore minerals from solutions.

Basic and ultrabasic igneous rocks, as well as their metamorphic analogs, are low in alkalis, but are relatively rich in calcium, which is a precipitator for some acid anions (F^-, CO_3^{-2}) and for some trace elements (W, Nb, and Ta); thus calcium has a disintegration effect on the corresponding complex compounds. Basic and ultrabasic rocks generally cause sharp changes in composition of natural solutions interacting with them, and are responsible for various mineral-forming reactions, sometimes resulting in accumulations of ore minerals.

Carbonate rocks which are rich in calcium promote important changes in the composition of solutions and lead to the neutralization of the acid mineral-forming solutions interacting with them, especially if these solutions contain the fluorine ion. The zone of interaction of hydrothermal solutions with carbonate rocks is a well defined reaction barrier where numerous mobile compounds of the ore trace elements decompose; this is accompanied by crystallization of the corresponding minerals.

Thus, the chemical environment in which an element migrates is determined by a combination of the nature of the internal changes which take place in the composition of natural solutions with decreasing temperature, on the one hand, and by changes in the composition of the solution effected by the surrounding rocks, on the other.

Temperature

Temperature is of major significance in the migration of the chemical elements. It enters, as an independent thermodynamic parameter, into many crucial equations describing the state of a system under particular physicochemical conditions. Its role in the evolution of the chemical composition of a cooling, high-temperature aqueous solution, was briefly discussed above where the chemical environments of migration were considered. However, this factor has much wider implications. Solubility of many compounds in aqueous solutions is known to increase appreciably with higher temperature; the rates of chemical reactions also change with the temperature. A drop in temperature leads to a decrease in the solubility and stability of some high-temperature complexes of ore elements, which may result in large-scale precipitation of the corresponding ore minerals from solutions. For instance, a decrease in temperature causes crystallization of galena from the alkaline hydrothermal solution rich in chlorides (Helgeson, 1967). The temperature factor in magmatic processes plays a major role in element migration, since the formation and subsequent behavior of igneous melts depend on the melting points of chemical compounds and their mixtures.

Pressure

Pressure is another major thermodynamic parameter affecting the state of natural systems and tendencies in their behaviour. Thermodynamic equations exist which relate the activity and chemical potentials of specific components with changes in the pressure and volume taking place during a reaction. We have already discussed the effect of pressure on complex formation in supercritical aqueous solutions. From the point of view of the principle of the shift in equilibrium (Le Chatlier's principle) the quantitative ratios of the ions bound in complexes, to ions which exist in the dissociated state in the supercritical solution, as well as the stabilities of the complex associations, are directly proportional to pressure. This implies that an abrupt decrease of pressure in a system (for instance, in the case of formation of fissure zones during tectonic processes), must result in a large-scale decomposition of acidocomplex groups, and this is accompanied by an abrupt increase in the acidity of the solutions and the overall degree of dissociation. Consequently, a sudden decrease in pressure in a system with a high-temperature supercritical solution may serve as an impetus for the deposition of ore minerals.

The effect of pressure on complexing in a hydrothermal solution must be much less than it is in supercritical solutions, and this has been proved experimentally (Helgeson, 1967). It is, therefore, assumed that in those regions where water is in the liquid phase, the effect of pressure on the instability constant of a complex, or on the activity ratios of the components, may be neglected.

The clues to an understanding and interpretation of the external factors affecting geochemical migration of the elements in natural processes should be sought within the framework of chemical thermodynamics. It is impossible to use geochemical distribution patterns properly for practical situations unless one understands these factors. Even though thermodynamics at today's level of knowledge may, in some cases, only help to outline the most general laws which control the course of the natural processes, an understanding of these general patterns is a valuable tool in the hands of the geochemist engaged in prospecting for mineral deposits.

Factors Affecting the Migration of the Chemical Elements in the Supergene Zone

The factors influencing the migration of the chemical elements in the supergene zone are determined by the thermodynamic conditions on the surface of the Earth's crust, which are different in many ways from those prevailing in the deep zones of the Earth. Migration in the supergene zone occurs at essentially a constant pressure (about 1 kg/cm^2) and with essentially insignificant variations in temperature. The chemical environment on the surface of the crust may be studied directly by means of geochemical investigations of the processes involving the supergene alteration of rocks, and the chemical composition of natural waters, soils, air, etc. *Biological factors,* as well as the transport of substances in the colloidal or adsorbed forms, play a major role in the migration of chemical elements in the supergene zone.

Chemical environment

The chemical environment of supergene migration of the elements is determined primarily by the pH and Eh* values of the natural surface systems. These have been discussed in great detail by Perel'man (1968).

*The oxidizing-reducing potential Eh (or the redox potential), determines the ability of the system to supply electrons for the components being reduced in the reaction $R - ne^{-} = R^{+n}$, where e^{-} is the number of electrons given up by the reduced element to the oxidizer. Eh is measured in volts. The zero level of the potential is assumed to be the transition of the gaseous molecule of hydrogen into the ionized state. $H_2\ (gas) \rightarrow 2H^{+1} + 2e^{-} = \pm 0.000$ volts. The oxidizing-reducing potential of this reaction, in which the activities of the oxidized and reduced forms in the solution are equal, is called the *standard* and is designated as E_0.

The pH value in the supergene zone, in most cases, ranges from 3 to 9. Meteoric and river waters on the Earth's surface are neutral or close to neutral. Waters in some alkaline types of soils, as well as in sea water, show the highest alkalinity among waters found in the surface environments. These waters are enriched with strong bases; their pH reaches 8 or 9. Maximum acidity (from 3 to 5) is characteristic of natural waters surrounding sulfide deposits, as well as of waters in peat bogs. The higher acidity of waters from deep thermal springs is due to deep-seated processes.

The pH and the oxidizing - reducing environments of the supergene zone are largely responsible for the mobility of the elements, their differentiation along their flow paths, as well as for their precipitation (concentration) at certain values of pH and Eh.

The role of pH is expressed primarily through the different effects that solutions, with different values of pH, have on rocks during the processes of weathering. The effect of natural solutions on the surface of rocks increases as the pH becomes lower. Acidic or weakly acidic waters with a pH of less than 6 are particularly favorable for the migration of the majority of the trace elements found in ores. At the same time, an increase in the pH of surface waters may cause the precipitation of certain elements in the form of hydroxides. This pH-dependent precipitation is controlled by the solubility products of the corresponding hydroxides and, therefore, by the contents of the corresponding elements which may possibly be in the surface solutions. In this connection, A. I. Perel'man stresses that the low contents of certain cations, such as Ni^{2+}, Co^{2+}, Zn^{2+}, Mn^{2+}, Ag^{2+}, Pb^{2+}, as well as numerous other trace elements, in waters in the supergene zone, confirms the possibility of their presence in solution even where the pH values are relatively large. Therefore, the pH of surface solutions does not directly affect the supergene migration of these elements. Conversely, cations, such as Cr^{3+}, Ti^{4+}, Th^{4+}, Sn^{2+}, and Zr^{4+}, can only be transported by strongly acidic solutions, and they readily precipitate as the alkalinity of the solutions increases. These strongly acidic solutions form, for example, during the oxidation of ore deposits containing sulfides. They are capable of transporting numerous cations, which are normally weakly mobile, in the form of sulfates or halogen compounds, over certain distances away from the deposit. The distances depend on the rates at which the pH of the solutions increase in the process of its neutralization. However, even alkaline solutions, whose pH is larger than 7, can transport several high-valence ions (generally in a complexed form). These include, in particular, hexavalent chromium and molybdenum, and pentavalent arsenic and vanadium.

Similar to what occurs in endogenic solutions, solutions in the supergene zone exhibit widespread complexing. Consequently, supergene migration of the chemical elements in natural waters can be regarded as the migration of simple ions to a first approximation only, because this does not always represent the actual form in which the elements are present in solution. For the majority of the metals, the formation of complex ions causes an increase in the pH of hydroxide precipitation, and generally favors the solubility of the metals. In this connection, the experimentally determined values of the pH of hydroxide precipitation (tables of which are given in many reference books on geochemistry) can only be of limited significance in the assessment of natural phenomena.

The complex compounds in which ore trace elements occur in natural solutions at the surface, obey the same laws which apply to the endogenic solutions considered above. The main factors responsible for dissociation of the complexes, in this case too, are changes in the pH of the solutions, or precipitations of the dissociated ions of the complex compound. The forms in which insoluble minerals typically precipitate during supergene processes are controlled by the precipitating agents. These precipitating agents include S^{2-} CO_3^{2-}, or, to a lesser extent, PO_4^{3-} and SO_4^{2-}. Hydrogen sulfide is the most active precipitator of the ore trace elements in alkaline solutions. Even small amounts of the S^{2-} ion in a solution can cause complete isolation of the metals present in the solution into a solid phase (assuming the natural fresh waters contain approximately $1 \cdot 10^{-8}$ to $1 \cdot 10^{-6}$g/liter of Cu, Zn, Ni, Co, and Hg).

Hydrolysis of the complexes, due to variations in the pH of the solutions (provided the total content of the trace elements in the solution is sufficient), ultimately leads to the precipitation of metal hydroxides. It should be remembered that many hydroxides have a reasonably high solubility and, therefore, it is possible to regard hydroxides as likely forms in which metals can be transported in supergene solutions. The same applies to the carbonates of nickel, copper, zinc, and silver.

A specific geochemical feature of the supergene zone is the presence of organo-metallic complexes, called *chelates,* many of which are water-soluble and can, therefore, be forms in which metals are transported in surface and ground waters. The chelates which migrate in soils are readily assimilated by plants, thus contributing to the geochemical exchange between the supergene solutions and the biosphere.

The majority of chemical reactions which take place in the supergene zone are accompanied by the exchange of electrons i.e., they can be regarded as oxidizing-reducing reactions. As mentioned previously, the possibility of losing or acquiring an electron by an atom is controlled by the electronegativity of the element. Consequently, electronegative elements, such as oxygen, the halogens, etc. are also the strongest oxidizers. Oxidizers in the supergene zone also include Fe^{3+}, Mn^{4+}, Mo^{6+}, Cr^{6+}, As^{5+}, V^{5+}, S^{6+}, etc, all of which are capable of receiving electrons in oxidizing-reducing reactions. The components of such reactions which lose electrons are reducing agents. They become oxidized as they reduce electronegative elements, i.e., the oxidizing agents. Numerous organic compounds, Fe^{2+}, Mn^{2+}, Cr^{3+}, S^{2-}, etc., are the most active reducing agents under conditions characteristic of supergene migration.

The redox potential is associated with the activity of the reduced (a_{red}) and oxidized (a_{ox}) forms of the reaction components, according to the following relation:

$$Eh = E_0 + 2 \cdot 10^{-4} \frac{T}{n} \log \frac{a_{ox}}{a_{red}}$$

where n is the number of electrons participating in the reaction and T is the absolute temperature.

The value of the redox potential is an important geochemical constant of any natural solution, and it determines the ability of a given solution to oxidize or to reduce ions. In the supergene zone this ability generally depends on the presence of one or several components which control the course of the oxidizing-reducing reactions in a given solution. These components, called *potential governing* components, primarily include free oxygen, organic compounds, hydrogen sulfide, ions of ferric and ferrous iron, and, to a lesser extent, ions of bivalent and tetravalent manganese. Iron is the most common indicator of the oxidizing-reducing environments found in the supergene zone. The presence of supergene ferrous iron compounds (vivianite and melanterite) indicate reducing environments of migration, whereas the corresponding oxide compounds, primarily brown and red iron hydroxides (limonite, goethite), are indicators of oxidizing environments.

A knowledge of redox potentials of the migration environments enables one to judge the possible forms in which ore trace elements are transported, and to draw sound conclusions as to the possibility of transport or precipitation of certain metals under specific geochemical conditions. It is also necessary to take into account the pH of the environment. This is because many oxidizing-reducing reactions which tend to oxidize a cation, require much smaller Eh values in alkaline environments. This was illustrated by Perel'man with the following example based on iron. The boundary at which ferrous converts to ferric iron ranges from Eh = (+0.4) to (+0.6) in strongly acidic environments (which exist in the zones of oxidation of sulfide deposits, and in acidic swamps of the taiga). However, this boundary occurs below zero (Eh has a negative value) in alkaline environments (in soils, and in weathering crusts in deserts).

Waters in the supergene zone, which contain free oxygen in the dissolved form, are usually characterized by variations in Eh of from +0.150 to +0.700. As one goes deeper, the Eh in groundwater decreases from values typical of surface oxygenated waters, to a thousandth of a volt at depths greater than 100 to 200 meters. The lowest values of Eh (less than zero) are characteristic of the reducing waters found in petroleum pools.

The conversion of iron which migrates in supergene waters from the ferrous to the ferric form, is accompanied by the precipitation of iron hydroxides, and is of major significance in the aqueous migration of the ore trace elements, This is especially true in view of the phenomena of adsorption of trace elements by the precipitating colloidal iron oxides. Therefore, an understanding of the geochemical environments which control the oxidation of iron in natural waters is helpful in studies of the migration features of the ore trace elements.

Colloids and sorption

Colloidal forms of migration and the phenomena of sorption play an important, if not a leading, role in the migration and concentration of iron, aluminum, and manganese, as well as numerous trace elements, in the supergene zone. The colloidal (sol) form of migration of the elements is of particular significance under humid climatic conditions, where the majority of the amphoteric trace elements found in ores migrate, in all probability, in the colloidal form.

However, in rivers flowing in temperate climatic zones, the bulk of the iron, manganese, phosphorus, and metallic trace elements is also transported in the form of colloidal solutions or mechanical (finely dispersed) suspensions.

In view of the fact that colloids are capable of sorbing ions and molecules from aqueous solutions, even if their content is negligible and does not reach the solubility product, it is advisable to recognize that a specific *colloidal-sorption form of migration of the trace elements* can exist. Finely dispersed suspensions of clay materials, humic organic colloids, hydroxides of iron, aluminum, manganese, etc., can be sorbents, that is, carriers of trace elements in surface and ground waters. Consequently the precipitation of a trace element along the paths of its aqueous migration does not depend on its solubility product or on its activity in the solution. Rather it is controlled by the precipitation of a sorbent which is capable of being a carrier of this trace element. For the overwhelming majority of trace elements whose concentration in waters in the supergene zone is very low, and which almost never reaches values necessary to ensure precipitation from the solution, the sorption precipitation (more specifically coprecipitation) together with the sorbent-carrier is the principal mode of precipitation. The concentration of trace elements from percolating aqueous solutions by colloids in soils or in clay sediments, is also an important mechanism by which trace elements are accumulated in certain types of geological formations.

The colloidal-sorption form of trace-element transport in waters of the supergene zone must be taken into account in all geochemical conclusions which attempt to use the distribution patterns of particular elements in recent sediments or soils.

The trace elements sorbed by finely dispersed precipitators in soils and sediments of various types can be extracted into solution. This is accomplished by treating the rocks or soils with a weak (2%) hydrochloric acid solution at low temperatures or during heating. This hydrochloric acid extract, which does not attack the silicate fraction of the material being studied, makes it possible to judge the relative proportion of the trace elements transported in solution and those adsorbed within the sediments or soils.

Biological migration

Biological migration of the chemical elements is an exclusive feature of the supergene zone. In 1940 Vernadskii, the founder of the theory of the geochemical role played by living organisms on the Earth's surface, emphasized that "taken as a whole, there is no chemical force on the Earth's surface which is more permanent and which is, therefore, more powerful in its ultimate results, than living organisms."

Three major stages in the biogeochemical migration cycle have been identified: 1) disintegration of rocks under the effect of biogenic factors, resulting in the formation of soluble compounds of some elements; 2) extraction of biogenic elements and trace element impurities from the air and from aqueous solutions, and their accumulation in organisms; and 3) accumulation, decomposition, and mineralization of organic remains. Each of the above stages is characterized by specific features of migration, concentration, and dispersion of the chemical elements, which are the subject matter of biogeochemistry.

Living organisms (which participate in a constant exchange with the atmosphere and hydrosphere and, therefore, also with the lithosphere), in addition to the common biogenic elements (O, C, H, N, S, P, K, Fe, etc.), also accumulate various trace elements. This accumulation depends both on the requirements of the organisms, and on the geochemical characteristics of the environment in which the exchange takes place. Thus living organisms can, to a certain extent, be used as indicators of the geochemical features of the environment. From this point of view, plants which develop a close geochemical relationship with soils and underlying rocks, play a special role. By extracting nutrient solutions from relatively deep soil horizons, plants accumulate certain trace elements and then, after decay, they enrich the surface soil horizons with these elements. Repeated biogenic cycles can result in considerable accumulation of trace elements in the upper horizons of soils. This makes it possible, in some cases, to use their distribution patterns for reaching conclusions about specific features of the composition of the underlying rocks, in particular, about the presence within such rocks of unusually high concentrations of certain ore elements.

The following chapters of this book will show that these features can sometimes be successfully used for the solution of practical problems associated with prospecting for mineral deposits.

Geochemical barriers

Perel'man in characterizing the phenomena of the supergene concentration of elements, introduced the concept of *geochemical barriers* (Table 15). This term implies an abrupt change in the physicochemical environment in the paths of migration of the elements, causing precipitation of certain elements from solution. This concept can be applied successfully to the processes of endogenic migration, in which *temperature and decompression*

Table 15. Geochemical barriers

Barrier type	Characteristics
Temperature	This is very important for migration in endogenic processes (see External Factors of Migration). The role of this barrier in exogenic processes is insignificant.
Decompression	In endogenic processes an abrupt decrease in the pressure within the system plays a major role in the processes of mineral formation. It is much less significant in exogenic processes.
Acid-alkaline	The effect of changes in the acidity-alkalinity regime of a solution during endogenic processes is sometimes a decisive factor in the separation of many components in the solid phase, and in the concentration of ore substances. It is of less significance in exogenic processes; however, the alkaline barrier is responsible for the precipitation of iron, nickel, and other metals from solution when the solution comes into contact with limestones at the boundary of weakly acidic soil horizons and deeper levels rich in carbonate materials, etc.
Oxidizing-reducing	In endogenic, as well as in exogenic processes, a sudden change in the oxidizing-reducing environments in the paths of migration has a decisive effect on the precipitation of some metals;
a) Oxidizing	takes place as juvenile or ground waters, low in oxygen, come into contact with surface waters rich in oxygen. It is very important in the precipitation of the oxides of iron and manganese in surface waters;
b) reducing hydrogen sulfide	causes precipitation of the majority of metals in the form of sulfides;
c) reducing gley	causes precipitation of some anion-producing metals, such as uranium, vanadium, and molybdenum
Sulfate and carbonate	Occurs at the initial interaction of sulfate and carbonate waters with other types of waters rich in calcium, strontium, and barium. Gypsum and celestine are formed.
Adsorption	Typically an exogenic geochemical barrier. It is of great importance in the precipitation of trace elements from surface and ground waters.
Evaporation	Occurs in regions of rapid evaporation of ground waters. It is accompanied by salinization, the formation of gypsum, etc.
Mechanical	Results from changes in the velocity of water flow (or air movement) and is responsible for the precipitation of heavy minerals. it plays a major role in the formation of placer deposits.

barriers caused by abrupt changes in the temperature and pressure, play a major role in the processes of mineral formation.

The *adsorption and oxidizing-reducing barriers,* whose effects are often combined, are of prime significance during the supergene migration of elements. These barriers have also been recognized by Perel'man. The adsorption barrier is responsible for the precipitation, at the surface, of various finely dispersed particles of positively charged (precipitators: clay particles, organic colloids, peat, etc.) and negatively charged (precipitators: hydroxides of iron and aluminum, etc.) ions. In particular, precipitation of the finely dispersed iron hydroxide phase at the oxidizing barrier will also cause sorption and precipitation of the negatively charged complex ions of arsenic, phosphorus, uranium, vanadium, etc. In turn, coagulation of organic colloids is accompanied by sorption precipitation of the cations of copper, lead, zinc, etc. Concentration of the above groups of cations and anions at the sorption barrier also takes place during the processes of percolation of solutions through clayey and organic media, or through soils and unconsolidated deposits rich in humus, as well as on hydroxides of iron and aluminum.

The reducing hydrogen sulfide barrier is of prime importance in the processes of endogenic concentration of the sulfophile elements in the form of sulfides. In the supergene zone, this barrier is encountered in regions where oxidizing surface or ground water, or weakly reducing waters, come into contact with decaying organic matter (for example, the widespread occurrence of concentrations of pyrite and marcasite in shells of fossil mollusks). It also occurs when such waters encounter oil pools, etc. Deposition of heavy metals sulfides, primarily pyrite, is a typical indication of the reducing hydrogen sulfide barrier.

3

General Concepts Used in Applied Geochemistry

General Information

Geochemical methods of prospecting for mineral deposits make use of the characteristics of the migration and concentration of ore elements and associated elements during the processes of the formation, and supergene destruction, of mineral deposits. The major concepts of basic geochemistry, as well as the theory of ore deposits, provide the scientific basis for these methods.

Geochemical exploration is carried out by systematically sampling appropriate natural materials, and this is followed by laboratory studies which include: a) determining the contents of some chemical elements which are indicators of ore mineralization; this is achieved by using various highly sensitive analytical techniques; b) processing of the resulting analytical data with the help of statistics; and c) interpreting the resulting information using all available geochemical and geological knowledge. The last part is, as a rule, the most complicated.

Depending on the type of natural materials sampled, geochemical studies are classified as *lithogeochemical, hydrogeochemical, biogeochemical,* and *atmogeochemical.*

The lithogeochemical method is based on studies of the distribution of elements in rocks, unconsolidated materials, and soils, and is of major importance in exploration for mineral deposits. [In the Western literature, the term lithogeochemical is restricted to the study of rock material. Editor] The hydrogeochemical (sampling of natural waters), biogeochemical (sampling of living substances, mainly plants), and atmogeochemical (sampling of air) methods are, in most cases, auxiliary and are used in situations where lithogeochemical methods are not sufficiently effective. They are also used in solving certain specific exploration problems.

In practice, all geochemical criteria for exploration are based on the appreciable differences in the distribution of individual chemical elements (or groups of elements) in natural materials associated with mineral deposits, by comparison with those materials lacking this association. The comparisons are based on the local distribution characteristics of chemical elements in bedrock unaffected by any mineralization processes, as well as in unconsolidated materials, soils, waters, and plants, whose composition is similarly unaffected by endogenic or exogenic ore concentrations. These characteristics, which may be described mathematically with the aid of statistical estimates of the local distribution parameters of the chemical elements (see Chapter 1), determine the *local geochemical background* — a concept commonly used in geochemical exploration.

If one knows the distribution parameters of the contents of the indicator elements within a geochemical background (or logarithms of the contents, depending on the distribution law), it is easy to estimate the range of variations of the contents of these elements with any desired probability for the boundary values. In actual geochemical practice, the ranges corresponding to the probabilities of the following boundary values are most commonly used: 0.05; 0.01; and less frequently 0.003. Maximum deviations from the background contents in exploration are called *minimum-anomalous (threshold) contents* and are determined with a specified level of significance (respectively, 5; 1; 0.3%; etc.). The boundary values with a relatively low probability of appearing in the background population, are used as the criteria for recognizing abnormal contents of the individual indicator elements revealed during sampling. Consequently, by using statistical criteria, which characterize the background geochemical population, it may be possible to recognize various contents or groups of contents not belonging to the geochemical background. These abnormal contents result from various processes, including the processes of concentration of the ore material. Statistical procedures for determining the parameters of the geochemical background, and the minimum-anomalous content of the indicator elements, will be discussed in detail below.

The formation of ore and large non-economic concentrations of chemical elements compared with their regular background distribution in common rocks of the lithosphere is a rare event, whose probability of occurrence is generally very small. Therefore, all types of elemental distributions in natural materials associated directly or indirectly with such concentrations, and differing appreciably from the distributions characteristic of the geochemical background, are classified as *geochemical anomalies*. Different types of geochemical anomalies exist, depending on the geological nature of the environment.

For example, local geological processes mainly related to magmatic and hydrothermal activity, in particular when rocks interact with endogenic solutions of various types, bring about the redistribution of elements, with some chemical elements being introduced and others removed. These zones of rocks, characterized by anomalous distributions of elements in comparison with the geochemical background, form the so-called *endogenic geochemical anomalies*.

Local areas in which unconsolidated deposits or soil occur, and which are characterized, for some reason or other, by anomalous distributions of certain chemical elements as compared with the geochemical background, form *supergene lithogeochemical anomalies*. Such anomalies are often associated with the phenomenon of supergene destruction of ore concentrations which were produced by various processes.

The abnormally high contents of trace elements (including ore elements) within the region of an endogenic or supergene lithogeochemical anomaly usually cause anomalous contents of these elements in plants growing in the region, compared with the geochemical background. This results in the formation of *biogeochemical (more specifically phytogeochemical) anomalies* within the boundaries of which all plants, or their individual species, are characterized by increased contents of certain trace elements.

Hydrogeochemical anomalies are also recognized, and these are determined by changes in the elemental composition of natural waters in certain parts of the region being studied, in comparison with the geochemical background. Finally, *atmogeochemical anomalies* characterize areas of the atmosphere whose composition differs from those of the atmogeochemical background, due to the anomalous content of one or several gaseous components (SO_2, CO_2, vapors of Hg, radon, etc.).

The *concentration ratio* (CR) of a given chemical element within the anomalous zone, as compared with the average value of the geochemical background (x_{an}/x_b = CR), is a numerical index of the intensity (contrast) of a geochemical anomaly.

Geochemical methods of exploration for mineral deposits are based on the principle of detecting geochemical anomalies caused by the processes of primary concentration, or secondary dispersion, of the ore-forming and associated chemical elements in various natural materials.

After a geochemical anomaly is detected, one of the major problems in interpreting the results is determining whether or not the anomaly represents a geochemical halo of a mineral deposit.

A geochemical halo (aureole) of a deposit is the region in the enclosing rocks, unconsolidated formations, vegetation, as well as in surface and ground waters, in which an occurrence of an anomalous distribution of the ore elements or their associates has been revealed. The origin of a geochemical halo is associated with the processes of the formation of a mineral deposit (primary geochemical halo), or its supergene destruction (secondary geochemical halo).

Primary geochemical halos resulting from the processes of the formation of endogenic mineral deposits are called *endogenic geochemical halos*. These should not be confused with primary geochemical halos associated with the formation of exogenic mineral deposits.

Secondary geochemical halos form as the result of the supergene destruction of mineral deposits and their primary halos. The formation of secondary geochemical halos is associated with the phenomena of supergene dispersion of material which had primarily been concentrated in mineral deposits; it is for this reason that these halos are generally called *secondary dispersion halos**. Depending on the type of the geochemical anomalies which secondary dispersion halos represent, it is possible to identify: secondary lithogeochemical halos in unconsolidated deposits of the supergene zone (in some classifications lithogeochemical also includes geochemical halos in soils); lithogeochemical dispersion halos in bottom sediments of drainage basins; hydrogeochemical halos and dispersion trains in surface and ground waters; biogeochemical halos in plants; and atmogeochemical halos in the atmosphere and soil gas.

As pointed out above, the criteria for the recognition of a geochemical anomaly lie in the distribution patterns of one or several trace elements which characterize a certain type of metallic or non-metallic mineral deposit. The trace elements whose distribution in various types of natural materials may be used as criteria in the search for mineral deposits, are called *indicators* of mineralization. In geochemical exploration for ore deposits, the role of indicator is usually played by those elements which concentrate in ores characteristic of a given type of deposit, as well as by the most typical trace elements which accompany ore mineralization. Those elements which concentrate in ores are *direct indicators* of

*The term *dispersion halo* is, in some cases [particularly in the English literature; Editor], erroneously used to characterize primary geochemical halos. In fact, however, the phenomena of dispersion are *not* involved in the formation of primary halos; rather, primary geochemical halos are intimately connected with the processes of *concentration* of material in mineral deposits.

mineralization, whereas the associated elements are classed as *indirect indicators*.

The choice of indicator elements for geochemical exploration is based on both geochemical and economic considerations; the principle of maximum effectiveness is the major factor upon which the choice depends. The best results are usually achieved by using direct indicator elements which may point to the possible occurrence of a particular type of ore mineralization.

For example, copper is the best indicator of copper deposits, just as tin is the best indicator of tin deposits. However, optimal effectiveness will be achieved only if the element used as a direct indicator is very mobile, so that its abnormally high concentrations may be detected at considerable distances from the mineral deposit. Also, the quantitative analytical method used for the determination of this element must be rapid, sensitive, and inexpensive. If these conditions are not met, then it is sometimes better practice to use indirect indicators which have a definite correlation with the major ore components. Tantalum is an example. No efficient analytical methods are presently available to ensure rapid, and sufficiently sensitive, analysis of geochemical samples for this element. Therefore, in geochemical exploration for tantalum deposits a group of elements are used, which are indirect indicators of tantalum mineralization. These elements include, in particular, lithium which is relatively mobile and can be detected easily by inexpensive analytical methods.

The following practical problems may be solved by geochemical exploration methods:

1) regional metallogenic zoning and geochemical prediction based on the distribution patterns of chemical elements — this involves the search for indicators of ore in the stream sediments of drainage basins, as well as in igneous, metamorphic, and sedimentary geological complexes;

2) the search for mineral deposits exposed by erosion and later covered by unconsolidated sediments, based on secondary geochemical dispersion halos in stream sediments (dispersion trains), eluvial-talus deposits, soils, vegetation, natural waters, and in the atmosphere;

3) the search for mineral deposits and ore bodies which have not been exposed by erosion (hidden), and the assessment of ore occurrence at deep levels and within the hidden flanks of mineral deposits which have already been explored or exploited, using indications obtained from primary geochemical halos;

4) the exploration for oil and gas fields (this subject is not discussed in this book).

Geochemical methods used in exploration may successfully be applied to the solution of numerous purely geological problems, such as: determining the consanguinity of igneous rocks within intrusive and volcanic series; ascertaining the sequence of sedimentation, and establishing stratigraphic correlations in sedimentary suites; and, studying the characteristics of metasomatic processes in various rocks. Analysis of these and other common geological and geochemical problems which are dealt with during the study of the distribution patterns of chemical elements, require special treatment and are beyond the scope of this book.

4

The Use of Geochemical Specialization of Rocks in Geochemical Exploration

General Information

Rocks of the lithosphere which belong to the same specific rock type are, as a rule, characterized by a very similar chemical composition of their major components. It is also possible to identify a sequential series of trace elements for each group of rocks, in which each particular trace element occupies a definite place, in accordance with its average content in the given rock type (see Table 10). However, in contrast to the major elements, the content of trace elements in each rock type is likely to vary appreciably, due to various factors. As a result, the average contents of individual trace elements in rocks with similar abundances of the major components may vary by several times in either direction, or even by and an order of magnitude, deviating sharply from the regularity characteristic of the standard rock series.

This geochemical peculiarity in the distribution of trace elements is, as a rule, a distinctive characteristic of each specific rock type and, consequently, results of studies of the distribution of trace elements in rocks are now widely used in geology. The problem of *geochemical specialization* of geological complexes has become especially important from this point of view during the last decade. This term implies the *specific features in the distribution of one or several trace elements in rocks, expressed in appreciably higher or lower contents of these elements, or in the anomalous values of variance unusual for the given rock type.* There may be noticeable deviations from the typical sequential series of trace elements for given rocks. These geochemical features may be used as: indications of similarity or dissimilarity of rocks classified as one type; in the determination of their genetic affiliation; in the determination of their ore producing potential; etc.

When geochemical specialization is used for estimating the ore producing potential of rocks, the term usually implies the totality of all geochemical features which make this ore-bearing igneous, metamorphic or sedimentary complex different from rocks similar in composition and appearance, but which are barren.

The distinctive features of geochemical specialization of igneous rocks are determined by a range of factors, including the geochemical nature of the magma, the geological characteristics of its crystallization (including the effect of the enclosing rocks), as well as the character of postmagmatic alteration of the parent rocks.

Geochemical specialization in sedimentary rocks is the result of the geochemical differentiation of material during the processes of supergenesis and sedimentation followed by diagenetic transformation of the sediments.

Geochemical specialization of geological complexes may have regional or local significance, depending on the size of the area covered by this phenomenon. For instance, well known regional examples include: the pronounced specialization for copper and molybdenum of diorite-granodiorite rocks in the Pacific mobile zone which extends for thousands of kilometers along the western coast of North and South America; the presence of tin in Cretaceous granites of the Chukchi Peninsula; and the rare-metal specialization of the Jurassic granites in Eastern Transbaikalia, the Hercynian granites of Central Kazakhstan, the "young" granites of Northern Nigeria, and so forth.

In all these cases, geochemical specialization is determined by the general geological conditions related to the development of igneous activity in a particular geochemical province, and in particular, by the regional conditions of formation and emplacement of igneous complexes in a certain tectono-igneous cycle. The regional geochemical specialization of sedimentary complexes is, on the other hand, a reflection of specific features of the chemical composition of the erosional sources, and is the result of stable depositional environments. Thus, regional geochemical specialization in sedimentary rocks depends largely on the tectonic and climatic regime of the sedimentation process, and on the accompanying diagenetic alteration of the sediments. Typical examples are provided by the copper-bearing sedimentary complexes associated with continental nearshore marine facies rich in several ore elements, etc.

Local geochemical specialization of individual intrusive bodies and sedimentary suites, which is confined to an area much smaller than that of a region, is in most cases due to the specific conditions of formation and geological evolution of a given complex. These conditions are often determined by specific structural environments (for instance, the occurrence of an igneous complex in or near a deep fault zone, or the presence of an impervious layer screening the top of a magmatic body), by an unusual composition of the host rocks, etc. Local geochemical specialization of sedimentary rocks may be due to the small size of the primary rocks with anomalous compositions within the erosion source (for example, small ore deposits), by unusual physicochemical conditions in localized parts of the terminal reservoir of the runoff, as well as by some other local factors.

The term "specialization" is sometimes used in a more narrow sense to designate the relative enrichment of individual rock varieties with a particular metallic trace element. Such "specialized" rocks are often regarded as potentially metalliferous. It will be shown below, that the geochemical indications of an ore producing potential in a geological complex are by no means constrained by either the presence or the absence in the rocks of an abnormally high content of the elements accumulating in ores. On the other hand, it should be stressed that igneous, metamorphic, and sedimentary complexes with which economic concentrations of chemical elements are genetically or paragenetically associated, generally do exhibit specific geochemical features. These very features determine the possibility of differentiating between potentially metalliferous and barren geological formations, using typical associations of trace elements and their distribution patterns in rocks. The formations are called potentially metalliferous because geochemical factors alone, even though they may favor the concentration of ore, are not sufficient for the formation of economic mineralization. The accumulation of large amounts of ore materials which may eventually be of economic importance, depends on a large number of interrelated geochemical and geological factors.

For example, in the case of the formation of endogenic ore concentrations associated with the development of a particular igneous complex, the geochemical factors are responsible for: the behavior of the chemical elements making up the ore in the primary magmatic melt; the specific features of their separation from the magma chamber during the magmatic or post-magmatic stage of its development; the transport mechanisms of the ore elements; and the chemical reactions leading to the concentration of these elements in ores. The geological factors, in turn, control the separation and movement of the ore-forming melts and solutions by creating conditions favorable, or unfavorable, for the formation of economic concentrations of ore minerals. Mineral deposits of economic significance are produced only by a favorable combination of geochemical and geological factors.

The same applies to exogenic ore concentrations whose formation and size of accumulation are distinctly dependent on the combination, in time and space, of the geochemical characteristics of the weathering or sedimentation processes, and on the nature of the geological evolution of the given area during the period of ore accumulation.

Thus, the geochemical ore producing potential in a geological complex, may or may not develop into an economic concentration of ore, depending on the specific geological conditions. Nevertheless, in all cases the possibility of identifying potentially metalliferous geological formations, based on data from geochemical surveys, enables more accurate delineation of the area to be recommended for more detailed prospecting, and such data also provides the basis for scientific predictions of possible mineralization in poorly studied regions. It may also assist in reevaluation of the ore potential of those regions which are already being exploited.

The main criteria for the assessment of the geochemical specialization of igneous, metamorphic, and sedimentary rocks are based on: 1) the distribution of trace elements in rocks; 2) the distribution of trace elements in the rock-forming, minor, and associated minerals; 3) the relationships between pairs of isomorphic elements in rocks.

Criteria for the Assessment of Geochemical Specialization

Criteria based on trace element distribution patterns in rocks.

Specific features of the distribution of trace elements in rocks may be expressed primarily on the basis of appreciably higher or lower contents of a particular element (or group of elements), as compared with the regional or global average, and also on the basis of noticeable increases (or decreases) in the variance of the distributions. These features provide solutions to: nearly all geological problems which usually arise during geochemical studies of rocks within the framework of geological mapping, including subdivision of petrographically homogeneous igneous series, and establishment of their consanguinity; subdivision and correlation of barren sedimentary series; assessment of the areal extent of metamorphic and metasomatic processes within igneous and metamorphic complexes; etc. Abnormally high contents of an ore element, or an appreciably increased variance of its distribution, are regarded as favorable indications of the ore producing potential in the massif or rock series. Although these indications are neither the only ones nor unequivocal, they are undoubtedly among the most important geochemical criteria to be considered in exploration work.

Beus (1968) showed, from a general theoretical standpoint, that during crystallization of magmas containing increased concentrations of certain ore elements, the solid products of the crystallization (rocks), as well as the liquid and gaseous separations which form during the process of evolution of a magma chamber, must have relatively high contents of the same ore elements. During the process of fractional crystallization, the largest concentrations of ore trace elements whose properties differ rather markedly from those of the rock-forming elements, will appear in the latest (residual) portion of the melt. Accordingly, the maximum contents of these trace elements (taking into account the crystallochemical factors), should be expected in minerals formed in the final crystallization stage of an igneous rock.

However, some researchers (Tauson and others) maintain that there is no correlation between the content of ore trace elements in unaltered igneous rocks and the ore producing potential of the latter. These investigators stress the relationship between the increased contents of trace elements in rocks with the phenomena of their postmagmatic alteration, and this is accompanied by the redistribution and, in some cases, the accumulation of individual ore elements. This redistribution, under the effect of postmagmatic solutions, usually accompanies the processes of ore formation and is, therefore, a favorable indication during the assessment of the ore producing potential in the igneous complex being studied. Redistribution of ore elements following postmagmatic alteration of rocks, is especially well illustrated by the resulting greater irregularity in the ore element distribution, which is characterized by statistical estimates of variance and its derivatives: standard deviation and variation coefficient. In addition to rock varieties which have appreciably lower contents of certain trace elements because of their removal, redistribution also produces facies of altered rocks with high contents of the same elements.

The phenomena of postmagmatic removal and redeposition of ore elements as a result of action by supercritical and hydrothermal solutions, have been used as the basis for the so-called mobilization hypothesis which explains the formation of certain types of postmagmatic ore deposits. The formation of economic ore concentrations as a result of the removal of ore elements from rocks and their subsequent redeposition (mobilization) within altered varieties of the parent rocks, or at some distance beyond the source of their removal, has been studied for such trace elements as tin, beryllium, tungsten, molybdenum, tantalum, etc. It should be stressed that, in all cases, the above redistribution processes of ore substances were clearly indicated by an appreciably larger variance of the element distributions and the occurrence of local manifestations of ore trace elements in the parent rocks.

It follows that, regardless whether the higher contents of ore elements or their associates in rocks, were generated during the magmatic stage or they are related to the phenomena of their postmagmatic alteration, the characteristics of distribution of the trace elements may, in general, serve as criteria for ascertaining the ore producing potential in igneous rocks with which the processes of ore accumulation are genetically or paragenetically associated.

Tables 16 to 18 present data on the contents of tin and lithium in granitoids, as well as nickel, copper, chromium, and sulfur in basic and ultrabasic rocks. These data illustrate differences in the average contents of ore trace elements and their associates in ore-bearing and barren igneous complexes. It is, in some cases, convenient to use simple or multiplicative ratios of concentration of the indicator elements, rather than their absolute average contents. Table 18 shows examples of the differences between concentration ratios of the indicator elements in nickel-bearing and barren basic and ultrabasic rocks.

Table 16. Average contents of tin in granites from tin-bearing and barren complexes (ppm).

Region	Barren granites	Tin-bearing granites	Reference
Kalba, Gorny and Rudny Altai	5	16-30	Barsukov and Pavlenko (1956)
Chukchi Peninsula	7-8	10-16	Lugov (1964)
Transbaikalia, Far East, Kazakhstan, Urals, Middle Asia	5	15	Beus (1966)
Caucasus, Ukraine	2±0.7	—	Beus (1966)
USA, Canada, Japan	3	—	Onishi and Sandell (1957)
Great Britain	—	27	Butler (1953)
Yugoslavia and France	—	37	Jedwab (1955)

Table 17. Distribution of lithium in metalliferous and barren granitoids.

Granites (regions of occurrence)	Arithmetic mean		Standard deviation of logarithms of contents	Range of variation of contents, ppm (with probability 0.01 for boundary values)	Estimate of probability of a lithium content equal to or higher than 100 ppm
	of contents x, (ppm)	of logarithms of contents, log x			
Granites of the Earth (global parameters of distribution)	38	1.504	0.252	8-130	0.024
Barren for tin, tungsten, beryllium, and tantalum (Ukraine, Caucasus, Urals, Kazakhstan, Transbaikalia)	37.5	1.500	0.256	8-130	0.026
Parent rocks for pegmatite deposits of lithium, beryllium, and tantalum (Kazakhstan, Middle Asia, Transbaikalia)	100	1.946	0.221	25-300	0.405
Parent rocks for greisen and quartz-vein deposits of tin, tungsten, and beryllium (Kazakhstan, Transbaikalia, West Siberia)	80	1.683	0.436	4-540	0.236
Parent rocks for tantalum-bearing apogranites (Transbaikalia)	130	2.080	0.229	23-440	0.637

Table 18. Concentration ratios of ore trace elements in ore-bearing and barren (for Ni and Cu) basic and ultrabasic rocks* (from Polferov et al., 1968).

Region	Type of rock	Character of mineralization	Ni	Co	Cu	Cr	Ti	S	CCNi/CCS	H_2O %
Kola Peninsula	Basic	Absent	0,8	0,6	0,8	1,2	0,05	0,05	16,0	0,09
		Sulfide	2,1	1,3	0,8	1,2	0,3	1,1	1,1	2,13
	Ultrabasic	Absent	1,0	0,6	1,7	1,3	0,05	0,06	16,0	2,38
		Sulfide	2,3	1,0	5,1	0,9	0,3	0,8	2,9	4,03
Northern Baikal	Basic	Absent	1,3	0,4	0,9	1,3	0,02	0,07	18,5	1,16
		Sulfide	5,9	2,5	0,9	2,7	0,03	0,10	59,0	3,55
	Ultrabasic	Absent	0,6	0,5	1,1	1,4	0,15	0,01	60,0	0,72
		Sulfide	1,4	0,7	2,1	0,6	0,08	0,06	23,4	2,33
Norilsk	Basic	Absent	1,2	1,2	0,9	1,3	1,4	0,3	4,0	—
		Sulfide	1,9	1,1	2,7	3,9	0,8	1,0	1,9	—
Central Russia	Strongly metamorphosed ultrabasic rocks	Absent	0,12	0,57	2,31	0,84	0,02	2,30	0,05	—
		Sulfide	0,43	0,50	1,89	0,80	0,06	0,76	0,56	—
Northern Baikal	Strongly metamorphosed ultrabasic rocks	Absent	0,40	0,10	0,07	0,30	0,007	0,02	20	—
		Sulfide	0,95	0,46	1,60	0,60	0,064	0,06	16	—

*Concentration ratios have been calculated relative to the clarkes determined by Vinogradov for the corresponding groups of rocks.

However, the presence of ore in an igneous complex is, by no means, always expressed by high contents of ore elements in parent rocks. This may be illustrated by the ordinary distribution of the majority of sulfophile elements in igneous rocks which are considered to be the source of hydrothermal sulfide deposits of these elements, as well as by the distribution pattern of beryllium in pegmatite-bearing granites. In the latter case Beus (1960) noted that parent granites of even very large pegmatite deposits of beryllium usually do not contain more beryllium than is expected on the basis of the global average content of this element in granites. This may be explained by the widespread phenomena of microclinization of plagioclase in granites with which beryllium-bearing pegmatite occurrences are genetically related. The process by which plagioclase (the main carrier of beryllium in biotite granites) is replaced by potassium feldspar, results in the removal of beryllium and, consequently, the parent rocks are relatively low in this element. Thus, the presence or absence of increased contents of the corresponding elements in the parent rocks of ore deposits, is always determined by the specific processes of mineral formation in these rocks, either prior to mineralization or following ore accumulation.

Abnormally high contents of ore elements and their associates, as well as the increased variance in their distribution in rocks, probably may also be used as criteria for assessing the ore producing potential in some sedimentary formations. According to Strakhov, areas with increased contents of ore elements develop in sedimentary series containing economic ore concentrations of sedimentary origin, by comparison with the general background of the enclosing rocks. Redistribution of ore elements in sedimentary rocks also occurs when mineral deposits are formed during the process of diagenetic transformation of sediments. This is accompanied by the removal of ore elements from sedimentary rock types unfavorable for the concentration of ore minerals, followed by enrichment of these elements in rocks whose structural features (porosity, jointing, etc.) favor accumulation of the ore material. All this takes place within an ore-bearing suite. Diagenetic redistribution of trace elements in sediments usually leads to a greater variance in their distribution, and to an appearance of markedly higher contents of certain elements compared to the general relatively low background in rocks surrounding the ore-bearing horizons. Sedimentary deposits of copper, uranium, and some other elements associated with ore-bearing sandstones, exemplify formations of this type.

Criteria based on the distribution patterns of ore-element admixtures in rock-forming minerals.

In general, the extent of the ismorphic incorporation of a trace element into the crystal structure of a host mineral is determined by the crystallochemical features of the given structure, as well as by the ratio of activities of the trace element and the element it replaces in solutions or melts during the process of mineral formation. An increased activity of an ore element in a solution or melt during the formation of a mineral(s) which may incorporate the given ore element as an isomorphic admixture in its crystal structure, must find expression in the higher contents of this element in one or several minerals representing a certain stage of mineral formation. The distribution of the trace element contents in minerals makes it possible, in some cases, to judge the conditions under which the minerals formed, the geochemical environment which existed at the time of crystallization of the indicator mineral, as well as the relative activities of ore elements in the solution or melt from which the mineral crystallized.

The possibility of using the features of the chemical composition of typomorphic minerals (from the point of view of their trace-element contents) was first suggested by Fersman in the middle 1930's. Subsequently, micas have received special attention from researchers. The first feature to be noticed was that a relationship exists between the content of tin in pegmatitic muscovites and the presence of cassiterite in pegmatite bodies (Ahrens; Ginzburg). An increased content of cesium in lepidolites was suggested as a positive indication in the search for pollucite (Ginzburg). Heinrich and Beus, independently, discovered that the contents of beryllium, tantalum, and niobium in the muscovites from pegmatite bodies may indicate the presence of beryl and tantalite-columbite in these pegmatites.

A significant advance in the utilization of the trace element distribution in individual indicator minerals was made in 1956 when Barsukov explained the relationship between the increased content of tin in biotites from granites, and the presence of tin in the granites. This extended the possibility of using geochemical data on the distribution of trace elements in rock-forming and accessory minerals for exploration purposes to rocks which are associated with ore concentrations by more distant links than in the case of pegmatites. Trace element admixtures in some accessory minerals found in rocks (cassiterite and zircon) have been shown to give valuable information. In the case of cassiterite, a study of these admixtures by Dudykina made it possible to develop a genetic scheme characterizing the typomorphic features of this mineral, and also to use the tantalum content in cassiterite as an indirect criterion in exploration for tantalum deposits.

In some cases, practically negligible differences between the distribution of indicator elements in parent rocks of ore deposits and in similar yet barren rocks, become more distinct if the contents of these elements are compared in individual minerals which are the concentrators or bearers of these elements in the rock. For example, in the above-mentioned case of the parent granites of beryl-bearing pegmatite deposits, even though there may be no difference in the content of beryllium in the pegmatite-producing and barren granites, there may be quite distinct differences in the content of this element in plagioclase (0 to 15 ppm in barren granites, as compared to 15 - 42 ppm in pegmatite-bearing granites). For lead, which does not form any conspicuous concentrations in igneous rocks which are considered as the source of lead deposits, higher contents of this element have been detected in potassium feldspars from granitoids from ore-bearing regions by comparison with potassium feldspars from barren granitoids (Slawson and Nackowski, 1958; Beus, 1966, 1968).

Table 19 illustrates differences in the distribution of elements in minerals which are indicators of ore, from metalliferous and non-metalliferous rocks. It shows the distribution of tantalum in muscovites from pegmatites containing tantalite-columbite, in comparison with muscovites from micaceous and ceramic pegmatites not containing this mineral.

Table 19. Distribution of tantalum in muscovites from granite pegmatites.

Type of pegmatite field (region; number of samples)	Arithmetic mean $\bar{x}$ of contents (ppm)	Arithmetic mean $\bar{x}$ of logarithms of contents	Standard deviation of logarithms of contents s	Coefficient of variation v, %
Deposits from micaceous and ceramic pegmatites (Karelia, Mama, Biryusa, Urals, Ukraine, Baikal Region; 20 samples)	4	0,490	0,255	50
Deposits of rare-metal pegmatites containing tantalite-columbite (Transbaikalia, East Kazakhstan, Central Asia; 120 samples)	35	1,486	0,245	54

Criteria based on the ratios of contents of geochemically related elements in rocks and minerals.

These criteria are based on differences in relations between isomorphic elements in rocks, which are considered a result of the different behavior of elements in various environments of migration.

For example, the migration characteristics and isomorphism for the majority of chemical elements are substantially different during the magmatic process by comparison with the postmagmatic stage when the elements migrate in supercritical or hydrothermal aqueous solutions. Isomorphic relations between chemical elements in magmatic melts are determined largely by a similarity of ionic radii and, consequently, by the possibility of forming coordination structures with oxygen which are close in parameters and energy. Minor, and certainly appreciable, differences in chemical properties of the rock-forming and isomorphic trace element make isomorphism more difficult, but do not rule it out altogether. For a given massif or complex, relatively stable thermodynamic conditions of magmatic crystallization are responsible for the stability of the relative activity of isomorphous pairs of elements in the melt. The latter circumstance, in turn, is the cause of the relative stability of certain relationships between isomorphic elements in the products of magmatic crystallization in a genetically related igneous rocks series. This pattern was comprehensively studied by Ahrens (1952) who proposed that the constant ratios of isomorphic geochemically related elements can be used for solving various geological problems, in particular, the determination of the consanguinity of intrusive and volcanic series. The very pronounced isomorphic relationships between pairs of rock-forming elements and geochemically similar trace elements, are used for the above-mentioned purposes. These include K - Rb, Al - Ga, and others, or the isomorphic relationships between such trace elements as Rb - Tl, the rare-earth group, etc.

Criteria based on differences, rather than similarities, in the geochemical properties of isomorphic elements are of the greatest practical significance, since they enable definite conclusions to be made about changes in the conditions of mineral formation. In particular, during the processes of mineral formation associated with postmagmatic solutions, the outcome of the "struggle" for a site in the crystal structure of a newly formed mineral between the major host element and the trace element admixture depends, to a large extent, on changes in the relative activities of these elements which, in turn, are determined by the acidity-alkalinity regime of the solutions. The differences, however insignificant they may be in the chemical properties of the isomorphic elements, become a major factor favoring or preventing an isomorphic incorporation of a certain element into the mineral structure.

Let us discuss a typical isomorphic pair: aluminum - gallium. Gallium, which is a typically dispersed element, similar to aluminum in its chemical properties and ionic radius, is very closely associated with aluminum during the processes of magmatic crystallization. The overwhelming majority of the gallium atoms in the Earth's crust are dispersed in the structures of the aluminum-bearing minerals. The gallium-aluminum ratio in a genetically related series of igneous rocks is extremely stable, and there is a pronounced positive correlation between the contents of aluminum and gallium. That this ratio varies very little within different types of igneous rocks was shown quite convincingly by Nockolds and Allan (1958) who analyzed a large number of samples. Gallium, however, has somewhat more basic properties than aluminum. Therefore, it can hardly be expected that a direct correlation between the contents of these two elements should be maintained in the reaction products of aqueous solutions formed under significantly different conditions in the acidity-alkalinity regime, despite the similarity in ionic radii, and only very slight differences in the chemical properties of aluminium and gallium. This feature is well illustrated during the greisenization process in which paragenetic associations formed under the conditions of strongly acidic solutions, and containing the high-alumina mineral topaz (48.2 to 62% to Al_2O_3), contain very little gallium, with the resulting very low gallium to aluminum ratio (Beus, 1960).

Consequently, quantitative relationships between isomorphic trace and major elements, must be different in the same minerals of igneous origin, by comparison with those produced from supercritical or hydrothermal aqueous solutions. These differences may prove very helpful in recognizing metasomatically altered varieties of igneous rocks whose detection by petrographic means is difficult in some cases.

In regional exploration for oxyphile ore elements such as beryllium, tin, tungsten, and tantalum, the recognition of metasomatically altered granitoid facies makes it possible to predict deposits of the above-mentioned elements in albitite and quartz-greisen formations. The accuracy of the prediction is, in this case, increased by the distinctive behavior of magnesium and lithium in the parent granites of these deposits. The tendency of lithium to concentrate, and magnesium to disperse, in later facies of ore-bearing granites in which the processes of high-temperature postmagmatic metasomatism* are more or less distinctly expressed, allows the magnesium to lithium ratio in granites to be used successfully as a criterion in the assessment of their ore producing potential. The potassium to rubidium ratio is a much less sensitive, although it is a characteristic indicator in this case (Table 20).

Table 20. Some indicator ratios in granitoids.

Granitoid type	K/Rb	Mg/Li	Zr/Sn	V/Nb
Average for granitoids of the lithosphere	170	370	80	4,4
Average for granites of the lithosphere	170	90	60	2,1
Granites with marked expression of the processes:				
of postmagmatic metasomatism	About 100			
of microclinization, albitization and muscovitization	150	20		
Average for granitoids not related to Li, Be, Sn, W, or Ta deposits	170	270	76	
Average for granitoids related to Li, Be, Sn, W, and Ta deposits	130	75	30	
a) including parent biotite granites of pegmatite deposits of Li, Be, Ta and Cs	160	40	12	0,8
b) the same for pegmatite and apogranite deposits of Ta	126	30	14	0,6

The ratios of all other elements (not only isomorphic elements) whose migration features in a given process differ sufficiently (for instance, Zr/Sn, V/Nb in granitoids, etc., see Table 20), also can be used as indicators. Kvyatkovskii et al. (1972) proposed that the average contents and dispersion patterns of the ratio S/P_2O_5 can be used as an indicator for the identification of potential nickel-bearing massifs in basic and ultrabasic rocks. Polferov (1962) had suggested earlier that the concentration ratios of nickel and sulfur in rocks could be used for the same purpose (this means that ratios of contents of these elements may also be indicative).**

*A pronounced decline in the role of magnesium in granitoid facies which have undergone rather strong postmagmatic alterations is due, in all probability, to an increase in the relative activity of iron in relation to magnesium in solutions characterized by a somewhat increased overall acidity (activity of acids). This results in the replacement of magnesium by iron in biotites. Since biotite is the major carrier of magnesium in granite, the removal of magnesium from this mineral results in its low content in the rock as a whole.

**It should be remembered that the use of concentration ratios, in place of estimates of the average contents (or logarithms of the contents), makes the results more graphic; however, this frequently lowers their accuracy. This is because the final result, besides incorporating an unknown error in the estimated averages, also incorporates an unknown (sometimes appreciable) error in the global average. It is, therefore, recommended that only statistically processed estimates of the concentration ratios be used as criteria.

Statistical processing of quantitative information on the distribution of indicator elements and indicator ratios in rocks, makes it possible to describe mathematically the geochemical specialization of geological complexes, and to substantitate criteria for recognizing promising regions for certain types of ores. Geochemical information, obtained as a result of analytical studies is systematically arranged in accordance with the petrographic characteristics of the sampled rocks. The resulting sets of geochemical data are then processed using conventional statistical techniques in order to estimate the distribution parameters of the contents, or logarithms thereof (depending on the distribution law). At the same time, ratios of typical indicator elements are calculated and processed statistically. The most informative ratios are usually those of the indicator elements (for example, Mg/Li, and others) whose migration tendencies have opposing directions. Depending on the geochemical characteristics of the rocks being studied, as well as on the nature of the problem being undertaken, it is possible to calculate quotient, additive, or multiplicative concentration ratios for various indicator elements, or groups of elements, relative to the global or (what is much more effective) regional averages.**

When placing a certain type of rock in a particular group one can use (in addition to the estimates of the arithmetic means and standard deviations) assessments of the probabilities that certain contents or ratios will be encountered. These are chosen by means of comparing the distribution parameters of the indicator elements and indicator ratios in known ore-bearing and barren geological complexes. This technique makes it possible, in some cases, to reduce considerably (to 10 - 30) the number of the samples necessary for an objective consideration of whether or not the rock being studied belongs to a certain type of geochemical specialization (Table 21).

Table 21. Geochemical criteria for recognizing granitoid complexes parent to deposits of rare metals (Li, Be, Sn, W, Ta)

Indicator element or indicator ratio	Estimates of distribution parameters in granitoids*					Content (ppm) or ratio used as criterion	Probability of indicator value being in the population of granitoid samples	
	barren			productive			barren	productive
	x, ppm	s	s_{log}	x, ppm	s_{log}			
Li	37	—	0,26	80	0,43	100	≤0,02	≥0,23
Sn	5	2,98	—	15	0,35	20	≤0,01	≥0,16
Mg/Li	270	—	0,34	75	0,40	30	≤0,001	≥0,27
						100	≥0,98	≤0,38
Zr/Sn	76	—	0,39	30	0,36	30	≤0,07	≥0,46

* Estimates of logarithmic distribution parameters are taken from Table 17. In the case of normal distribution, estimates of content distributions are used.

Below, using lithium as an example, one of the simplest methods for substantiating evaluation criteria is illustrated. Tables 17 and 21 list estimates of distribution parameters for lithium in granites which are related to deposits of the so-called rare elements (Li, Be,

**In making use of the multiplicative and additive indexes, one should bear in mind that the grouping and summation of the chemical elements for such transformations must have a definite geochemical meaning. It is essential, in particular, that all the elements being combined have unidirectional correlations.

Sn, Ta, and W) as well as the estimates of the parameters for granitoids which have no relation to deposits of these elements. The estimates, calculated from representative sets of data, may be processed by simple statistical procedures (see below) in order to establish statistical limits of variation in the contents within the populations being compared (in the next to the last column of Table 17 they are estimated with the probability 0.01 for the boundary values). In making the comparison between these variations, it is necessary to select one or several values of the contents which are typical of one set, and rare (close to the critical extreme values) for the other. In the case illustrated, lithium contents in the range of 25 and 100 ppm may be selected. The lower value is close to the global average, but is not typical of the parent granites of rare-metal deposits; the higher value is characteristic of the latter, but close to the critical value for barren granites. In order to quantitatively estimate the possibility of using the selected values as the criterion for exploration, we shall calculate, with the help of the Laplace integral function (Φz), the probability of lithium contents, higher than the chosen value, being in the population.

I. We shall make calculations for the value 25 ppm. $a = 25$; $\log a = 1.398$.

For barren granites:

$$z = \frac{\lg a - \lg x}{S_{\lg}} = \frac{1,398 - 1,500^*}{0,256} = -0,400. \qquad \Phi_z = -0,155,$$

(*$\lg x$ and $S_{\lg}$ values from Table 17)

where a is the selected value of the content;

z is the argument of the Laplace integral function;

Φz is the Laplace function (taken from tables).

The probability $P_{x \geqslant 25} = 0.5 - (-0.155) = 0.655$.

Consequently, the probability of samples having contents higher than 25 ppm in the barren granites is 0.655, or approximately 66%. Accordingly, 44% of the samples will have contents lower than the chosen value.

For productive granites related to pegmatite deposits:

$$z = \frac{1,398 - 1,946}{0,221} = 2,48; \qquad \Phi_z = -0,493;$$

$$P_{x \geqslant 25} = 0,5 - (-0,493) = 0,993,$$

i.e., 99% of the lithium contents in samples from this population are higher than 25 ppm, and only 1% are smaller than this value.

For productive granites related to greisen and quartz-vein deposits:

$$z = \frac{1,398 - 1,683}{0,486} = -0,586; \qquad \Phi_z = -0,224;$$

$$P_{x \geqslant 25} = 0,5 - (-0,224) = 0,724,$$

i.e., the probability of having samples with lithium contents less than 25 ppm within this population is about 28%.

II. Analagous calculations will be made for 100 ppm of lithium. $a = 100$ and $\log a = 2.000$.

For barren granites:

$$z = \frac{2{,}000 - 1{,}500}{0{,}256} = 1{,}95; \qquad \Phi_z = 0{,}474;$$

$$P_{x \geqslant 100} = 0{,}5 - 0{,}474 = 0{,}026,$$

i.e., the probability of having samples with such contents in the population being considered is quite small (about three percent).

For productive granites related to pegmatite deposits:

$$z = \frac{2{,}000 - 1{,}946}{0{,}221} = 0{,}22; \qquad \Phi_z = 0{,}087;$$

$$P_{x \geqslant 100} = 0{,}5 - 0{,}087 = 0{,}413,$$

i.e., nearly half of all the samples within this population of productive granites exhibit contents greater than 100 ppm.

For productive granites related to greisen and quartz-vein deposits:

$$z = \frac{2{,}000 - 1{,}683}{0{,}436} = 0{,}72; \qquad \Phi_z = 0{,}264;$$

$$P_{x \geqslant 100} = 0{,}5 - 0{,}264 = 0{,}236.$$

The calculated probabilities may be summarized as follows (Table 22).

Table 22. Indicator probabilities based on lithium contents in certain granites

Granite	Indicator contents, ppm	
	25	100
Barren granites:	0,44	0,03
Parent of pegmatite deposits	0,01	0,41
of greisen and quartz-vein deposits	0,28	0,24

A comparison of the data shows that a lithium content equal to 100 ppm may be used as a satisfactory criterion for grouping granitoids into productive and non-productive categories for the given group of elements. At the same time a lithium content of 25 ppm may be used only for identifying "pegmatite-bearing" granites, because the probabilities of contents smaller than 25 ppm being found in barren granites and in parent granites of greisen and quartz-vein deposits are not sufficiently distinctive.

The probabilities derived above may be used to determine the optimal size of set (i.e., the number of samples required) which will be sufficient for a reliable classification of the sample set. The tables prepared by Chernitskii (1957) may be used for the purpose. These tables relate the number of trials, ensuring an outcome of a certain random event, with a guaranteed probability of its occurrence at least a certain specified number of times, given the probability of such an event in one trial. A minimum of three to five samples with the chosen indicator value (with a probability of 0.95), could be conditionally considered optimal for the smallest possible set of samples. Table 23 presents the number of samples which should be taken for the above examples in order to have at least three or five samples with critical values of x within the set.

Table 23. Minimum size of the sample set which will contain one, three, or five samples with critical (indicator) lithium values, the probability being 0.95

Type of sample set and indicator values	Number of critical samples in the set		
	1	3	5
Barren granites:			
a) 25 ppm	5	12	20
b) 100 ppm	98	206	304
Parent granites of pegmatite deposits:			
a) 25 ppm	300	620	910
b) 100 ppm	6	13	19
Parent granites of greisen and quartz-vein deposits:			
a) 25 ppm	9	20	30
b) 100 ppm	11	24	35

It follows from the data of Table 23, that if it is found possible to have at least three samples in the population with indicator contents of lithium equal to 100 ppm or higher, then the optimal size of the geochemical sample set which is necessary to ensure the reliable identification of potentially metalliferous or barren granitoids, is 24 samples. The probability of two samples showing 100 ppm or more lithium in the set of samples from barren granitoids is only 0.015, and for three samples it is less than 0.005. In order to have at least five critical values in the sample set representing productive granites, the size of the set in this case should be increased to 35 samples.

Small-Scale (Regional) Geochemical Mapping as a Method for Studying Geochemical Specialization During Exploration

Geochemical specialization of geological complexes is studied during geochemical exploration with a view toward solving specific problems which usually arise during the early reconnaisance stage. As previously mentioned, an assessment of the ore producing potential in igneous, metamorphic, or sedimentary complexes makes it possible to delineate the most promising areas in which to conduct further, more detailed, surveys. However, this method of prediction cannot be considered a direct method of prospecting. Conditions which determine its application, in combination with other specific exploration techniques, will be considered below. It should be emphasized that methods of this type are especially important for studies concerned with the assessment of ore potential in areas covered with thick sediments transported from distant sources, which is the case within many platformal regions. In such cases, conclusions which determine the direction of subsequent geological prospecting are often based on core samples from isolated single boreholes; therefore, the proper geochemical evaluation of the ore potential with respect to a certain massif or complex intersected under a thick sedimentary series, by one or several boreholes, is of great economic significance.

Geochemical specialization of geological complexes is usually studied during exploration by means of geochemical mapping of bedrock at a scale of from 1:200,000 to 1:50,000. It is most desirable to do geochemical mapping concurrently with geological surveys. In those regions which are already mapped, readily available geological information may be used. On the basis of small-scale geochemical mapping of the bedrock and on geological maps of appropriate scale, "geologic-geochemical prediction" maps are produced.

Subdivision of geological complexes in the area being studied, based on the different types of geochemical specialization, is the main objective of the small-scale geochemical mapping of the bedrock. The specific exploration problems which may be solved can be summarized as follows:

1) recognition of potentially metalliferous geological complexes, with the objective of conducting more detailed exploration work towards finding particular mineral deposits;

2) recognition within the area being mapped of potentially metalliferous geological features (zones and facies of metasomatically altered rocks, tectonic dislocations, zones of jointing, halos beyond the contacts of intrusive bodies, etc.), and among them zones especially recommended for further studies;

3) evaluation of the distribution parameters of those elements which are indicators of mineralization in rocks, with the objective of analyzing changes in the geochemical background, and subsequently using these data in detailed exploration for mineral deposits based on primary geochemical halos.

The results of small-scale (regional) geochemical mapping of bedrock may also be used for solving the various geological problems described briefly above.

The work carried out during the small-scale geochemical mapping intended for studying geochemical specialization of rocks, includes the following procedures (Beus, 1966); a) geological description of the geological sampling sites with a very thorough investigation of the geological factors affecting the distribution parameters of the chemical elements (grade of metamorphism, intensity of superimposed processes, extent and character of the supergene alteration, etc.); b) geochemical sampling of the bedrock; c) analytical study of the selected samples; and d) processing of the results of the analyses, and the preparation of a geologic-geochemical map.

The detail of geochemical mapping depends on the scale of the geological or topographic map, and on the accuracy necessary for the evaluation of the distribution parameters of the chemical elements which characterize each type of rock noted on the geological map. Therefore, the direct correlation between the size of an outcrop and the number of geochemical samples collected from it, is not a decisive principle of sampling. Regardless of the area it occupies, each rock type indicated on the map, taking into account the scale of mapping, must be characterized with the same degree of accuracy in terms of the distribution parameters of the chemical elements which may be used as indicators of the presence of ore. The major factor in determining the quantity of samples collected from a given type of rock during small-scale (regional) geochemical mapping is the degree of irregularity in the distribution of the indicator elements. This can be evaluated by the variance of the distribution of the element contents, or by the coefficient of variation.

Generally, average estimates having an accuracy equal to $\pm 20\%$, and determined at the

5% significance level, are considered sufficient for the purposes of small-scale geochemical mapping. The number of samples required for this accuracy depends on the coefficient of variation of the contents of the corresponding elements, and is expressed by the following formulas: for an accuracy $\pm 20\%$, $n20\% = v^2/100$; for an accuracy $\pm 30\%$, $n30\% = v^2/200$, where n is the number of samples sufficient for the specified accuracy, and v is the coefficient of variation of the indicator element contents.

Geochemical sampling with an accuracy equal to $\pm$ 30% of the arithmetic mean is permissible in some cases when semi-quantitative spectrographic analytical methods are used. For those indicator elements whose coefficients of variation of the contents in a given rock type do not exceed 60%, 30 samples are usually sufficient to provide the required accuracy of $\pm$ 20% (this category includes the majority of trace elements in common rock types). When semi-quantitative spectrographic analytical methods are used, the coefficient of variation sometimes increases to 100% due to random errors introduced by the analytical procedure. In such cases, the number of geochemical samples taken from the given rock types is increased to 50. Based on available information, the number of samples indicated by the above formulas is valid for areas as large as 50 km^2 in the case of igneous rocks, and 200 km^2 in the case of sedimentary rocks. If the area of exposure of the given rock type is larger, then the number of the samples collected should also be increased.

During geochemical mapping conducted concurrently with a geological survey, field observations and geochemical sampling are performed on essentially the same traverses, cross sections, pits and observation sites.

The bedrock is sampled along traverses oriented in the direction of maximum variability of the complex or rock type being studied (usually across the strike of the massif, suite, etc.). It is recommended that geochemical profiles be selected in such a way that they intersect all facies, or phase varieties, of the rock type being studied, even including those which are not indicated on the geological map at the particular scale. This should not affect the uniform nature of the sampling.

During small-scale (regional) geochemical mapping, samples are also collected from any zones of altered rocks, zones of tectonic dislocations, veins, and other geological formations which may be encountered even though they may not have been included in the mapping scale. This is done because such rocks may be of interest from the point of view of revealing ore concentrations. Three to five geochemical samples are taken from each of these features. They are subsequently studied separately and are not included in the set of samples characterizing the rock type being mapped.

It is now appropriate to discuss the weight of the specimens collected, and the method of geochemical sampling used in small-scale geochemical mapping. Experimental studies have shown that the optional weight of a geochemical sample is 100 to 150 grams.* The method of sampling is based on a grid system.

In order to select the indicator elements, all samples are analyzed by the semi-quantitative emission spectrographic method which determines a reasonably large number of trace elements. The solution to any problem which may arise from the geochemical mapping usually requires a specific combination of various indicator elements. There are two limit-

*The weight of a geochemical sample from which accessory minerals are to be studied, may reach 1 kg. In this case the portion to be analyzed, weighing 100 g, is quartered after the sample is ground to a 1 mm size.

ing cases which determine the methods used to process the results of the semi-quantitative emission spectrographic analysis:

1) previous studies in the area have established those elements which indicate the presence of ore, the age, etc. (depending on the nature of the problem);

2) indicator elements necessary for the solution of problems which arise during geochemical mapping of a given area are unknown.

The first case is the simplest. Using all available data, mathematical processing of the analytical results is performed for the particular indicator elements, in order to answer unequivocally the question which has arisen. This is done by means of comparing the distribution parameters of these elements with the known standards. Sometimes precise determination of the parameters may require additional spectrographic, spectrophotometric, or even chemical analyses.

In the second case, the semi-quantitative emission spectrographic determinations of all trace elements are processed mathematically. Then, several elements whose distribution parameters in the several rock types differ most, are chosen. If more accurate data on the indicator elements are necessary, then the geochemical samples are analyzed for these elements quantitatively. Usually two to four indicator elements are sufficient for unequivocally solving the problem of similarity, or dissimilarity, of rock types.

It should be stressed that the significance of the indicators increases greatly if there is some correlation between them. This is why the selection of correlatable elements as indicators, is a major problem in the processing of the analytical data. In this case, the classification of rocks may, if necessary, be based on the additive and multiplicative indexes derived by means of grouping those elements which have a positive correlation.

In the final stage of geochemical mapping, the geochemical parameters characterizing the distribution of the chemical elements in rocks are plotted on a structural-geological basis, i.e., geologic-geochemical maps are prepared. Geologic-geochemical maps are expected to contribute to the solution of very specific geological problems, the major one being the prediction of the presence of ore in the geological complexes within the given area. It is for this reason that geologic-geochemical maps are prepared which show the distribution of those elements in the rocks which may be used as indicators of the ore producing potential in the geological complexes.

The following types of practical geologic-geochemical maps may be prepared from the results of small-scale (regional) geochemical mapping of bedrock.

1. Geochemical specialization maps of geological complexes:

   a) monoelement;

   b) multielement.

2. Geochemical specialization maps of tectonic zones and postmagmatic formations.

Monoelement geochemical specialization maps of geological complexes are usually transparent overlays, drawn with dashed lines, placed on top of structural-lithologic maps. For each of these overlays, colors or legends within the rock types delineated on the map, illustrate the distribution parameters of a typical chemical element chosen as an indicator of the ore potential.* Indicator criteria for the legend may be: a) the average contents of the

*Maps of geochemical specialization may also be drawn on photocopies or blueprints with the geological data shown by dashed lines.

element; b) the range of variation of the contents determined with a specified probability; c) the concentration ratios relative to the average value of the geochemical background; d) the coefficients of variation of the contents; etc. The criterion depicted on the geochemical specialization map is expected to advantageously describe specific features of the geological formations.

Multielement geochemical specialization maps show the distribution in rocks of geochemically related pairs of elements, their ratio (for example, for granitoids: magnesium - lithium, potassium - rubidium, zirconium - tin; for basic and ultrabasic rocks: nickel - sulfur, etc.), or correlation coefficients of the indicator elements, etc. For those elements whose contents are characterized by direct correlation, multicomponent maps may be prepared, showing the products of the logarithms of the contents of the correlatable elements, or the products of the concentration ratios of these elements, and so on. Multielement geochemical specialization maps are usually prepared in those cases where geochemical mapping problems cannot be solved using monoelement maps.

Geochemical specialization maps of tectonic zones and postmagmatic formations are prepared on a transparent overlay, or directly on structural-lithologic base maps. The base maps, in addition to tectonic zones, also show postmagmatic features (vein lodes, zones of hydrothermal alteration, etc.), even though they would not normally appear at the scale of mapping.

Symbols for those trace elements whose contents in the given formation differ considerably from the geochemical background, are indicated in colored ink next to the tectonic zones and postmagmatic features.

The types of practical geologic-geochemical maps listed above together with maps of the geochemical dispersion trains, form the basis for the *metallogenic prediction map*. The latter is the main objective of geochemical studies.

During the preliminary preparation stages of the metallogenic prediction maps, complexes and rock types are identified on the geologic map. These may be assessed as promising indications of the presence of ore, based on the element distribution patterns. This makes it possible, with the help of structural criteria, to delineate areas to be recommended for detailed prospecting with the objective of locating specific mineral deposits.

Two different cases may arise during the preparation of the geologic-geochemical (metallogenic prediction) map:

1) The geochemical criteria used for the identification of potentially metalliferous geological complexes are known from exploration experiences in other regions (when mapping at a 1:50,000 scale, such criteria may be established based on results from mapping at scales of 1:200,000 or 1:100,000);

2) The geochemical criteria for recognizing potentially metalliferous geological complexes are not known.

An example. One of the problems to be solved by the metallogenic prediction map is the recognition of promising areas from the point of view of exploration for tin deposits.

The following geochemical criteria for recognizing tin-bearing granitoid intrusive complexes are known from previous studies:

a) the probability of tin contents being equal to 10 ppm or higher is 0.82;

b) the probability of tin contents being equal to, or less than, 5 ppm is 0.03; the arithmetic mean of the tin content in a tin-bearing massif is 15 ppm or higher.

Using the distribution parameters of tin derived from the results of sampling granites in the region covered by mapping, it is possible to calculate from the available geochemical samples the probabilities of the tin contents being greater than 10 ppm and smaller than 5 ppm. Samples corresponding to potential tin-bearing granites are identified, and then the promising massifs are delineated on the prediction map by appropriate colors or cross-hatching.

If the geochemical criteria for the presence of ore are not known, then it is necessary, during the preparation of the prediction map, to determine the geochemical parameters (criteria) which may be helpful in revealing potential metalliferous geological complexes.

To accomplish this, typical geological complexes in which ore deposits are known to occur in genetic or paragenetic relationship within the territory being mapping, are studied by geochemical techniques. Results of the geochemical sampling are used for estimating the distribution parameters of the ore elements and their associates. The estimates serve, subsequently, as standard values for comparison with the distribution parameters of the indicator elements in other complexes of similar composition located throughout the mapping area. Complexes or rock types with geochemical characteristics close to above-determined standard, are classified as promising.

The data obtained from studies of geochemical specialization in tectonic zones are an important constituent part of the total information used for constructing the geochemical foundation of metallogenic prediction maps.

Analysis of these data makes it possible to distinguish, within each specific tectonic zone, a typical complex of ore elements and their associates which are indicative of the ore producing potential within a given structure. Plotting these ore associations on the prediction map permits the possible assessment of tectonic zones from the point of view of directing detailed prospecting to particular types of mineral deposits.

Consequently, the geologic-geochemical basis for the metallogenic prediction maps is the structural geologic map on which special symbols are shown for:

a) geological complexes recognized as potentially promising for specific mineral deposits, based on explicit geochemical criteria;

b) promising tectonic zones which have indications of ore elements, and in which economic concentrations may be encountered.

An essential component of geologic-geochemical prediction maps is information on the anomalous contents of ore elements in stream sediments obtained from drainage basin surveys (dispersion trains of ore elements).

The use of geochemical information in the preparation of metallogenic prediction maps contributes to a better justification for any prediction. A geologist who makes the most of this information is able to support his conclusions with a rigorous quantitative foundation, proceeding from the objective distribution laws which explain the concentrations of chemical elements within the Earth's crust.

5

Primary Geochemical Halos of Mineral Deposits: Their Use in Prospecting

Introduction

Primary geochemical halos are zones surrounding metalliferous deposits or ore bodies, which are enriched or depleted in several chemical elements, as a result of the introduction or redistribution of these elements during the process of ore formation.

Both primary geochemical halos and zones of rock alteration surrounding mineral deposits are genetically related to the phenomena of ore formation, and it is not always possible to draw a sharp line of demarcation between them. For example, zones of sericitization of ore-bearing rocks may also be regarded as primary geochemical halos of potassium, and zones of albitization may be regarded as primary geochemical halos of sodium. This dual classification has long been recognized and it has acquired a certain practical significance. Primary geochemical halos which are genetically related to ore bodies, are expressions of the basic geochemical principles underlying the processes of rock alteration in the zones surrounding mineral deposits. However, in many cases, these processes do not result in any appreciable alteration in the mineral composition. On the other hand, zones of altered rocks usually are characterized by a particular type of mineralogical alteration caused by the metalliferous solutions.

As mentioned previously, ore bodies or deposits usually are surrounded by halos which represent either the introduction or the removal of chemical elements. Many studies have shown that the former type of halo is much more common. Besides, they are usually large and are, consequently, of greater practical significance. A great practical significance of those halos, which results from the introduction of elements, is that they are composed of the elements which are typomorphic for the ore bodies. Therefore, they are direct indicators of the ore bodies. On the other hand, the removal of elements could have taken place either in the stage of ore formation, or during the processes of alteration of the enclosing rocks prior to (or following) the mineralization. At present, there are no reliable criteria by which the relationship between halos characterized by the removal of chemical elements and their associated ore bodies can be utilized and this, naturally, reduces their practical value. We will, therefore, discuss mainly the characteristics, formation, and problems involving the practical utilization of halos formed by the introduction of chemical elements. The criteria and the methods for the interpretation of primary geochemical halos are determined by the distribution patterns of the chemical elements in the regions surrounding ore bodies. For this reason, we will consider the main laws governing the composition and structure of primary geochemical halos based on examples from several mineral deposits.

Procedure for Studying Primary Halos

Sampling, Processing, and Analysis of the Samples

The results obtained from studying the distribution patterns of the chemical elements around known ore bodies or mineral deposits are the basis for the development of the techniques which use geochemical halos in the search for ore deposits. These systematic studies usually consist of geochemically sampling ore bodies and their enclosing rocks on the surface and in cross section. The results obtained from processing the analytical data are used for determining the compositional and structural characteristics of the geochemical halos. These are subsequently used in interpreting other geochemical anomalies discovered during geological exploration.

Primary geochemical halos are studied by means of sampling selected cross sections or profiles, usually oriented across the strike, of both the enclosing rock and the ore. The sampling grid is planned in such a way that each potential geochemical anomaly is crossed by at least two profiles. The profiles may have different lengths, however, they should extend beyond the areas of any altered rock. Geochemical samples are taken along the profile or along the wall of a working section, or from cores obtained during exploratory drilling, at an interval of 5 to 10 meters. Geochemical sampling of bedrock must be accompanied by a detailed geological description of all the intervals sampled.

Exploration for ore deposits using primary geochemical halos, or any other type of halo, requires the selection and the analysis of a great number of samples. Therefore, the amount of labor involved in the sampling must be considered in the choice of the sampling technique.

The so-called *chip-channel* method is the main sampling technique. It consists of selecting, at a specific interval, 5 or 6 small broken fragments which are combined into one sample (each fragment is approximately 3 to 4 cm^2 across). This method is not only fast, but it has another significant advantage. It is much more sensitive than channel or bulk sampling to anomalous contents of indicator elements, i.e., halos delineated by this method are much more extensive.

All the samples chosen are ground to 1-mm particles by means of a jawbreaker. One fourth of the material, after grinding and quartering, is pulverized and analyzed. It is essential to ensure that the samples are not contaminated during the processing. The samples are analyzed by the semi-quantitative emission spectrographic technique in order to determine a wide range of chemical elements.

In order to ensure reliable results from the semi-quantitative spectrographic analysis, it is necessary, in addition to both external and internal laboratory controls, to check the reproducibility of the results by means of the repeated analysis of 10% of the samples from each batch. The analytical data are then further processed only if the results of the reproducibility tests are satisfactory; the procedure is explained in the "instructions" (*Instructions On Geochemical Exploration Techniques,* 1965).

Methods of Detecting Primary Halos

The basis of geochemical exploration for ore deposits consists of discovering, and subsequently interpreting, geochemical anomalies (including primary halos in bedrock) which are the result of the processes of ore formation.

Primary geochemical halos are detected and delineated with the help of conventional techniques which are based on the comparison between the areas being studied, and the background distribution parameters of the elements. In order to calculate these parameters and to determine the minimum anomalous (threshold) contents of the chemical elements for each variety of enclosing rock, the results of geochemical bedrock sampling in specially selected background areas are generally used. These background areas must be far removed from the ore bodies or deposits and must not be affected by mineralization.

Primary halos of the chemical elements are accurately determined both in plan, and in cross section, on the basis of the minimum anomalous contents calculated at the 5% significance level.

Study of many ore deposits has shown that indicator elements form primary halos of different size, depending on their composition and other characteristics of the mineralization. For example, the vertical extent of primary halos of certain elements in polymetallic skarn deposits exceeds hundreds of meters, whereas in some different types of mineral deposits (such as gold ores), primary halos of the same elements are so small and weak that it is difficult to recognize them during geochemical exploration. The development of special methods for the reliable detection of weak geochemical halos is, therefore, a very pressing problem.

Of particular interest are the factors which control the size of primary geochemical halos. Let us first consider the distribution characteristics of the elements in primary halos. Morris and Lovering (1952) established that as one goes away from an ore body, the concentrations of lead and zinc in primary halos decrease very close to exponentially, based on a study of lead and zinc deposits in the Tintic district of Utah. Graphs of the distribution of the elements, plotted in a semilogarithmic coordinate system, approached straight lines.

However, this pattern in primary halos associated with hydrothermal deposits is, in most cases, masked by the irregular distribution of the elements. It becomes reasonably distinct only after the data are properly smoothed.

Figs. 4 and 5 illustrate the distribution of several elements around ore bodies, plotted in a semilogarithmic coordinate system. The "smoothed" graphs are very close to straight lines (smoothing was performed by the "sliding" window (moving average) technique, the window consisting of three samples).

Analysis of the plots of elemental distributions (Fig. 5) shows that the size of a halo is directly related to the minimum anomalous content (threshold) and to the gradient of the concentration. The *gradient of concentration* is equal to the tangent of an acute angle, α, formed by the distribution graph and the x-axis, and it is determined by the ability of the elements to impregnate rocks enclosing ore bodies (Grigorian, 1963).

Solovov (1966) later suggested that the term *mobility of an element* be used. It is characterized quantitatively by $1/\lambda$, where $\lambda = \tan \alpha$.

It follows from the above that the only way to enhance weak halos is by reducing the interfering effect of the background content, as this interfering effect limits the size of the halo.

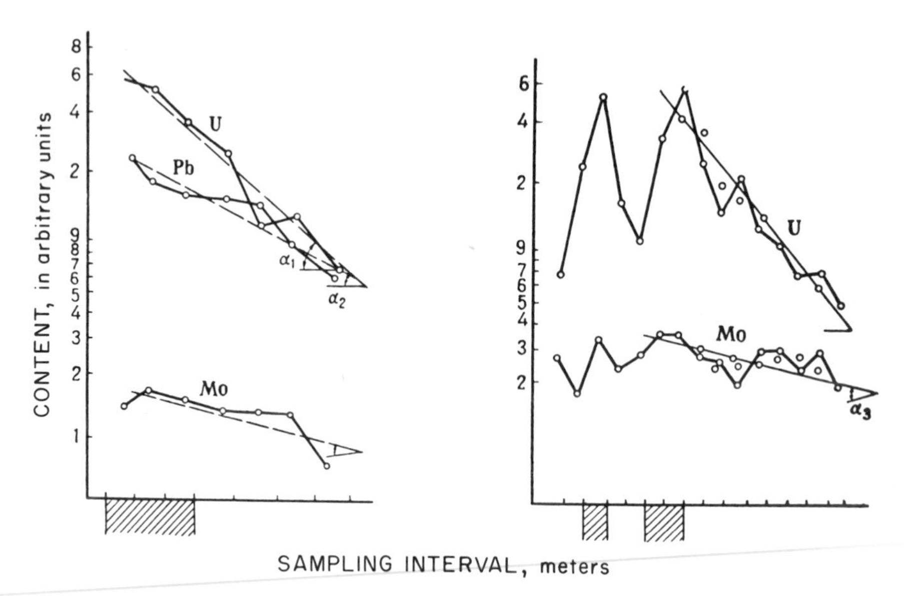

Fig. 4. Empirical and smoothed graphs of element distributions around ore bodies. (Note: Vertical stripped zones are mineralized areas).

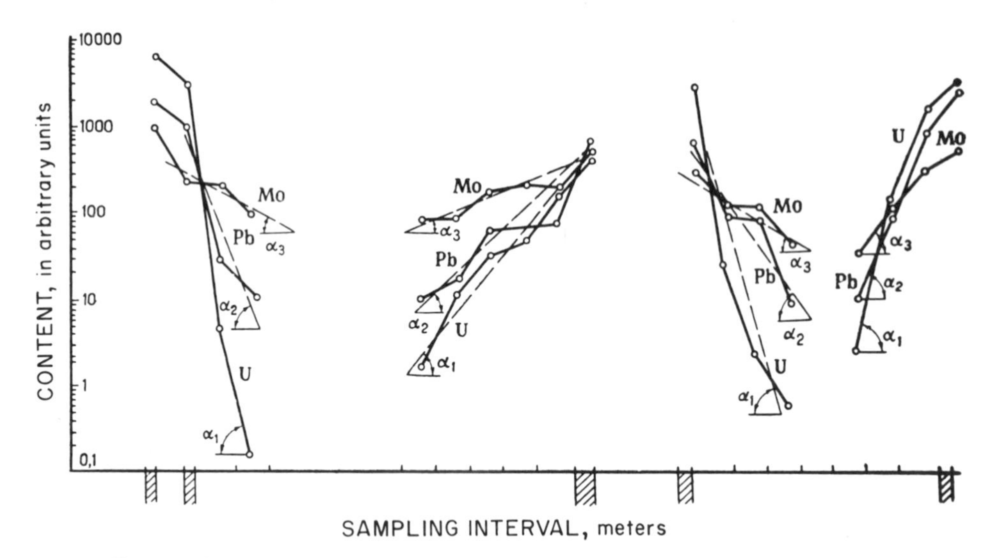

Fig. 5. Empirical and smoothed graphs of element distributions around ore bodies. (Note: Vertical stripped zones are mineralized areas).

Methods of Enhancing Primary Halos

We will now describe four methods of enhancing primary halos, based on reducing the interfering effect of the background distribution of the indicator elements in the enclosing rocks.

1. Method based on the quantitative analysis of geochemical samples.

The value of the minimum anomalous content (threshold), on the basis of which a halo is delineated, depends on the *major distribution parameters* of the chemical elements in the enclosing rocks. *The major distribution parameters are the arithmetic mean, the variance, and the standard deviation.* The average content (the arithmetic mean in the case of normal distribution, or the arithmetic mean of logarithims in the case of lognormal distribution) is usually calculated using a sufficiently large number of samples, and it is a value very close to the actual mean for the population. The standard deviation characterizes variations in the contents of the elements. These variations are due both to the non-uniform distribution of the elements in the rocks being studied, and to errors introduced by sampling, processing, and analysis of the geochemical samples. Thus:

$$S^2 = S^2_{ntr} + S^2_{smpl} + S^2_{prc} + S^2_{an}$$

Where S^2 is the total variance of the contents;

S^2_{ntr} is the natural variance caused by the non-uniform distribution of the element content;

S^2_{smpl} is the varance due to sampling;

S^2_{prc} is the variance due to the processing of the samples;

S^2_{an} is the analytical variance.

Emission (semi-quantitative) spectrographic analytical methods are usually used for geochemical samples and these methods may introduce a large random error, sometimes reaching 60% or more. Because of this, the halos being delineated turn out to be smaller than they really are, as a result of the fact that the relatively large value of the analytical variance results in higher minimum anomalous contents. Quantitative analytical methods, in comparison with semi-quantitative methods, have a much lower analytical variance. Thus, the use of quantitative methods for the analysis of geochemical samples might be expected to help in detecting larger primary halos of the indicator elements, and to reveal deeply hidden ore bodies. However, experience has shown that the additional size of the halo which results when quantitative methods are used, is not great enough to justify the use of these more costly techniques. Fig. 6 compares the size of a primary lead halo detected by both semi-quantitative and quantitative analytical methods. The quantitative method increases the size of the halo by only 15%.

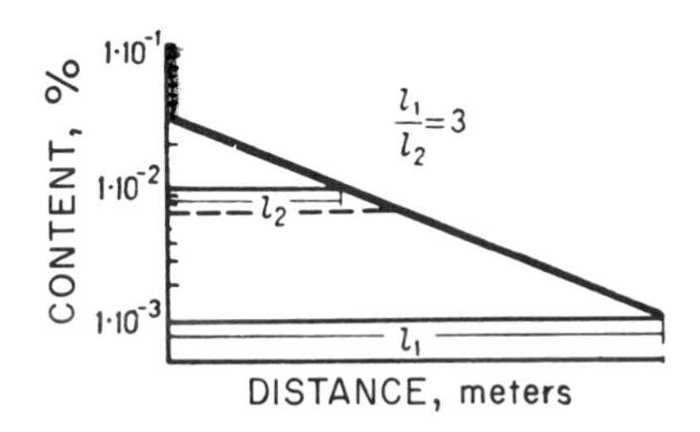

Fig. 6. Comparison between the sizes of lead halos detected by partial chemical analysis (l_1), semi-quantitative emission spectrographic analysis (l_2), and quantitative emission spectrographic analysis (dashed line).

2. Techniques using partial chemical analysis of geochemical samples.

In those cases where the distribution patterns of the indicator elements in the enclosing rocks (syngenetic dispersion) and halos (epigenetic dispersion) differ, it is possible to enhance the size of a geochemical anomaly. To accomplish this, the effect of the syngenetic dispersion (the background) of the element, which makes the accurate representation of the halo difficult, can be reduced to a minimum with the help of partial chemical analysis of the geochemical samples. The results obtained by using these methods to achieve a better delineation of the primary halos are illustrated below by examples of uranium and lead, which are the main indicators in hydrothermal uranium deposits.

Uranium. The partial analysis method for uranium takes into account the form in which primary uranium halos occur in enclosing rocks (in the background regions). It has been found that uranium introduced into enclosing rocks as a result of the formation of ore bodies, generally occurs in the form of its own minerals, that is, in the form of uraninite and sooty uraninite (pitchblende). These minerals are readily soluble when samples containing them are treated with a 2% soda (Na_2CO_3) solution (with some added hydrogen peroxide to oxidize the tetravalent uranium). This method makes it possible to determine the content of the so-called mobile uranium in the samples, and it has been described in detail in several publications (e.g., Grigorian and Yanishevskii, 1968). It has been shown experimentally that only insignificant amounts of syngenetically dispersed uranium are extractable by this soda solution from granite samples collected beyond the uranium halos. This means that primary dispersed uranium was present in the granites mainly in forms which are insoluble in a soda solution (such as in accessory minerals, etc.). Because of this, the minimum anomalous contents of the mobile uranium ($2 \cdot 10^{-4}\% = 0.0002\% = 2$ ppm) was found to be appreciably smaller than the minimum anomalous content deduced from the total content of uranium in the samples ($1.6 \cdot 10^{-3}\% = 0.0016\% = 16$ ppm). (The distribution of uranium in both cases obeyed the lognormal distribution law. The minimum anomalous content was determined at the 5% significance level.)

Plots have shown that the use of partial chemical analysis helps to reveal halos of mobile uranium, and these halos are much larger than those based on the total uranium content in the samples. In the case illustrated in Fig. 7, the halo occurs over a hidden ore body with a

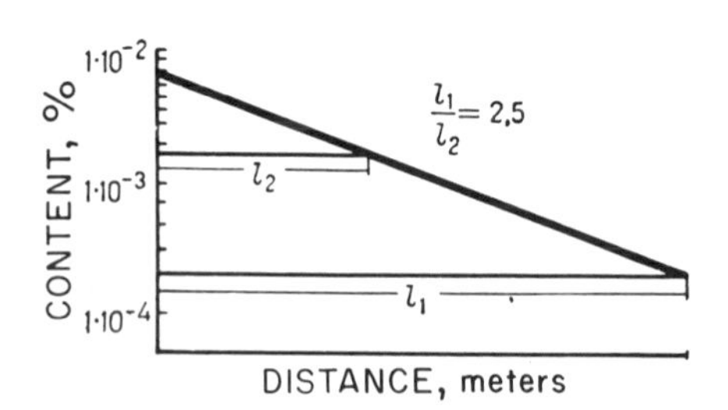

Fig. 7. Comparison between the sizes of uranium halos detected by the partial analysis technique (l_1), and by determining the total uranium content in the samples (l_2).

maximum anomalous uranium amounting content of 0.007% (in the center of the halo). Thus, the partial analysis method made it possible to delineate a halo of mobile uranium 2.5 times as large as that delineated by determining the total uranium content in the samples collected.

Fig. 8 shows a plot of the vertical variation of the linear productivity [for definition see Glossary] of the mobile uranium halo, which is plotted on the x-axis using a logarithmic scale. The y-axis represents distances from the hidden ore body using a linear scale. The graph is essentially a straight line. In order to determine the vertical extent of the uranium

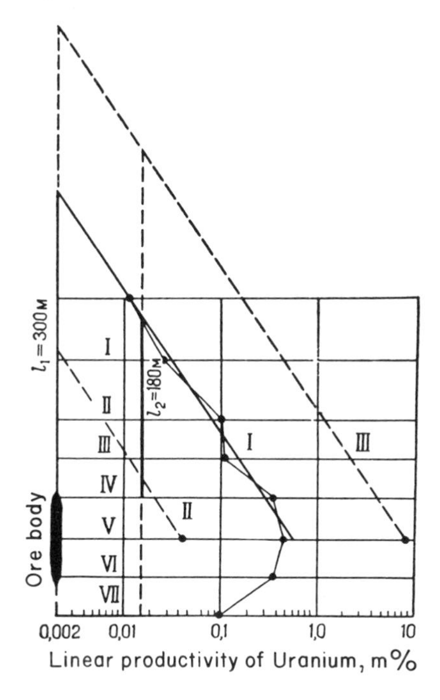

Fig. 8. Comparison between the vertical extents of uranium halos over a hidden ore body, detected by different methods (the ore body is shown by dark shading). l_1 is the extent of the mobile uranium halo; l_2 is the extent of the halo as determined by the total uranium content.
(The I, II . . . VII in the vertical direction are levels below the surface. The other I, II, and III are exploration or mining horizons.)

halo over an ore body, the plot of linear productivity of the halo is extended until it intersects with the line parallel to the y-axis, which corresponds to the minimum anomalous (threshold) productivity. A halo characterized by both a minimum width and a minimum content equal to a threshold content has such a productivity. The minimum width of a halo which can be detected during geochemical prospecting is equal to the sampling interval (5 meters in the given case). However, in order to avoid the possibility of error in this particular case, the minimum width selected was equal to 10 meters, because the halo must be intersected by at least two samples. If at least two samples do not intersect an anomaly, then high analytical data from any one sample, which may be due to random factors such as fluctuations in the background concentrations of the elements, will be erroneously considered as anomalous. Consequently, the minimum anomalous (threshold) linear productivity of the mobile uranium halo will be 0.0002 • 10 = 0.002 m% (meter percent), whereas for the halo detected on the basis of the total uranium content it will be 0.0016 • 10 = 0.016 m%.

It follows from Fig. 8 that the extent of the uranium halo (based on extrapolation of the total content), is equal to 180 meters (the halo does not reach the surface), whereas the extent of the mobile uranium halo is 300 meters. Thus, by using partial chemical analyses, it is possible, in this case, to recognize hidden ore bodies, with the help of primary uranium halos, at depths at least 1.5 times greater than before.

The above conclusion has been confirmed by the results obtained from drilling such halos. Fig. 9 illustrates the uranium halos obtained by drilling a hidden ore body. The halo around the ore body had been outlined in two ways: (1) by determining the mobile

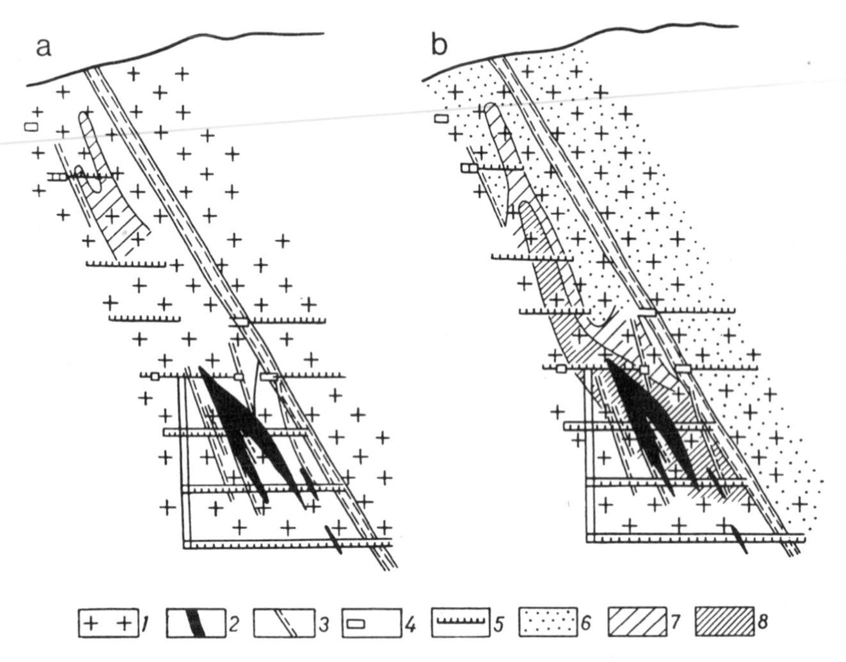

Fig. 9. Primary uranium halos around hidden ore bodies. *a* — as revealed by determining the total uranium content in the samples. *b* — as revealed by the partial analysis method.
1 - granite; *2* - ore body; *3* - dislocation (fracture); *4* - underground workings and boreholes; *5* - sampling interval; *6* to *8* - uranium contents, % (*6* - 0.0002 to 0.0005; *7* - 0.0005 to 0.005; *8* - more than 0.005).

uranium; and (2) by determining the total uranium in the samples. In order to reduce the number of analyses, the total uranium content was determined only on samples collected at the surface and along the first horizon (of the vertical profile).

Comparison shows that the mobile uranium halo is wider and longer than the halo which can be detected from the total uranium content. In all samples collected at the surface, the contents of the mobile uranium are anomalous, whereas the total uranium contents in the same samples are less than the minimum anomalous content. In other words, the uranium halo at the surface of the given cross section is not detectable by the total uranium content, and its width along the first horizon is extremely small. This implies that the hidden ore body shown in Fig. 9 could not be detected during exploration using uranium halos based solely on the total uranium content in the samples.

Fig. 10 illustrates uranium halos detected in the vicinity of two closely spaced hidden ore bodies which are represented at the surface by only two samples with minimum anomalous total uranium contents. At the same time, many more of the samples collected at the surface are from sites within the mobile uranium halo. In addition, a second anomalous zone of mobile uranium was detected on the surface directly over the ore bodies. Thus the mobile uranium zone is wider than the halo detected solely on the basis of the total uranium content.

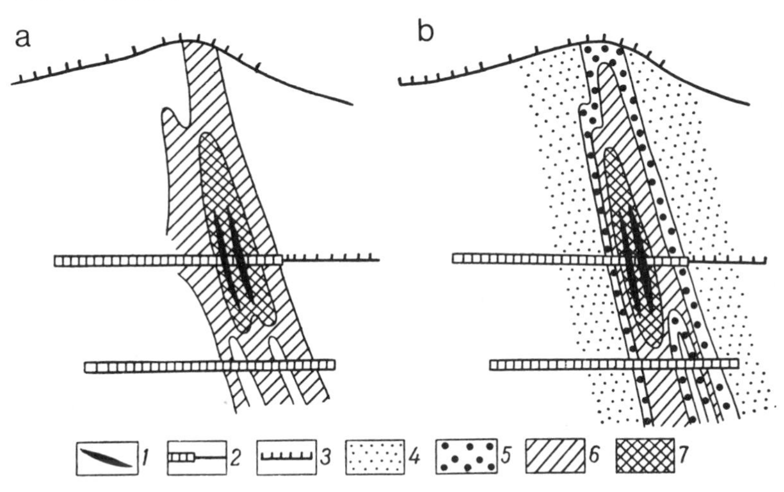

Fig. 10. Primary uranium halos around hidden ore bodies. *a* — as revealed by determining the total uranium content in the samples. *b* — as revealed by the partial analysis method.
1 - ore body; *2* - underground workings and boreholes; *3* - sampling interval; *4* to *7* - uranium contents, % (*4* - 0.0002 to 0.0005; *5* - 0.0005 to 0.0015; *6* - 0.0015 to 0.005; *7* - more than 0.005).

The partial analysis method for the chemical analysis of uranium is much simpler than the analytical method used for the determination of total uranium. This makes it possible to carry out many analyses for mobile uranium, these analyses being essential for geochemical exploration.

Lead. The lead halo outlined by means of emission spectrographic analysis based on the determination of the total lead content, was smaller than the mobile uranium halo. The lead halo was not detected at the surface. The use of lead has been suggested as an indicator of zonality for uranium in the type of deposit being described. It is supposed to enable the erosion level of the primary halo to be determined. The absence of lead in a surface halo is supposed to indicate a deep level of erosion, and the absence of hidden ore bodies at depth. However, this holds true only when lead and uranium halos have identical dimensions over the ore body. Thus, if the lead halos are smaller, then uranium halos alone, without any confirming anomalous contents of lead, will be detectable in parts of halos most distant from hidden ore bodies; this [the absence of lead] may result in erroneous conclusions about the absence of hidden mineralization at depth. It is, therefore, necessary to look for the lead halos in greater detail.

Galena is an important source of lead in endogenic ore halos. However, Tauson (1961) reported that the bulk of lead in granites is present as isomorphous admixtures in potassium feldspars. In order to determine how lead occurs in the samples being considered, the method recommended by Tauson for the processing of samples was used to selectively dissolve galena. The dissolution procedure involves the use of hydrochloric acid with a concentration of 1:50 (by volume), to which common salt (concentration 1 g/liter) is added. It has been shown experimentally that hydrochloric acid with this concentration does not destroy the crystal lattices of the rock-forming minerals, but at the same time, it is a good solvent for galena. The addition of salt to the solvent favors the retention of lead in solution owing to the formation of a complex compound of the Na $[PbCl_4]$ type, which is more soluble than $PbCl_2$. The content of lead was determined by the dithizone method.

Statistical processing of the analytical results obtained from samples outside the halos made it possible to determine the minimum anomalous contents of easily extractable lead (it is $1 \cdot 10^{-3}\% = 0.001\% = 10$ ppm). This is one order of magnitude less than the minimum anomalous content determined using the total lead content, which is equal to $1 \cdot 10^{-2}\%$ or 0.02% (the minimum anomalous contents were calculated at the 5% significance level).

The dimensions of the halos detected by the partial extraction method are compared for lead and for uranium in Fig. 11.

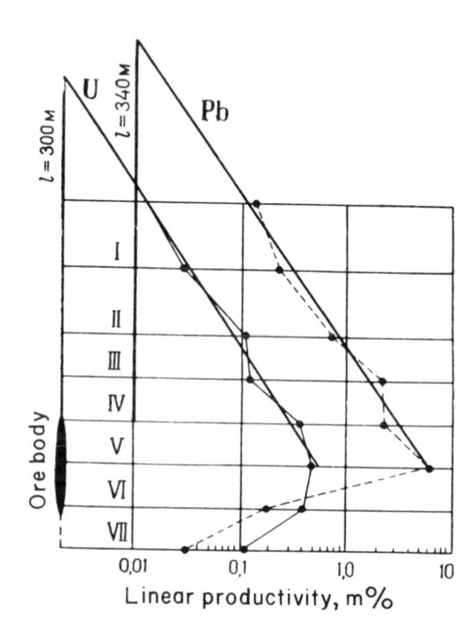

Fig. 11. Comparison between the sizes of the uranium and lead halos detected by partial analytical methods (the ore body is shown by dark shading).

In the case of the primary lead halo, the minimum anomalous content of lead equals $1 \cdot 10^{-2}\%$ (or 0.02%); thus, the application of partial analysis makes the halo 3 times as large as that delineated by the emission spectrographic (a total) analytical method (see Fig. 6).

The use of partial analyses is especially justified for those elements whose halos are neither sufficiently distinct nor large. Fig. 8 shows, for example, that the use of partial analysis makes it possible to extend the uranium halo, having a maximum linear productivity, by 1.2 times vertically. The vertical extent of the halo with a minimum linear productivity can be increased almost by a factor of 10.

3. Analysis of the heavy fractions from geochemical samples.

This method of enhancing anomalies is based on the selective concentration of the heavy fractions from geochemical samples. These samples are assumed to have come from primary halos which contain indicator elements introduced by metalliferous solutions, and which were deposited within the halos in the form of heavy minerals (sulfides, etc.). This is illustrated below by the results obtained from studies carried out at a uranium-sulfide deposit occurring in granites.

In order to obtain quantitative data on the heavy mineral concentrates, and to ascertain the optimal conditions for using this method to study halos, the distribution of elements accompanying uranium (i.e., lead, molybdenum, copper, and zinc) were compared in different size fractions from granite samples. The weight of each sample was about 2 kg. The samples were crushed to fragments smaller than 1 mm, and they were then sieved to various particle sizes ($-1+0.6$; $-0.6+0.4$; $-0.4+0.25$; $-0.25+0.1$; -0.1 mm). The resulting material was further divided into light and heavy fractions using bromoform which has a density of 2.8. A selective enrichment of the indicator elements was found in the heavy fractions. This enrichment was especially pronounced in those samples collected from geochemical halos, and particularly in the sample classes with the smallest particles.

In addition to data on the contents of the indicator elements, the yield of each individual fraction (expressed as the weight percent of the entire sample being analyzed), is another quantitative index of the accumulation of an element in a specific fraction.

The following formula can be used for a quantitative characterization of the elemental accumulation in a specific size fraction:

$$Q = C \cdot b,$$

where C is the content of the element in the fraction, (in percent);
b is the quantity of this fraction in the sample (in percent).

The quantity b may be calculated as follows:

$$b = \frac{d \cdot 100}{D},$$

where d is the weight of the fraction, g;
D is the total weight of the sample, g.

The parameter Q is the *enrichment index* of the indicator elements in the various size fractions. Calculations have shown that this parameter is much larger for indicator elements from anomalous samples, than it is for samples collected outside an anomaly.

Table 24 lists the data on the distribution of selected indicator elements in different fractions from two granite samples, one of which was collected within the halo of a uranium deposit, and the other outside. The first sample differs appreciably from the second in having increased contents of the indicator elements and a considerably larger per-

Table 24. Distribution of elements in different size fractions of geochemical samples.

Fraction	Size group, mm	Weight, g	Content, %				Enrichment index, $n \cdot 10^{-3}$			
			Pb	Mo	Cu	Zn	Pb	Mo	Cu	Zn
			General sample from within the halo							
Light		98	0,02	0,001	0,003	0,01	—	—	—	—
	−0,6+0,4	408	0,02	0,0006	0,001	0,01	—	—	—	—
	−0,4+0,25	320	0,01	Traces	0,0006	0,01	—	—	—	—
	−0,25+0,1	382	0,03	0,0006	0,002	0,01	—	—	—	—
	−0,1	125	0,03	0,002	0,003	0,02	—	—	—	—
Heavy	+0,6	19,1	0,5	0,005	0,1	0,15	710	7	140	210
	−0,6+0,4	7,5	0,3	0,003	0,03	0,1	167	1,67	16,7	55
	−0,4+0,25	7,45	0,3	0,008	0,1	0,1	165	1,65	55	55
	−0,25+0,1	3,67	1,0	0,008	0,2	0,3	271	2,11	54	81,3
	−0,1	0,45	1,0	0,03	0,2	0,3	34,8	1,04	7	10,4
			General sample from outside the halo							
Light		187	0,006	0,0002	0,0006	0,007	—	—	—	—
	−0,6+0,4	349	0,002	0,0002	0,0002	0,007	—	—	—	—
	−0,4+0,25	250	0,006	Traces	0,0003	0,007	—	—	—	—
	−0,25+0,1	325	0,006	»	0,0008	0,007	—	—	—	—
	−0,1	130	0,01	»	0,003	Traces	—	—	—	—
Heavy	+0,6	3,88	0,09	0,0008	0,0115	0,028	28	0,24	3,1	8,7
	−0,6+0,4	1,0	0,1	0,001	0,0030	0,03	8	0,08	0,24	2,4
	−0,4+0,25	0,98	0,06	0,002	0,003	0,02	4,73	0,158	0,236	1,58
	−0,25+0,1	1,85	0,1	0,0003	0,02	0,03	14,8	0,0446	3,0	4,5
	−0,1	0,05	0,1	0,0003	0,03	0,03	0,4	0,0012	0,12	0,12

Note: Sensitivity of the analysis: Pb $1 \cdot 10^{-3}$%; Mo $1 \cdot 10^{-4}$%; Cu $1 \cdot 10^{-4}$%; Zn $1 \cdot 10^{-2}$%.

centage of heavy mineral fractions. The maximum content of the indicator elements is usually detected in the heavy fractions with the smallest particle size (−0.1 mm). This same table gives the enrichment index values for the indicator elements in the heavy fractions. The comparison shows that the maximum enrichment indexes are characteristic of the heavy fractions obtained from within the halo. There is a pronounced dependence of the enrichment index on the particle size: the smaller the size of the heavy fraction, the larger the index.

All these data indicate that the samples can be differentiated better by using the enrichment index, rather than the total content of the indicator elements. For example, the total content of lead in the anomalous sample (0.02%) is only 3.3 times that of the background (0.006%). This means that the *anomaly ratio, K,* (the ratio of the anomalous to the average background content; Saukov), in this case is 3.3. By using the enrichment indexes, the anomaly ratio will be:

$$K = \frac{Q_{an,}}{Q_b}$$

where Q_{an} is the enrichment index for the sample collected from within the halo;
Q_b is the enrichment index for the sample collected outside the halo.

The anomaly ratios are listed in Table 25 [and are calculated from data in Table 24.]

The anomaly ratio for the enrichment indexes is always larger than the anomaly ratio calculated by using the total contents of the indicator elements.* This is illustrated by the figures in the last column of Table 25, which compare the anomaly ratios (A) determined on the basis of the formula:

$$A = \frac{K_{1,}}{K_2}$$

where K_1 is the anomaly ratio inferred from the enrichment indexes;
K_2 is the anomaly ratio determined from the total contents of the indicator elements.

Fractions with the smallest particle sizes (−0.1 mm) are characterized by the largest anomaly ratios. Copper is the only exception; its largest anomaly is in the medium particle size fraction (−0.4 + 0.25). It is obvious that the greater the contrast between the ratios, the larger the halos which will be detected (other conditions being equal).

Table 25. Anomaly coefficients.

Indicator element	General (bulk) sample	Size Group (heavy fraction), mm					Anomaly ratios compared
		+0,6	−0,6 +0,4	−0,4 +0,25	−0,25 +0,1	−0,1	
Lead	3,3	25	21	35	18	87	7,6 to 26
Molybdenum	5,0	6	21	10	47	865	1,2 to 173
Copper	5,0	9	70	230	18	57	1,8 to 46
Zinc	1,4	16	23	35	18	87	1,2 to 58

In order to obtain comparative data on the dimensions of the halos detected by the analysis of the heavy fractions, graphs of distribution based on the enrichment indexes were plotted on semilogarithmic coordinates. Graphs of distribution of the indicator element contents (total analysis of the bulk samples) were plotted for comparison on the same diagrams. In constructing the graphs it was assumed that: (1) the enrichment index would decrease exponentially; (2) the gradients of the curves for all elements are equal (including graphs of the element distribution); and, (3) the variations in the background contents of the elements, as well as the enrichment indexes, vary in a similar way. Figs. 12 and 13 show the graphs for copper and molybdenum.

*[The anomaly ratio can be defined and calculated two ways: (1) using the total contents of the indicator elements (this is the most common method); (2) using the enrichment indexes (A).
The anomaly ratio calculated by using the enrichment indexes is always larger because of larger differences in this parameter between anomalous and background samples. Editor.]

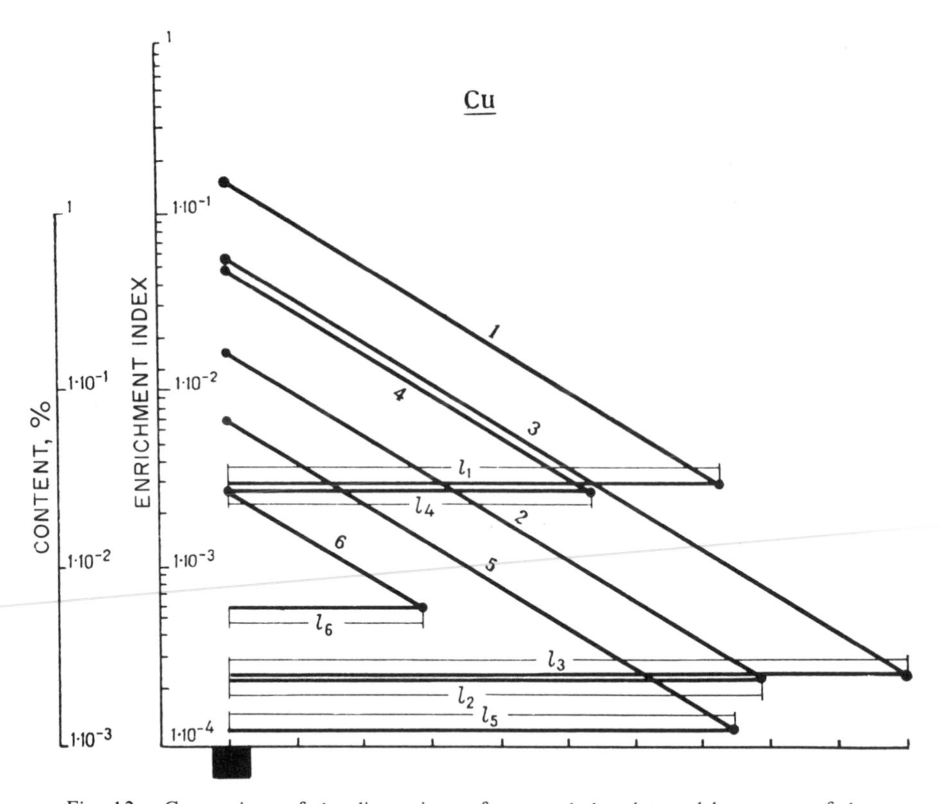

Fig. 12. Comparison of the dimensions of copper halos detected by means of the analysis of the heavy fractions from various size groups.
1 to *5* are graphs of the heavy fractions, mm (*1* - 1.0 to 0.6; *2* - 0.6 to 0.4; *3* - 0.4 to 0.25; *4* - 0.25 to 0.1; *5* - less than 0.1); *6* is the graph of the total contents.
[The enrichment index (Q) has been defined in the text as Q = C•b, where C is the content of the element in the particular fraction, and b is the quantity of this fraction in the sample (both in per cent). Thus, this diagram shows that the largest halos are revealed when heavy fractions (inclined lines) are used by comparison with halos which could be detected by analyzing the whole sample (horizontal lines). Editor.]

Six graphs have been plotted for each element, five of them for heavy fractions of various particle sizes, and the sixth is the graph of distribution of the total contents. The graphs are based on two points, and they correspond to the maximum (anomalous sample) and minimum (background sample) values of the parameters being studied. The distance between these two points on the abscissa (at the appropriate scale) will be equal to the half-width of the halo of a given element. (The section starts with a background sample just outside the halo and goes across the halo to the maximum anomalous sample which is supposedly situated at the center of the halo. Therefore, if an appropriate scale is used, it is possible to show the half-width dimension of the halo.)

It follows from Fig. 13 that the halos detected by means of the analysis of the heavy fractions (lines 1 to 5) are much larger than those detected with the help of analyses from the bulk sample (line 6). Table 26 lists the sizes of the halos detected by means of the analysis of

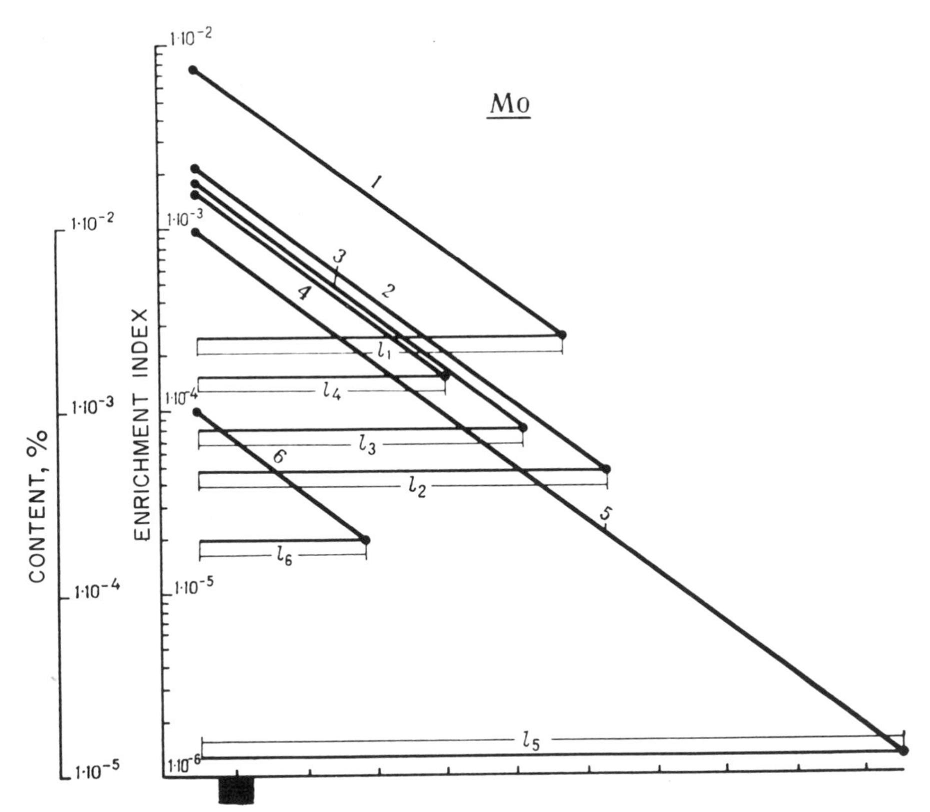

Fig. 13. Comparison of the dimensions of molybdenum halos detected by means of the analysis of the heavy fractions from different size groups.
1 to *5* are graphs of the heavy fractions, mm (*1* - 1.0 to 0.6; *2* - 0.6 to 0.4; *3* - 0.4 to 0.25; *4* - 0.25 to 0.1; *5* - less than 0.1); *6* is the graph of the total contents.
[The enrichment index (Q) has been defined in the text as Q = C•b, where C is the content of the element in the particular fraction, and b is the quantity of this fraction in the sample (both in per cent). Thus, this diagram shows that the largest halos are revealed when heavy fractions (inclined lines) are used by comparison with halos which could be detected by analyzing the whole sample (horizontal lines). Editor.]

Table 26. Comparison of the sizes of halos.

Indicator elements	Heavy fractions (mm)					Representative class*
	+0,6	−0,6 +0,4	−0,4 +0,25	−0,25 +0,1	−0,1	
Lead	2,5	2,4	2,6	2,2	3,8	−0,1
Molybdenum	2,1	2,3	1,8	1,5	4,1	−0,1
Copper	2,7	2,9	3,8	2,0	2,8	−0,4 + 0,25
Zinc	6,6	6,6	6,8	5,6	8,6	−0,1

* The class which helps reveal the largest halo is called the *representative class*.

the heavy fractions in units relative to the size (width) of the halo detected by means of the analysis of the bulk sample.

The conclusion concerning the larger dimensions of the halos delineated on the basis of

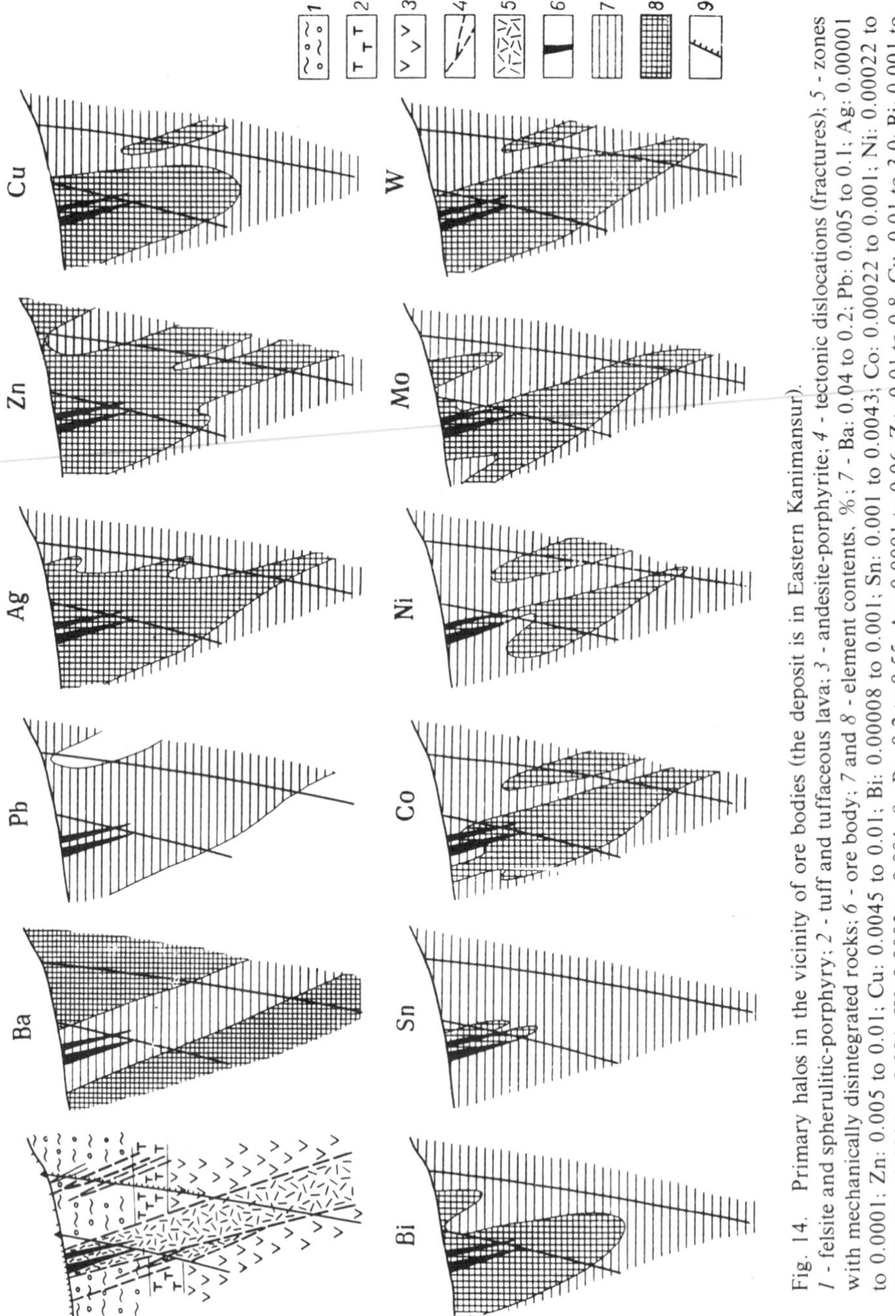

Fig. 14. Primary halos in the vicinity of ore bodies (the deposit is in Eastern Kanimansur). *1* - felsite and spherulitic-porphyry; *2* - tuff and tuffaceous lava; *3* - andesite-porphyrite; *4* - tectonic dislocations (fractures); *5* - zones with mechanically disintegrated rocks; *6* - ore body; *7* and *8* - element contents, %; *7* - Ba: 0.04 to 0.2; Pb: 0.005 to 0.1; Ag: 0.00001 to 0.0001; Zn: 0.005 to 0.01; Cu: 0.0045 to 0.01; Bi: 0.00008 to 0.001; Sn: 0.001 to 0.0043; Co: 0.00022 to 0.001; Ni: 0.00022 to 0.001; Mo: 0.0001 to 0.001; W: 0.0002 to 0.001; *8* - Ba: 0.2 to 0.55; Ag: 0.0001 to 0.06; Zn: 0.01 to 0.8; Cu: 0.01 to 3.0; Bi: 0.001 to 0.1; Sn: 0.001 to 0.002; Co: 0.001 to 0.0064; Ni: 0.001 to 0.0018; Mo: 0.001 to 0.0086; W: 0.001 to 0.1; *9* - sampling intervals.

analytical data from the heavy fractions, has been confirmed by results obtained from certain primary halos.

Figs. 14 and 15 show halos of certain elements around copper-bismuth ore bodies which wedge out at depth (in the Operyayushchii locality; the Kanimansur ore field). The halos

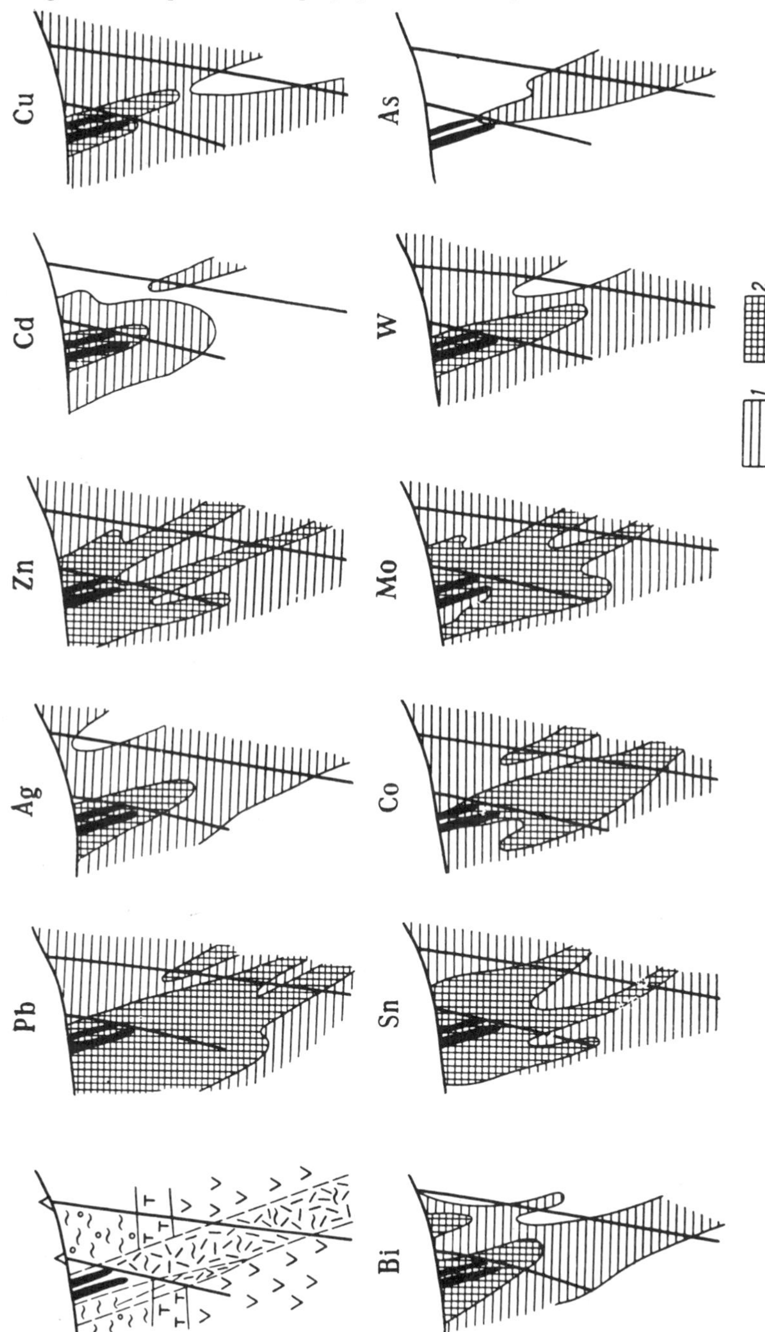

Fig. 15. Primary halos delineated on the basis of the enrichment index.
1 - primary halos; *2* - zones with increased values (background units greater than 100); for other symbols see Fig. 14.

are indicated in two different ways: first, on the basis of the data obtained from spectrographic analysis of the total samples (Fig. 14), and second, on the basis of the analysis of the heavy fractions (based on the enrichment indexes; Fig. 15). Anomalous enrichment index values were found in all the samples, and this complicates the comparison between the halos. Nevertheless, it is obvious that there is an appreciable enhancement of halos when the heavy-fractions are used. The enrichment index makes it possible to delineate additional anomalous fields with values exceeding 100 background units. By comparison, in halos based on the total contents, (which are delineated by means of a minimum anomalous content equal to 3 background units), those areas which do have contents larger than 100 background units do not form any distinct patterns because they are distributed sporadically.

The use of the heavy fractions obtained from geochemical samples has yet another advantage by comparison with the conventional method (analysis of the total sample). It is that the analytical detection limits for the indicator elements are much more easily attainable because of the selective enrichment of these elements in the heavy fractions (both anomalous and background). For example, the analytical methods need to detect 0.06, 0.003, and 0.02% lead, copper, and zinc, respectively, in the heavy fractions (see Table 24), and this is easily accomplished. In the case of the bulk samples, the analytical techniques must be able to detect as little as 0.006% lead, 0.0006% copper, and 0.005% zinc. Therefore, the analytical method employed (emission spectrography) will be able to reveal additional (peripheral) halos of the elements when the heavy fractions are analyzed. These additional halos will not be detected by analysis of the bulk samples, because of the lack of sensitivity of the analytical method at the very low abundance levels of the indicator elements.

4. Method based on composite halos.

It has been established that better defined geochemical halos can be revealed around ore bodies if the contents of a group of indicator elements are combined. There are two types of composite halos: (1) additive, and (2) multiplicative.

Additive halos are constructed by the simple addition of the contents of the indicator elements normalized against their average background contents in the enclosing rocks. The halos are, in this case, delineated by using the value of the total minimum anomalous (threshold) content of the indicator elements, which are determined by comparison with samples obtained from background regions.

Additive halos are larger and more distinct by comparison with monoelement halos. In addition, the effects of random errors are reduced to a minimum. Because of this, additive halos exhibit a closer relationship to the structural geological features associated with ore deposits. This, in turn, contributes significantly to the reliability of their interpretation.

Fig. 16 shows the primary halos of several chemical elements in cross section through a gold deposit. It is easy to see that the additive halos are larger and more pronounced than any of the monoelement halos.

Similar results are obtained if the contents of the indicator elements in each sample are multiplied (instead of being added), and multiplicative halos are constructed (see Fig. 16).

The multiplicative halo method is less time consuming compared to the additive method,

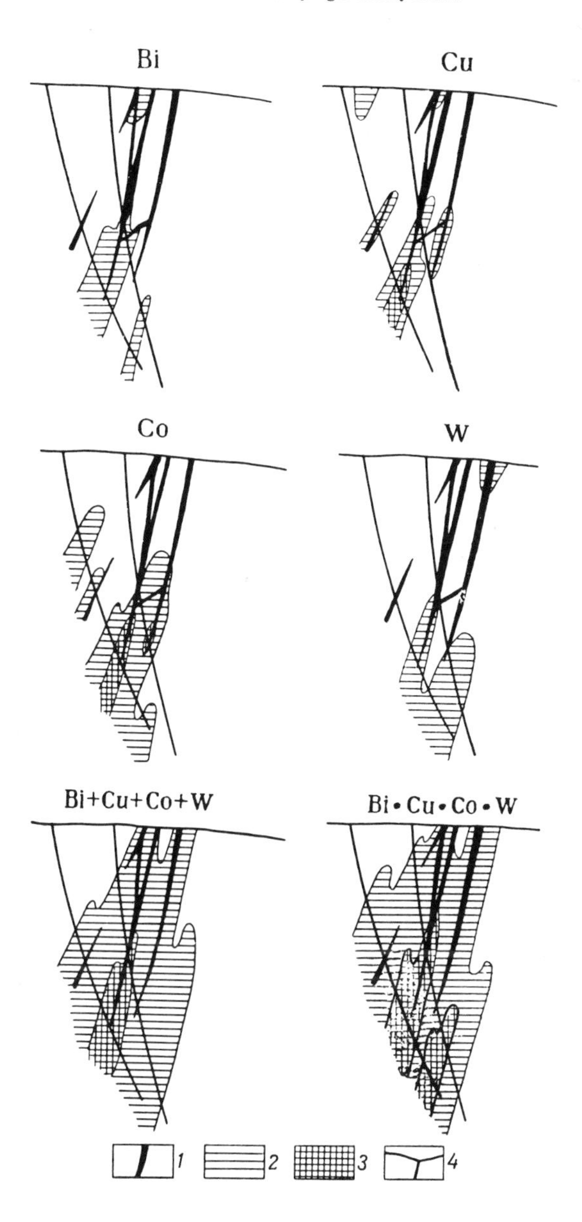

Fig. 16. Primary (monoelement, additive, and multiplicative) halos around gold ore deposits.

1 - quartz-gold veins; *2* and *3* - contents of elements within a halo, % (*2* - Bi, 0.0001 to 0.001; Cu, 0.01 to 0.1; Co, 0.0005 to 0.001; W, 0.0003 to 0.001; Σ, 10 to 100; Π, 1 to $10 \cdot 10^{-11}$; *3* - Cu, 0.1 to 0.5; Co, 0.001 to 0.008; Σ, 100 to 1500; Π, 10 to $720{,}000 \cdot 10^{-11}$); *4* - sampled cross sections.

[Σ is the total (sum) content of elements in the additive halo in "background units." Π is the product of element contents in the multiplicate halo. Editor.]

because it eliminates the necessity of determining the background contents of each indicator element and normalizing the values of their concentration in all the chosen samples.

In cases where the contents of certain elements are not detected because of the lack of sensitivity of the analytical technique, the element contents which can be used for the purpose of constructing composite halos, can arbitrarily be assigned a value equal to one half the limit of sensitivity of the analytical technique.

Methods of Studying the Zonality of Primary Halos

The zonality of primary geochemical halos is of great practical significance. Particular emphasis is placed on the vertical zonality of primary halos, because this can be used as a

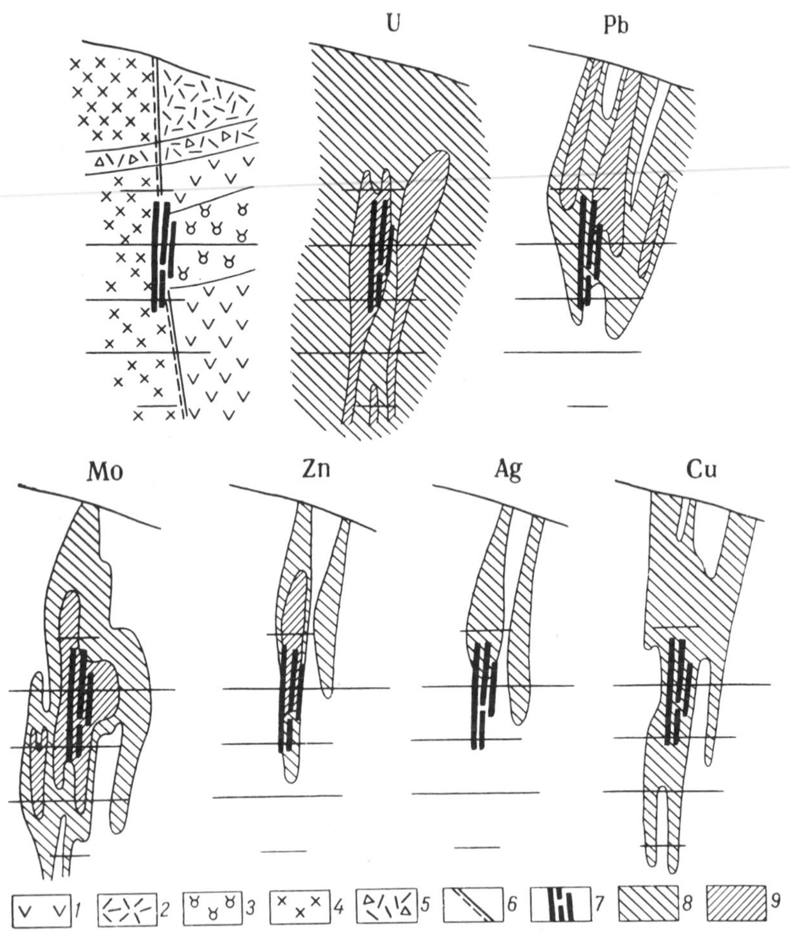

Fig. 17. Primary halos around uranium ore bodies (from Vertepov).
1 - quartz porphyry; *2* - tuffs of quartz porphyry composition; *3* - spherulite-porphyry; *4* - felsite-porphyry; *5* - tuffaceous breccia of quartz porphyry composition; *6* - fractures; *7* - ore bodies; *8* and *9* - element contents in primary halos (compared to geochemical background); (*8* - 2 to 10 units; *9* - more than 10 units).

criterion in assessing the exposed level of erosion of geochemical anomalies.

Vertical zonality in primary halos formed by hydrothermal mineralization was first recognized by Kablukov and Vertepov from uranium deposits (see Kablukov et al., 1964). These investigators observed that in a vertical section there is a very marked upward shift in the halos produced by elements which are associated with uranium, relative to the particular halo of uranium. This is illustrated in Fig. 17 by means of vertical cross sections. These scientists proposed that this zonality be evaluated quantitatively by using the average contents of the lead-uranium ratio. This ratio decreases regularly with depth, thus reflecting the zonal distribution of these elements in the halos. Fig. 18 illustrates this concept and shows the variations with depth of the ratios of the average contents of the indicator ele-

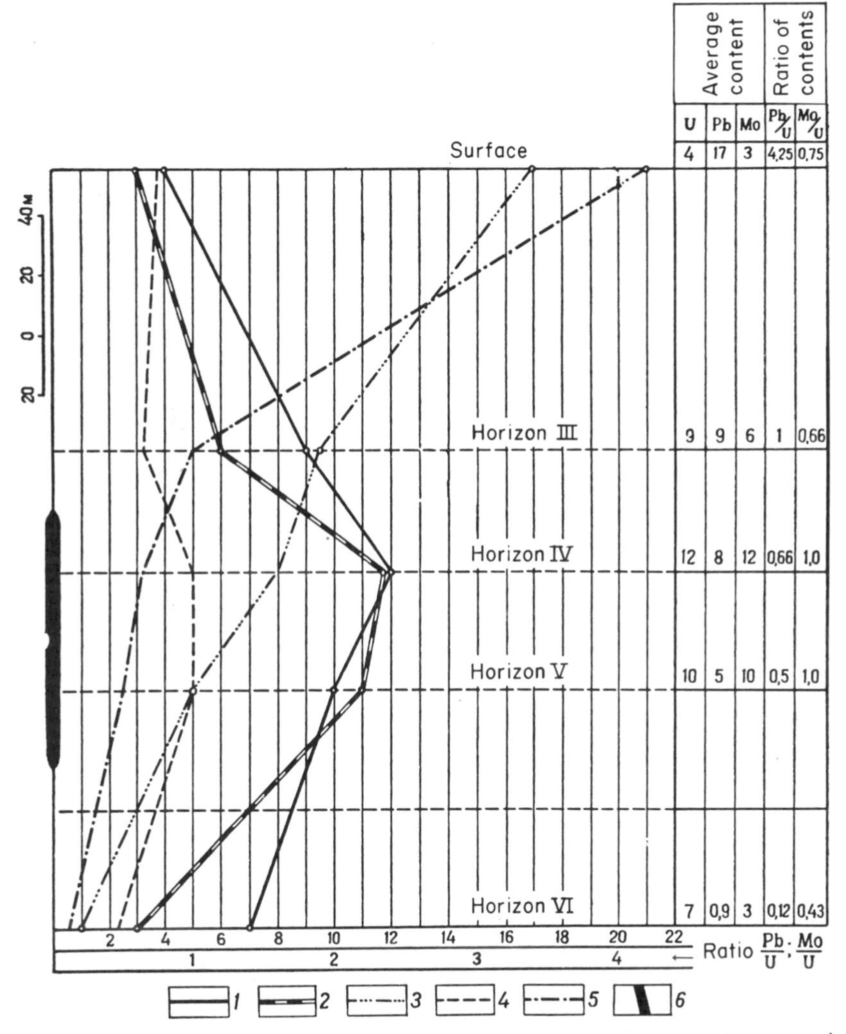

Fig. 18. Graphs showing variations in the average contents of indicator elements and ratios of indicator elements, with depth in halos.
1 - uranium; *2* - molybdenum; *3* - lead; *4* - Mo/U; *5* - Pb/U; *6* - ore body.

ment pairs in primary halos developed around a hidden uranium ore body within felsitic rocks. The graphs, which express the zonality by means of ratios of the average contents of the element pairs (lead-uranium and molybdenum-uranium), decrease monotonically with depth. The graphs also show distinct differences between the halos above and beneath the ore body.

Later Sochevanov (see, Kablukov et al., 1964) suggested that for the study of the zonality of halos in cross section, use should be made of the linear productivity of the halo (instead of the ratios of the average contents of the elements). *Linear productivity* is defined as the product of the average element content in the halo (in %), and the width of the halo (in meters). This results in a more distinctive and, therefore, more reliable, zoning of the halos.

Primary halos are multicomponent. Therefore, it is always necessary to compare zonalities of halos formed by different elements, so as to be able to choose the most effective indicator elements.

This can be done with the help of "sequences" which indicate the zonality of an element, and which is expressed by the *coefficient of contrast*. The coefficient of contrast is the ratio of certain parameters of the halo of a given element; it is determined in the upper (numerator) and lower (denominator) portions of the cross section being studied.

Table 27 lists the values of linear productivities of the halos, and the zonality of the coefficients of contrast, which have been calculated for a cross section of a uranium deposit in granites.

If elements are arranged in the decreasing order of their coefficients of contrast, the resulting sequence will illustrate (from left to right) the elements according to the zonality of their halos. The values for the case being considered are as follows (from Table 27): lead (46.0) - zinc (15.0) - copper (1.7) - uranium (0.6). The practical significance of these sequences is in that, with their help, it is possible to determine the most contrasting and, therefore, the most reliable indicator ratios to assess the exposed level of erosion. These ratios are based on pairs of elements at opposite ends of the sequence. For the given example, lead and uranium are such a pair.

The coefficient of contrast helps in obtaining unequivocal results only if the productivities of the halos show a monotonical vertical variation. However, this condition is often not encountered. Therefore, it is more reliable to use the *zonality index* of the indicator element, which is the ratio of the linear productivity of the halo formed by the given element, to the sum of the linear productivities of the halos of all indicator elements in the

Table 27. Linear productivities (in m%) and zonality based on the coefficient of contrast of halos.

Parameter	Indicator element			
	U	Cu	Zn	Pb
Linear productivity:				
Surface	0,3	1,2	3,0	12,0
Horizon V	0,45	1,3	1,6	5,0
Horizon VII	0,5	0,7	0,2	0,26
Coefficient of contrast	0,6	1,7	15,0	46,0

Note: The coefficient of contrast (defined in text) is obtained by dividing the surface content by the content in Horizon VII (Editor).

type of mineralization being studied. The linear productivities of the halos in these calculations are normalized in the following way: the maximum values of the linear productivities of the halos in the cross section are expressed in the same orders of magnitude, and then the rest of the values are defined more accurately.

This is shown (Table 28) by the results of the calculation of the zonality indexes for a cross section through the Aktash skarn-multicomponent metal deposit (Tadzhik, SSR).

Table 28. Linear productivities of halos (m %).

Level of halo	Indicator elements					
	Pb	As	Sb	Cu	Bi	Mo
Surface	1,5	0,17*	0,066*	0,96	0,07	0,00077
Horizon I	8,1*	0,006	0,006	0,75	0,03	0,0074
Horizon II	1,3	0,077	0,014	1,2*	0,16*	0,018
Horizon III	0,13	0,017	0,006	0,67	0,076	0,02

* Maximum values are indicated by asterisks.

Lead has the highest value in Table 28. The copper halo exhibits a similar maximum linear productivity value (on an order of magnitude basis). Consequently, the *normalization coefficient* (C_n) for copper (as well as for lead) will be unity. For other elements, the following normalization coefficients, which are equal to the differences between the maximum values of the lead and element in question, have been derived: arsenic, −10 (0.17-1.7); antimony, −100 (0.066-6.6); bismuth, −10 (0.16-1.6); molybdenum, −100 (0.02-2.0).

The values of the zonality indexes calculated using normalized linear productivity values of the halos are given in Table 29.

Table 29. Zonality indexes of elements.

Element	Normalized value of linear productivity					Zonality index			
	C_N	Surface	Borehole 407	Borehole 410	Borehole 411	Surface	Borehole 407	Borehole 410	Borehole 411
Lead	1	1,5	8,1	1,3	0,13	0,13	**0,764**	0,171	0,03
Arsenic	10	1,7	0,06	0,27	0,17	**0,148***	0,0056	0,036	0,04
Antimony	100	6,6	0,6	1,4	0,6	**0,574**	0,056	0,184	0,139
Copper	1	0,96	0,75	1,2	0,67	0,084	0,071	**0,158**	0,156
Bismuth	10	0,7	0,3	1,6	0,76	0,061	0,028	**0,211**	0,177
Molybdenum	100	0,07	0,74	1,8	2,0	0,0061	0,07	0,237	**0,465**
Σ Sum of linear productivities		11,5	10,6	7,6	4,3	—	—	—	—

*Maximum values for each element are indicated by bold numbers.

It is readily seen that the zonality index quantitatively expresses the relative accumulation of the elements in each horizon. It follows from Table 29 that the maximum relative accumulation of arsenic and antimony occurs at the upper horizon of the halos; lead, at the second horizon; copper and bismuth, at the third horizon; whereas the maximum relative accumulation of molybdenum, is at the lowest horizon. Based on the results of the calculations of the zonality indexes, the following sequence can be derived: (arsenic, antimony)— lead — (copper, bismuth) — molybdenum.

Those elements whose relationships are not clearly expressed, because the maximum values of their zonality indexes were found to be in the same horizons of the halos, are given above in parentheses. In order to further define the position of these particular elements in the zonality sequence, one can use the vertical variability of the zonality index. In those cases where comparison is made between elements whose maximum accumulation was found to be in the extreme levels of the halos (either upper or lower), the variability can be assessed with the help of the formula:

$$G = \sum_{1}^{n} \frac{D_{max}}{D_i},$$

where G is the variability index (or variability gradient):

D_{max} is the maximum value of the zonality index of a given element;

D_i is the value of the zonality index in the i-th horizon;

n is the number of horizons (exclusive of the horizons of maximum accumulation).

For arsenic and antimony (see Table 29):

$$G_{As} = \frac{0,148}{0,0056} + \frac{0,148}{0,036} + \frac{0,148}{0,04} = 26,4 + 4,14 + 3,7 = 34,22;$$

$$G_{Sb} = \frac{0,574}{0,056} + \frac{0,574}{0,184} + \frac{0,574}{0,139} = 10,2 + 3,1 + 4,1 = 17,4.$$

$G_{As} > G_{Sb}$, which suggests that antimony should follow arsenic in the zonality sequence. If the maximum values of the zonality index are typical for the lowermost horizon, then the elements in the sequence will be arranged in the order of increasing gradient.

If the maximum relative accumulation of several elements occurs at the level of the middle horizons, then the difference between gradients G_1 - G_2 can be used, where G_1 is the gradient in the upward direction, and G_2 is the gradient in the downward direction from the horizon of the maximum accumulation of the element. The order of the elements in the zonality sequence will depend on the difference between the gradients: the larger the difference, the farther down (to the right) in the zonality sequence the element is placed, and the converse also applies.

To illustrate the point, we show the relationship between copper and bismuth, whose maximum accumulation was found to be in the second horizon (see Table 29).

For copper:

$$G_1 = \frac{0,158}{0,071} + \frac{0,158}{0,084} = 2,23 + 1,88 = 4,11;$$

$$G_2 = \frac{0,158}{0,156} = 1,01; \qquad G_1 - G_2 = 4,11 - 1,01 = 3,1.$$

For bismuth:

$$G_1 = \frac{0,211}{0,028} + \frac{0,211}{0,061} = 7,54 + 3,46 = 10,0;$$

$$G_2 = \frac{0,211}{0,177} = 1,19; \qquad G_1 - G_2 = 10 - 1,19 = 8,81.$$

The difference in gradients of the zonality index for bismuth is larger, and thus it is placed in the zonality sequence to the right of copper.

Thus, the following sequence of indicator elements have been obtained to characterize the zonality of the halos: arsenic — antimony — lead — copper — bismuth — molybdenum.

The zonality index method is more accurate than the zonality sequence method based on the coefficient of contrast of the productivity ratios for element pairs. Owing to the summation of linear productivities of the halos produced by a wide range of elements, the effect of random errors in the determination of the initial data is reduced.

The sequences of the indicator elements characterizing the vertical zonality of primary halos are, as mentioned before, of great practical significance. They help in determining the most important indicator elements to be used for assessing the level of erosion of a geochemical anomaly. In the general case, these elements are the most distant from one another in the zonality sequence, because ratios of parameters of the halos formed by such pairs of elements vary more distinctly with depth and are, therefore, more reliable in practice.

Zonality in the structure of the above general composite halos may be revealed by constructing *partial* composite halos which, in contrast to general composite halos, are prepared for groups of elements with similar vertical distributions. The choice of these groups is based on the sequences of the indicator elements. Fig. 19 shows partial multiplicative

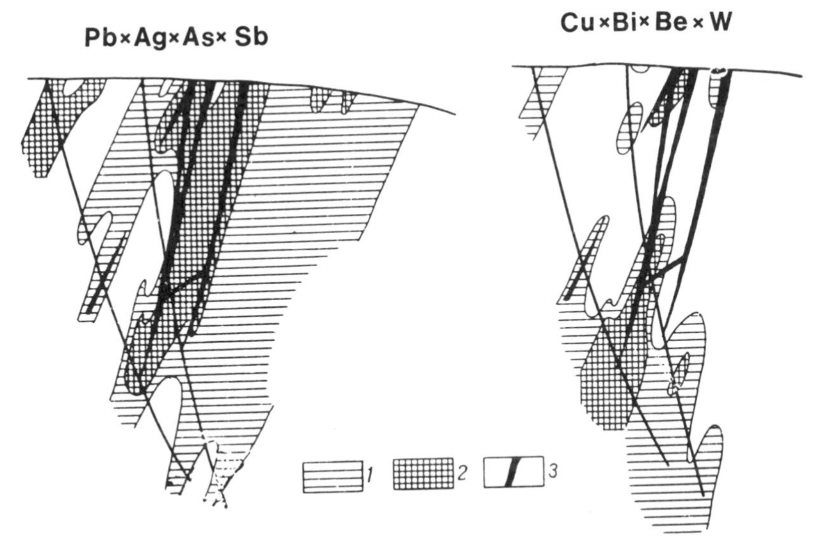

Fig. 19. Multiplicative halos of indicator elements in cross section (of gold deposits). *1* - Multiplicative halos; *2* - Areas with the maximum values of element content; *3* - ore bodies.

halos for a cross section of a gold deposit. These halos are characterized by a distinct vertical zonality owing to the opposing behavior of the halos formed by elements above (wedging out with depth) and below (sharp expansion with depth) the ore body. The ratio of linear productivities of these halos at a depth of 130 meters decreases by a factor of 1500.

Partial composite (multiplicative) halos of the groups of elements above and below ore bodies, chosen on the basis of their zonality, increase the contrast of the halo zonality. In addition, composite halos, owing to their larger size, furnish more reliable results in estimating the level of the erosion of geochemical anomalies.

A considerable increase in the zonality contrast can also be achieved by calculating ratios of the products of the average contents for two groups of indicator elements. This method was first proposed by Solovov et al., (1971). However, the method of composite halos is preferred, because composite halos are less susceptible to the effect of sampling or analytical errors, than are monoelement halos.

Primary Halos Associated With Steeply Dipping Ore Bodies

Below, based on examples from several deposits, we will briefly discuss the main characteristics of the composition and the structure of primary geochemical halos formed around steeply dipping metalliferous zones.

Bismuth deposit in skarns. *Central Chokalambulak.* The Chokalambulak ore deposit is situated in Western Karamazar (Tadzhik, SSR). The region is composed of limestones altered to marbles ($D_3 - C_1$), extrusive tuffaceous materials (P_1), and intrusive and contact metasomatic rocks. Limestones, in the form of isolated outcrops extending in an east-west direction, occupy the central part of the region, making up the southern limb of an anticline. They range from 5 to 200 m in thickness. The limestones are overlain by extrusive tuffaceous rocks with a sharp angular unconformity.

Intrusive rocks in the region are mainly acidic, and less frequently basic. The granodiorites of the Chokadambulak intrusive formation, constituting half the area of the deposit, are especially widespread. Granite porphyries occurring in stock-like granodiorite bodies are less common.

The metalliferous zone is confined to the limestone-granodiorite contact and it is represented by garnet and pyroxene skarns with magnetite. The magnetite generally occurs in massive or disseminated forms, less frequently it is banded, and magnetite ores result (Fig. 20). Bismuth metallization occurs in the magnetite skarns. The mineral composition of the ores is quite varied. The major ore minerals are: magnetite, muschketowite [magnetite pseudomorphous after hematite], pyrite, arsenopyrite, chalcopyrite, galena, sphalerite, cobaltite, sulfobismuthites of lead and copper, bismuthine, marcasite, native gold and scheelite.

The following stages of mineralization have been identified in the deposit: (1) early magnesian metasomatism; (2) early calcareous skarns; (3) skarn-magnetite; (4) actinolite-magnetite; (5) sulfide; (6) serpentine-chlorite-sulfide; (7) carbonate sulfide; and (8) quartz-carbonate. Bismuth metalization is associated with the serpentine-chlorite-sulfide stage.

Processing of the geochemical data obtained from samples around the ore bodies of the

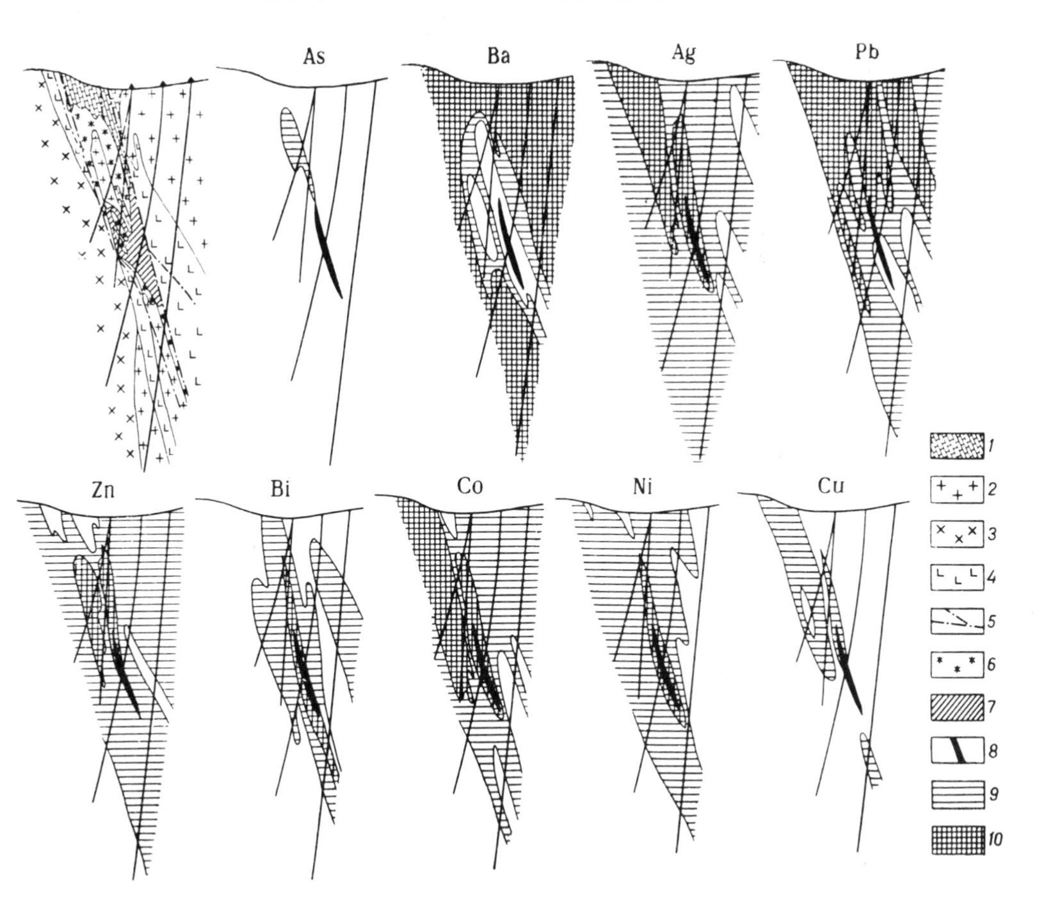

Fig. 20. Primary halos around ore bodies of the Chokadambulak deposit.
1 - limestones; *2* - granodiorites; *3* - skarned granodiorites; *4* - felsite-porphyries; *5* - fractures; *6* - skarning; *7* - magnetite skarns; *8* - bismuth ore body; *9* and *10* - element contents, % (*9* - 0.005 to 0.05% Zn; 0.0001 to 0.0005% Bi; 0.0005 to 0.001% Co; 0.0004 to 0.0005% Ni; 0.005 to 0.2% Cu; 0.0025 to 0.008% As; 0.0005 to 0.001% Ba; 0.00002 to 0.00005% Ag; 0.005 to 0.01% Pb; *10* - 0.001 to 0.07% Ba; 0.001 to 0.0004% Ag; 0.01 to 0.5% Pb; 0.1 to 0.3% Zn; 0.0001 to 0.007% Bi; 0.001 to 0.008% Co; 0.001 to 0.002% Ni).

deposit has revealed large and intensive halos of the elements (Fig. 20). Despite the considerable depth of occurrence of the blind ore body (nearly 200 m), there are appreciable halos of several elements at the surface, which indicate that the halos have a considerable vertical extent above the ore bodies. Arsenic appears to have the smallest halo, but this is probably due to the poor analytical sensitivity for this element.

The distribution of barium is quite unusual. This element has been removed from the ore body and adjacent regions (Fig. 20).

Polymetallic deposits in skarns. *Aktash.* The Aktash deposit is situated in the eastern part of the Kansai metalliferous area at the southern end of the Kuraminsky Range (Tadzhik, SSR). The area in which the deposit is found is characterized by an abundance of skarns formed at the contact of limestones and granodiorite porphyries. The limestones are sometimes serpentinized. The igneous rocks are silicified and chloritized. The skarn bodies are represented by veins and by lenticular and tubular bodies (Fig. 21). The skarns are composed of grossularite-andradite, diopside-sahlite, and vesuvianite.

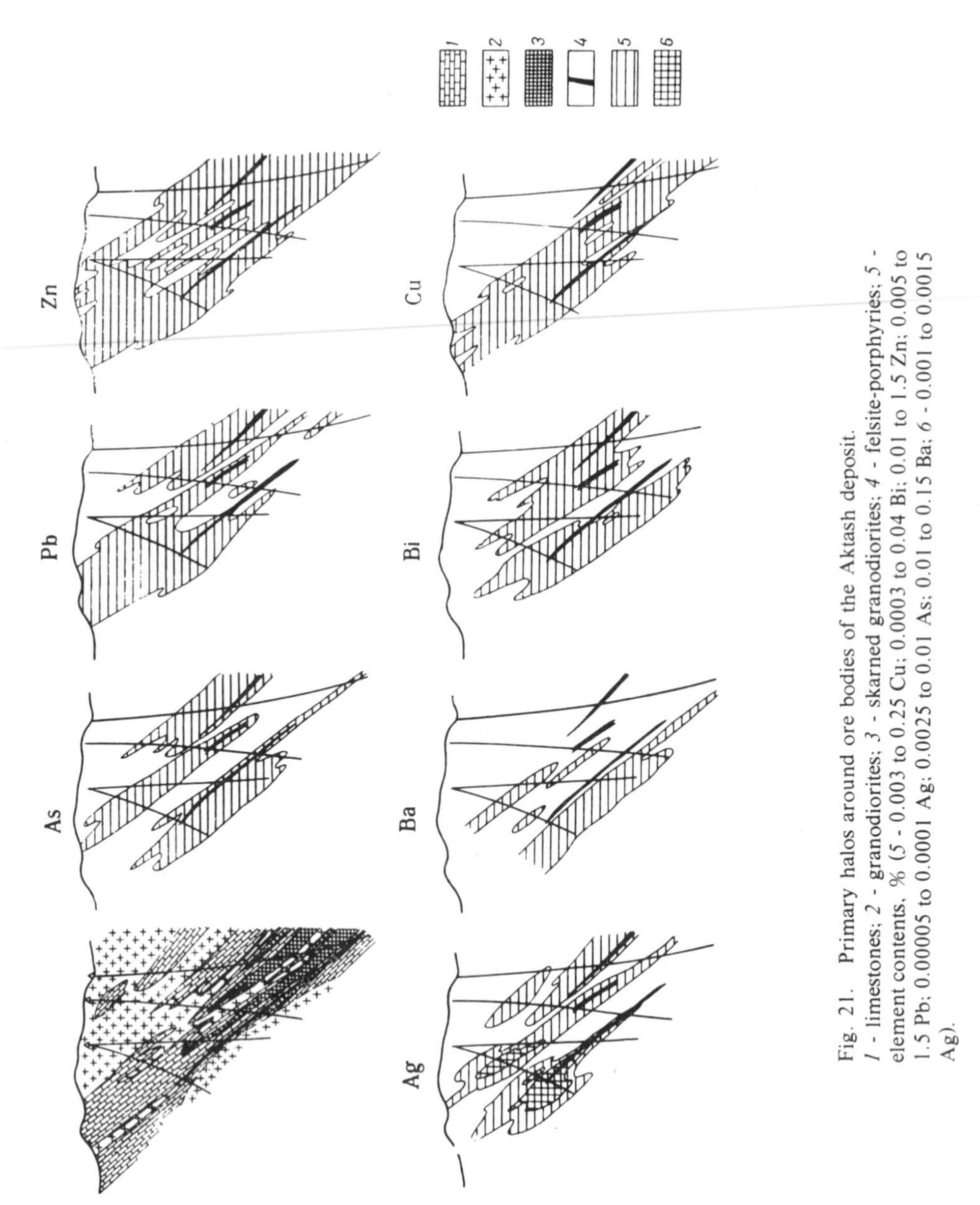

Fig. 21. Primary halos around ore bodies of the Aktash deposit.
1 - limestones; *2* - granodiorites; *3* - skarned granodiorites; *4* - felsite-porphyries; *5* - element contents, % (*5* - 0.003 to 0.25 Cu; 0.0003 to 0.04 Bi; 0.01 to 1.5 Zn; 0.005 to 1.5 Pb; 0.00005 to 0.0001 Ag; 0.0025 to 0.01 As; 0.01 to 0.15 Ba; *6* - 0.001 to 0.0015 Ag).

Localization of the mineralization is controlled by dislocations (fractures) which strike in approximately east-west and in northeast directions. Ores of the deposit are characterized by disseminated, fibrous-disseminated, and mottled-disseminated textures.

The deposit is high in magnetite, magnetite-sulfide ores in contact skarns, disseminated galena-sphalerite ores in the form of irregular bodies, and in high-grade tubular bodies of galena-sphalerite ores, where amphiboles, carbonates, and chlorites are abundant, along with the skarn minerals. Magnetite and sphalerite are the main ore minerals. Galena, chalcopyrite and bismuth minerals are present in minor amounts.

Fig. 21 shows the primary halos of the elements delineated around the galena-sphalerite ore bodies. Beside the elements shown, tin, cobalt, and molybdenum halos, varying in size and intensity, have also been revealed in the cross section. Samples collected at the surface above this cross section were not analyzed for barium, so that the distribution of this element was studied only on the basis of the results of core sampling from boreholes.

From Fig. 21, it can be seen that hidden ore bodies occurring at depths greater than 200 m are distinctly recorded at the surface by means of primary halos of several elements. The ore bodies are characterized by minimum contents of barium, as is also the case in the Chokadambulak deposit (Fig. 20).

Kurusai. The Kurusai deposit is situated at the northwestern end of the Kurusai Range (Tadzhik, SSR). Enclosing rocks are represented by Tournaisian stage marbles which are intruded by small dikes and stock-shaped bodies of quartzose porphyries, syenite porphyries, augite porphyrites, and other igneous rocks.

The deposit is confined to the region where steeply dipping short fissures and the major ore fault join. It is composed of tubular shaped, complexly branching mineralized skarn bodies. Two types of skarns have been recognized: (1) skarns formed at the contact of marbles and igneous rocks; and (2) an infiltration type of skarn in the marbles. Skarns of the second type are represented by andradite and andradite-grossular garnets, as well as by pyroxene (mangan-hedenbergite).

Fig. 22 shows the primary halos developed around blind (hidden) ore body No. 7. Primary halos formed by a wide range of elements have been detected in this, as well as in many other, cross sections studied in different portions of the deposit. In Fig. 22 only the halos of barium, lead, copper, and bismuth are illustrated, and these have been traced to a considerable depth (greater than 600 m). Zonality in the structure of the halos is expressed through the exclusive accumulation of barium in the upper part of the enclosing rocks, and of bismuth in the lower part, at the level where the ore body wedges out. The copper halo exhibits a distinct intensification and expansion with depth.

Another zone with anomalous element contents (not illustrated) was detected northeast of the halos around the known ore body. Within this band a broad and intensive barium anomaly (and to a lesser extent lead anomaly) developed, but there is essentially a total absence of copper and especially bismuth (the latter halos are indicators beneath ore bodies; see Fig. 22). This suggests that these anomalies are supraore halos probably from blind (hidden) mineralization northeast of the known ore body.

Harpenberg (Sweden). Results of geochemical sampling from this Swedish deposit are presented in order to compare primary halos from polymetallic deposits with different geological settings.

The Harpenberg deposit is situated in Central Sweden and is a skarn type of deposit. The

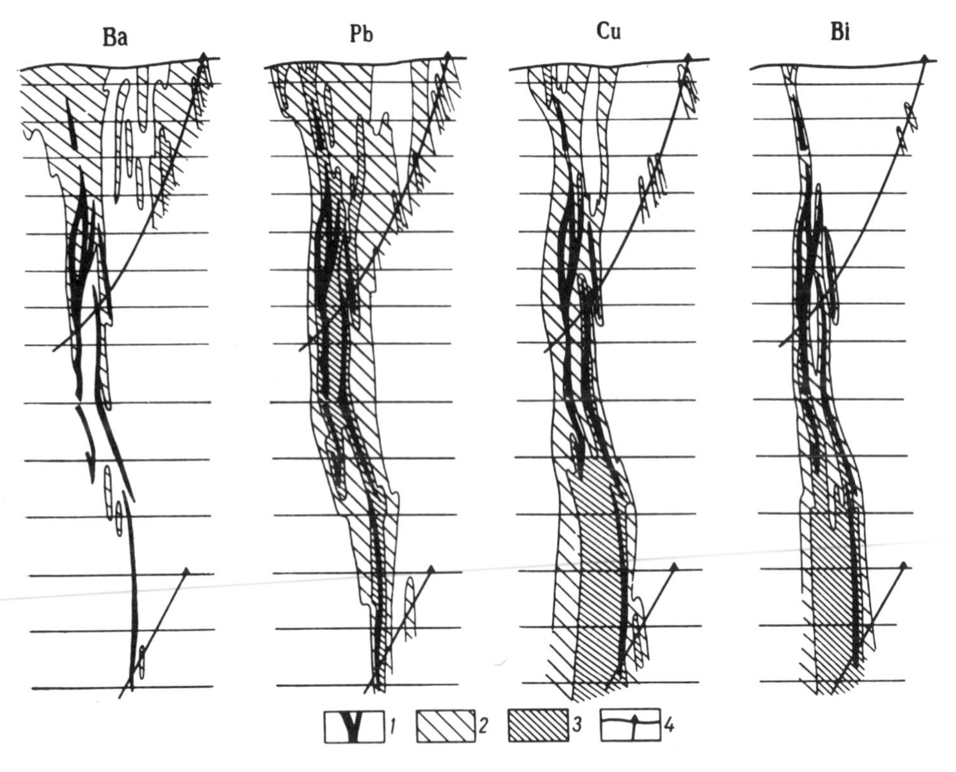

Fig. 22. Primary halos around an ore body in the Kurusai I deposit.
1 - ore body; *2* - and - element contents, % (*2* - 0.01 to 0.1 Ba; 0.01 to 0.1 Pb; 0.003 to 0.02 Cu; 0.0001 to 0.001 Bi; *3* - 0.1 to 0.6 Ba; 0.1 to 1.0 Pb; 0.02 to 0.3 Cu; 0.001 to 0.03 Bi); *4* - cross sections sampled.

skarns are magnesian, of the infiltration type, and are formed as a result of the intrusion of Precambrian granites into a complexly fractured leptite series.

Ore bodies occur in skarn-dolomite zones or in quartzites. Tremolite and talc are the common skarn minerals. The ore deposits contain galena, sphalerite, chalcopyrite, pyrite, pyrrhotite, molybdenite, tetrahedrite-tennantite, as well as stibnite, antimonite, and native silver. Gangue minerals are represented by quartz, calcite, and fluorite.

Fig. 23 shows the primary halos determined in cross section by means of sampling cores from horizontal underground boreholes along nine horizons and from two slanting deep boreholes. As a result of the processing of the geochemical data from the borehole cores, very distinct primary halos of numerous elements have been delineated. These halos extend along the pitch of the ore bodies. The halos produced by silver, lead, zinc, copper, and cadmium (at the surface) are the broadest.

A distinct vertical zonality has been revealed in the element distribution as a result of the development of the antimony halos only in the upper horizons, and of the tungsten, cobalt, nickel, and molybdenum halos at the root level of the ore bodies (see Fig. 23).

Despite distinct differences in the environments of mineralization, the Harpenberg deposit is practically an identical analog of the skarn multielement metal deposits considered above, based on the characteristics of the primary geochemical halos.

A very conspicuous feature in the distribution of the chemical elements in the above cross section is the intensification at depth (following a definite narrowing) of the silver, lead, and zinc halos, despite a complete wedging out of the ore bodies (note the deep slanting boreholes). Two parallel anomalous zones can be identified: the northern (at the extension of ore bodies), and the southern (at the limestone-quartzite contact).

These data made it possible to conclude that blind ore bodies probably exist at depth within the anomalous zones. Subsequent exploration conducted by the Boliden Mining Company within the northern anomaly revealed a large blind ore body.

Gold deposit in igneous rocks. The ore zone of the [unnamed] deposit being considered is composed mainly of igneous rocks of the Upper Akchinsky subsuite (group) of Middle Carboniferous age, represented by extrusive and pyroclastic material. These are generally agglomerate tuffs, andesite porphyrites, and lava breccias of andesite porphyrites.

The igneous rocks are intruded by dikes and subvolcanic bodies of various composition. These are dacite porphyries, syenite-diorite, andesite, and diabase porphyrites. Fractures, which are represented by zones of crushed rock and by brecciation, and which are sometimes filled with dikes or veins, are widespread. Propylitization is the most typical type of alteration of the enclosing rocks.

The morphology of the ore bodies is quite varied. Steeply and gently dipping ore bodies, the majority of which strike southward, have been identified in the deposit.

The formation of the ore bodies took place in several stages. The following stages have been recognized in the gold-bearing hydrothermal phase: silicate, huebnerite-quartz, gold-sulfide-quartz, and sulfide-quartz-carbonate. The gold-sulfide-quartz stage is the most strongly expressed. Typical of the gold deposits are: pyrite, galena, tetrahedrite-tennantite, chalcopyrite and sphalerite; native gold, silver, bismuth and silver; and gold and bismuth tellurides. These minerals form irregular, nodular, and streaky disseminations. Quartz is the main gangue mineral of the ore bodies.

The distribution of ore elements around the metalliferous zones was studied in several exploratory cross sections. Fig. 24 and 25 show the primary halos of numerous elements around the gold quartz veins. It can be seen from these figures that only lead, gold and silver halos are large and intensive, whereas the halos of other elements are weak and seldom extend beyond the metalliferous zone. However, this does not apply to antimony or arsenic, whose small halos can be explained by the low sensitivity of the analytical methods used to detect these elements.

The halos exhibit vertical zonality. The halos of antimony, lead, and silver are the widest and the strongest in the upper portion of the cross section and show a distinct tendency to wedge out at depth. The halos formed by tungsten, cobalt, nickel, and some other elements, on the other hand, are developed selectively in the lower portion of the cross section where the ore veins wedge out (see Fig. 25).

In contrast to the above elements, the arsenic halo has two "highs"; one in the upper part of the cross section, and the other where the gold veins wedge out (see Fig. 24).

As mentioned previously, most elements form halos which are small and weak. This appreciably reduces their effectiveness and the reliability of using these elements as indicators of the respective type of mineralization. Composite halos, such as shown in Fig. 19, may be

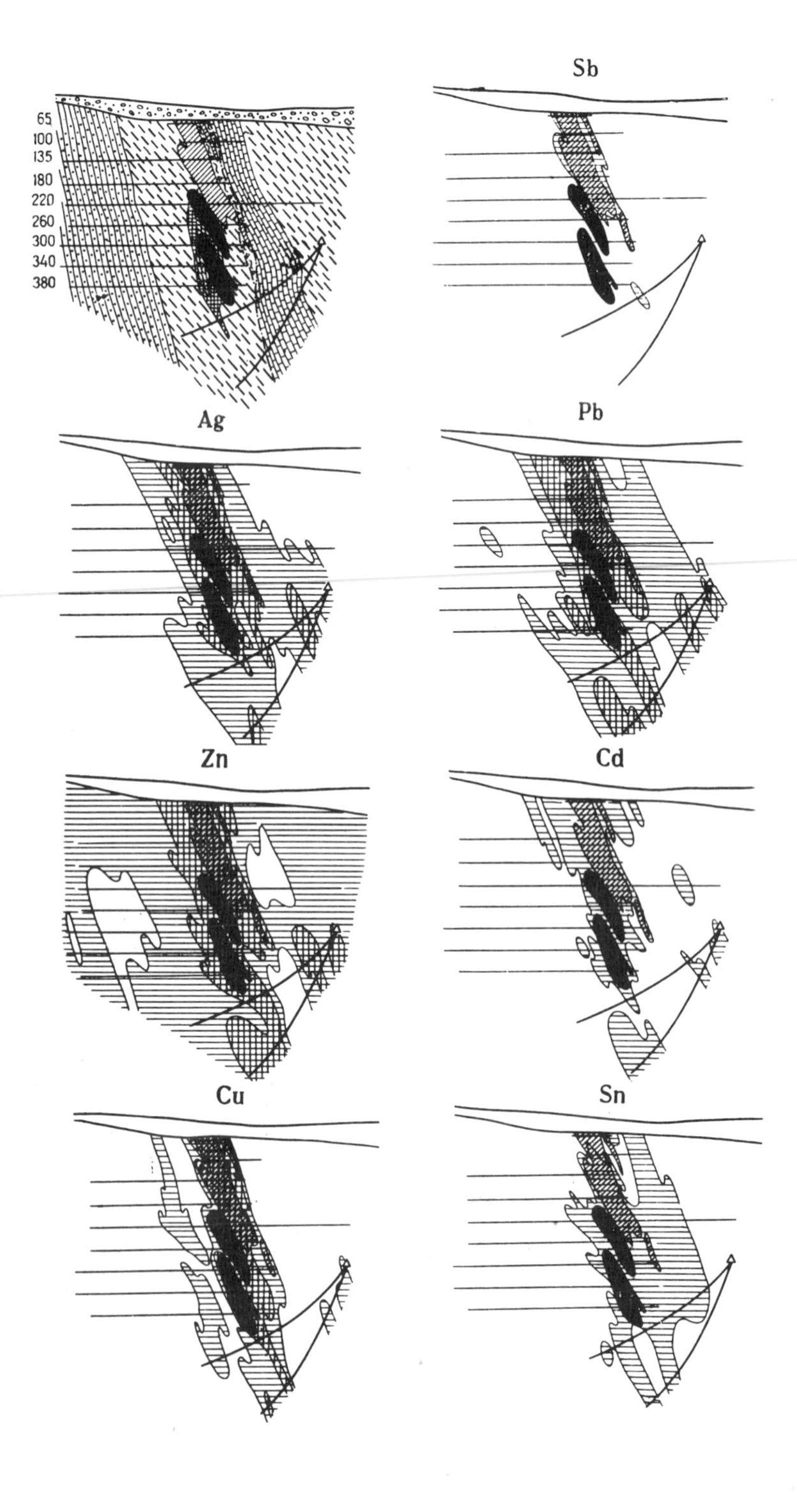
65
100
135
180
220
260
300
340
380
Sb
Ag
Pb
Zn
Cd
Cu
Sn

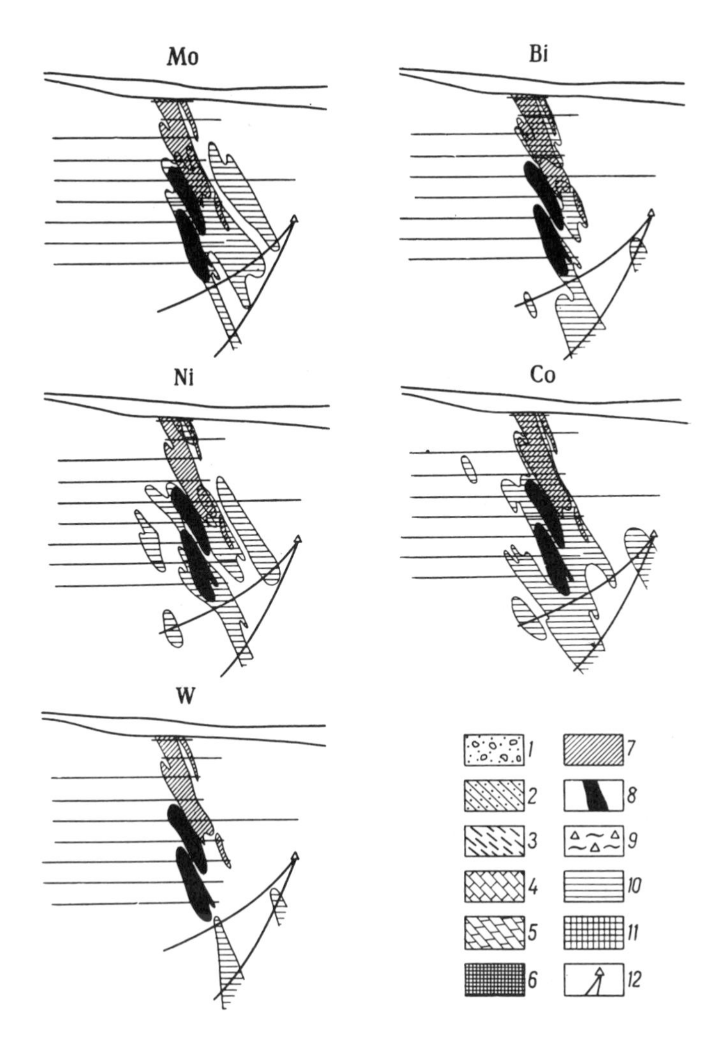

Fig. 23. Primary halos around ore bodies of the Harpenberg deposit (geological information from N. Pilava).
1 - alluvium; *2* - leptites; *3* - quartzites; *4* - limestones; *5* - dolomites; *6* - skarns; *7* and *8* - ore bodies (*7* - oxidized; *8* - unoxidized); *9* - fractures; *10* and *11* - element contents, % (*10* - 0.005 to 0.05 Sb; 0.00005 to 0.001 Ag; 0.005 to 0.05 Pb; 0.001 to 0.0025 Mo; 0.0003 to 0.013 Bi; 0.0005 to 0.001 Ni; 0.00025 to 0.001 Co; 0.003 to 0.01 Zn; 0.0005 to 0.001 Cd; 0.001 to 0.01 Cu; 0.001 to 0.005 Sn; *11* - 0.001 to 0.034 Ag; 0.1 to 1.5 Pb; 0.001 to 0.005 Ni; 0.001 to 0.017 Co; 0.01 to 1.5 Zn; 0.001 to 0.014 Cd; 0.01 to 1.7 Cu); *12* - cross sections sampled.

constructed in order to intensify weak halos and to increase their reliability and practical application.

Porphyry copper deposits (based on an example of deposits from the Almalyk group). The Almalyk metalliferous deposit (in the northwestern extension of the Kuraminsky Range) is composed of a variegated complex of sedimentary, extrusive, and intrusive rocks of

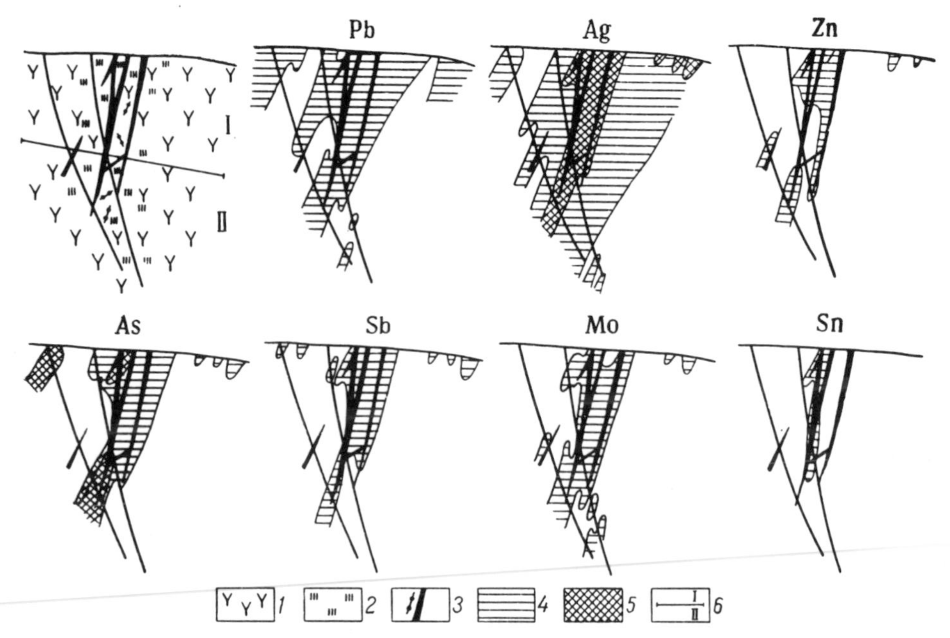

Fig. 24. Primary halos around ore bodies of a gold deposit in igneous rocks. *1* - andesite porphyries; *2* - altered rocks; *3* - gold-bearing veins; *4* and *5* - element contents, % (*4* - 0.001 to 0.05 Pb; 0.00001 to 0.0001 Ag; 0.005 to 0.05 Zn; 0.007 to 0.01 As; 0.003 to 0.01 Sb; 0.0001 to 0.001 Mo; 0.0001 to 0.005 Sn; *5* - 0.0001 to 0.001 Ag; 0.01 to 0.1 As); *6* - boundary of blocks.

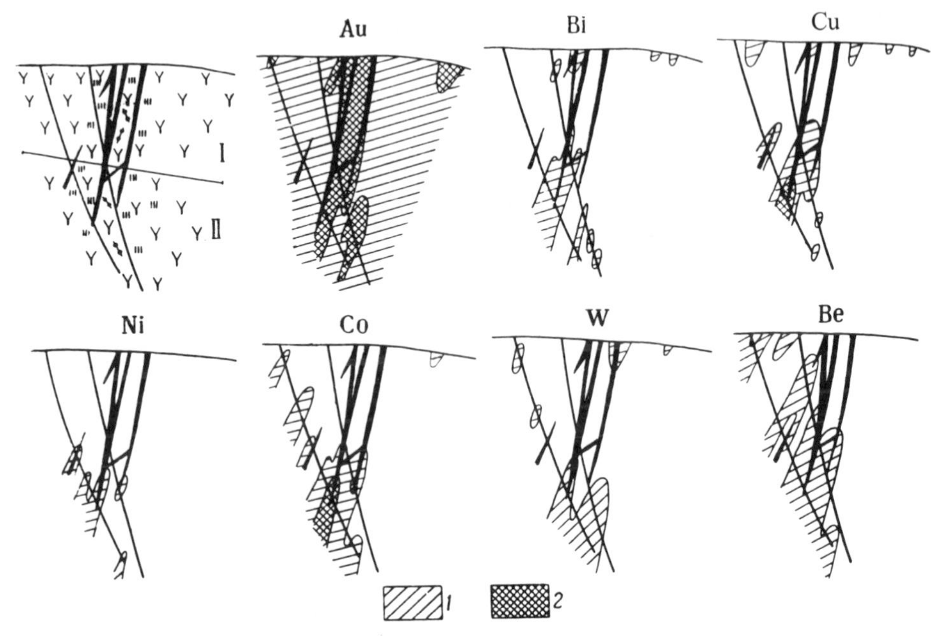

Fig. 25. Primary halos around ore bodies of the deposit (same deposit as Fig. 24). *1* and *2* - element contents, % (*1* - 0.000001 to 0.00001 Au; 0.0001 to 0.001 Bi; 0.01 to 0.1 Cu; 0.0005 to 0.001 Ni; 0.0005 to 0.001 Co; 0.0003 to 0.001 W; 0.0001 to 0.001 Be; *2* - 0.00001 to 0.001 Au; 0.1 to 0.5 Cu). Other designations as in Fig. 24.

different ages. The Lower Paleozoic is represented by Ordovician-Silurian metamorphic schists. The Middle Paleozoic is represented by Lower Devonian extrusives and by terrigenous carbonate deposits of the Middle Devonian to Lower Carboniferous. The Upper Paleozoic includes Carboniferous and Lower Permian sedimentary-igneous rocks which are widespread in the region.

Intrusive formations comprise about 70% of the region. Complexes of the Caledonian and Hercynian cycles are developed in the region. The Hercynian tectonic-magmatic cycle includes pre-batholith stages, batholith stages, and the stage characterized by minor porphyry intrusions. Dikes are very abundant, in particular dikes of granodiorite-porphyries, lamprophyres, diorite and diabase porphyrites.

The porphyry copper mineralization is concentrated near the stock-shaped granodiorite porphyry bodies which occur in fault zones on the northeastern and northwestern trends. Economic porphyry copper mineralization is usually confined to a gently sloping contact at the granodiorite porphyry intrusions. The enclosing rocks are propylitized, chloritized, and silicified.

Fig. 26 shows a cross section through a group of blind ore bodies at the northwestern Balykta deposits. A distinct vertical zonality was found in the structure of the halos. It is expressed in the selective accumulation of barium, lead, arsenic, and zinc in the supraore part of the halos, whereas molybdenum, copper, beryllium, and tin accumulate where the ore bodies wedge out.

Copper deposit. *Kafan.* Igneous and sedimentary-igneous formations of the Middle and Upper Jurassic occur in the geological structure of the Kafan metalliferous region situated in the southeastern part of Armenia. The Lower Bajocian epidotized andesite porphyrites and their lava breccias, which make up the lower igneous suite, are the oldest. The top portion is composed of andesite-dacite porphyrite flows up to 600 meters thick. They are conformably overlain by the Middle-Upper Bajocian igneous suite. The base is formed by a relatively persistent member of sandstones and tuffaceous sandstones, and then quartz porphyries, plagioclase porphyrites, and their tuffs and tuffaceous breccias follow. The Upper Jurassic suite is composed of tuffaceous breccias and conglomerates with thin flows of plagioclase and diabase porphyrites and rare lenticular bodies of limestones and tuffs. A thin layer of basal conglomerates lies at the bottom of the suite, which is 600 to 800 m thick.

Economically important copper mineralization at the Kafan ore field was found in a Middle Jurassic (Middle and Upper Bajocian) igneous-sedimentary rock series.

Rather extensive and strong primary halos formed by numerous elements have been detected in the vicinity of the ore bodies making up the deposit. Vertical zonality of the halos is expressed by the selective accumulation of barium, arsenic, and lead above the ore bodies, whereas bismuth, cobalt, and nickel show higher contents at the lower level of the ore bodies (Fig. 27).

Polymetallic vein deposits. *Eastern Kanimansur.* This deposit is located at the southern end of the Kuraminsky Range (Tadzhik, SSR). The ore field is composed mainly of Upper Paleozoic extrusive rocks.

The enclosing igneous rocks of Upper Paleozoic age are represented by tuffaceous lavas, tuffaceous breccias of quartz porphyries, felsite porphyries, and andesite porphyrites. There are numerous dikes of felsite porphyries, quartz porphyries, and diabase porphyrites within these rocks.

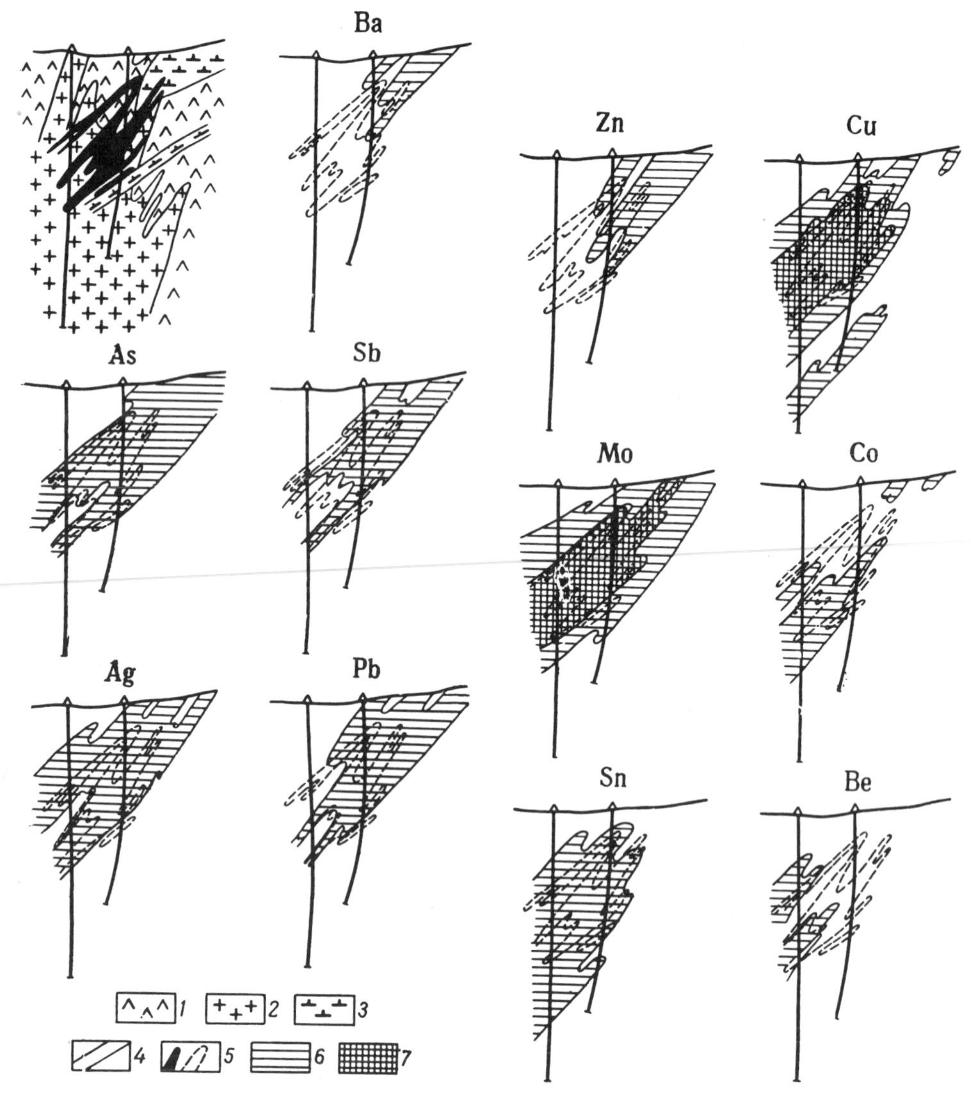

Fig. 26. Primary halos around blind ore bodies.
1 - syenite-diorites; *2* - granodiorite-porphyries; *3* - quartz porphyries; *4* - zones of rock disintegration; *5* - ore bodies; *6* and *7* - element contents, % (*6* - 0.1 to 0.5 Ba; 0.001 to 0.01 As; 0.001 to 0.07 Sb; 0.00001 to 0.0001 Ag; 0.01 to 0.07 Pb; 0.01 to 0.5 Zn; 0.1 to 0.3 Cu; 0.0001 to 0.001 Mo; 0.001 to 0.008 Co; 0.001 to 0.016 Sn; 0.0002 to 0.0017 Be; *7* - 0.3 to 1.0 Cu; 0.001 to 0.006 Mo).

This deposit is structurally associated with the southeastern limb of the Tavak anticline, and is complicated by a series of faults showing northeastern or easterly strikes and dipping in a south-southeast direction at angles of from 70 to 90°. The faults are thick zones of mechanically disintegrated and hydrothermally altered rocks, consisting in places of quartz, barite, fluorite, and sulfides.

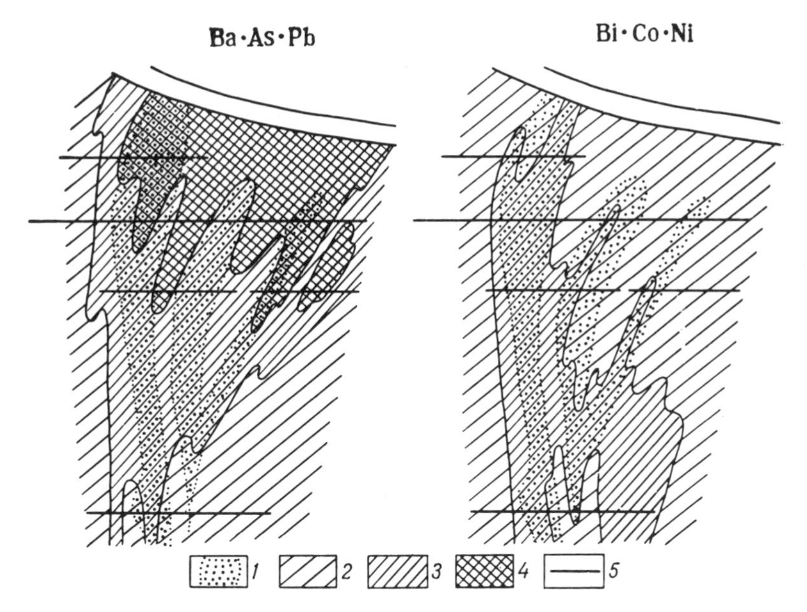

Fig. 27. Multiplicative halos of indicator elements in the vicinity of a stockwork zone.
1 - stockwork zone; *2* to *4* - product of contents (*2* - Ba·As·Pb, from $1 \cdot 10^{-11}$ to $1 \cdot 10^{-10}$; Bi·Co·Ni, from $1 \cdot 10^{-12}$ to $1 \cdot 10^{-10}$; *3* - Ba·As·Pb, from $1 \cdot 10^{-9}$ to $1 \cdot 10^{-8}$; Bi·Co·Ni, $1 \cdot 10^{-9}$ to $1 \cdot 10^{-8}$; *4* - Ba·As·Pb, from 1.10^{-7} to $1 \cdot 10^{-5}$); *5* cross sections sampled.

The above-mentioned faults are complicated by numerous fringing fissures, which, like the major structures, control the distribution of the ore bodies. The ore bodies in the deposit are confined to the Major Fault, to the Southern Branch Fault, and to the fringing fissures. They are lenticular in shape with numerous apophyses.

The following stages of mineralization have been recognized in the deposit: (1) quartz-hematite; (2) quartz-sulfide (copper-bismuth ores); (3) barite-fluorite-sulfide (lead-silver ores); (4) quartz-carbonate.

The copper-bismuth bodies have the following mineral composition: chalcopyrite, pyrite, bismuthinite, cosalite, and alkinite. Arsenopyrite, tetrahedrite, sphalerite, galena, and silver sulfosalts are present in lesser amounts, and the gangue minerals are represented by quartz and fluorite. The major minerals in the lead-silver ores are galena, sphalerite, pyrite, arsenopyrite, chalcopyrite, tetrahedrite, and stephanite. Quartz, barite, and fluorite are the principal gangue minerals.

Processing of the geochemical data has shown that halos of numerous chemical elements develop in the vicinity of the ore bodies occurring in the deposit. Figs. 28 and 29 show primary halos delineated around the Major zone. In this zone the ore bodies containing silver and lead occur close to the surface and are replaced by copper-bismuth bodies at greater depths.

Primary halos of most indicator elements are much wider than the ore bodies themselves. It is difficult to judge the vertical extent of the halos because of the shallow occurrence of the ore bodies. This necessitated studies of the distribution patterns exhibited by

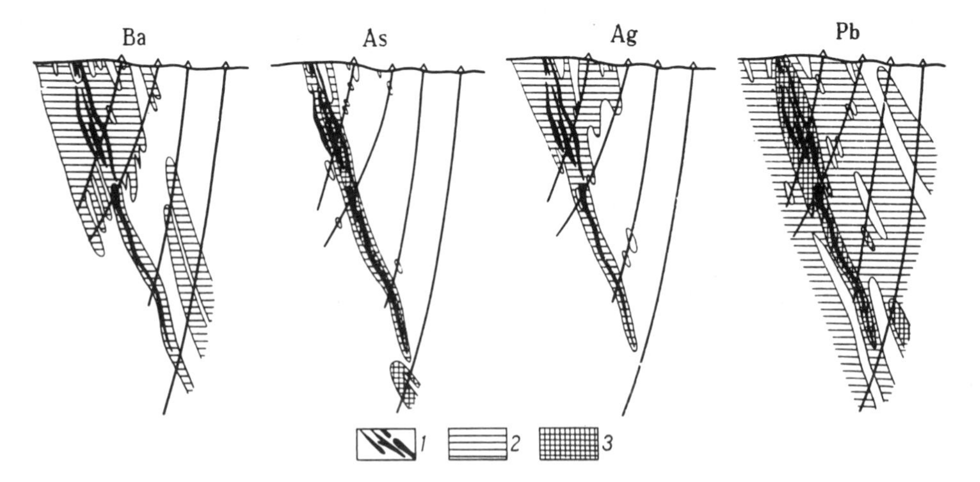

Fig. 28. Primary halos around ore bodies in the Major zone of the deposit (Eastern Kanimansur).
1 - ore body; *2* and *3* - element contents, % (*2* - Ba: 0.04 to 0.05; As: 0.0005 to 0.01; Ag: 0.00002 to 0.0001; Pb: 0.005 to 0.1; *3* - As: 0.01 to 0.1; Ag: 0.0001 to 0.06; Pb: 0.1 to 1.5).

the indicator elements only at the level of the ore bodies. Nevertheless, a distinct vertical zonality has been revealed in the structure of the primary halos. This zonality is expressed by a regular alternation of the halos formed by different elements. Certain elements, such as barium, silver, and lead (in part), form rather broad and distinct halos only at the upper levels of the cross section, more specifically, at the level at which the lead-silver ore bodies occur, and higher. The halos formed by these elements down dip from the ore bodies become narrow abruptly. This is especially true of the barium and silver halos. On the other hand, the halos formed by copper, bismuth, cobalt, molybdenum, tin, and tungsten (the latter two are not illustrated), are developed essentially at the lowest level of the cross section (see Figs. 28 and 29).

Vertical zonality in the element distribution is best exemplified by pairs of elements, such as barium (or silver) - cobalt (or tungsten). The halos formed by these elements are mirror images of one another.

From Fig. 28 it can be seen that the distribution of arsenic in the cross section differs from that of other elements. The arsenic halo is quite distinct at the level at which the lead-silver ores occur. It narrows down somewhat in the lower parts of the cross section, and then expands again, attaining a considerable width and intensity in that part of the cross section below the ore bodies. This behavior of arsenic is due, in all probability, to the mineral zonality resulting from the vertical alternation of the mineral forms in which this element occurs. It is most likely that in the upper horizons arsenic occurs mainly in the form of tetrahedrite, which is superseded by arsenopyrite at greater depths. This assumption is corroborated by data from mineralogical analyses. We will discuss this problem in greater detail below.

Vertical zonality exhibited by primary halos of the deposit being studied can be expressed through the following sequence of indicator elements: barium — arsenic$_1$ — silver — lead — zinc — copper — bismuth — cobalt — tin — arsenic$_2$ — tungsten. The place

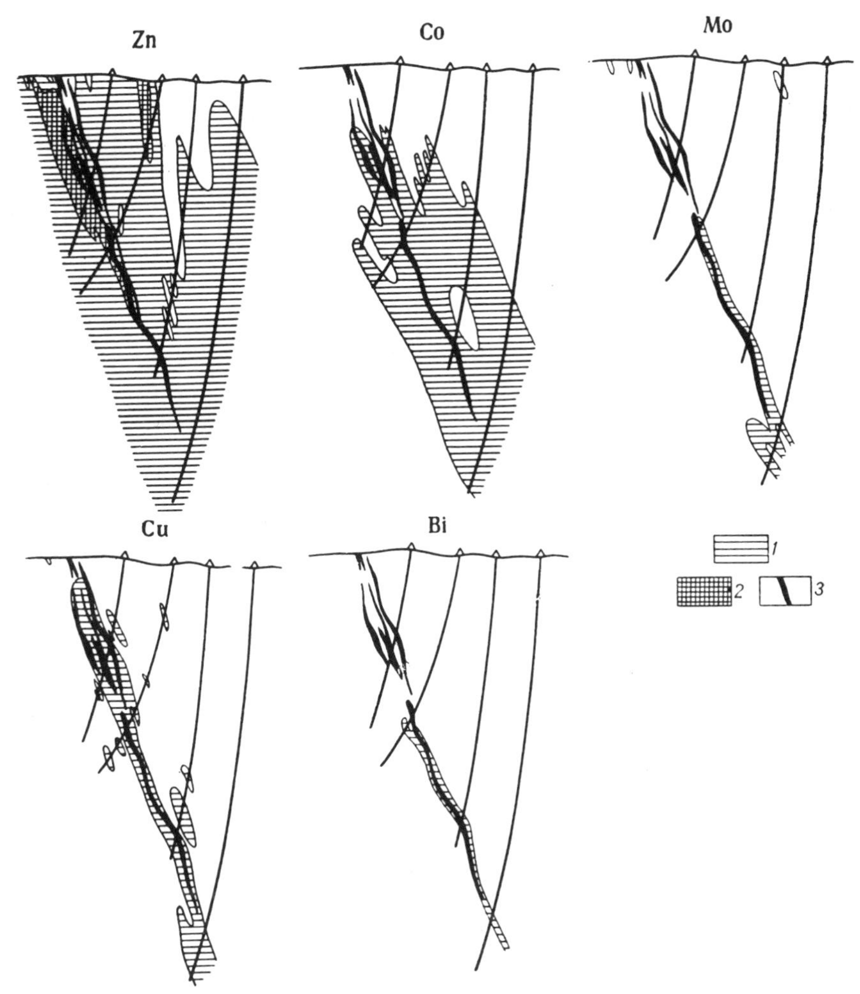

Fig. 29. Primary halos in the vicinity of ore bodies in the Major zone of the Eastern Kanimansur deposit.
Element contents, %: *1* - Zn: 0.005 to 0.01; Cu: 0.0005 to 0.01; Bi: 0.00008 to 0.001; Co: 0.0002 to 0.006; Mo: 0.0001 to 0.008; *2* - Zn: 0.01 to 0.8; *3* - ore body.

occupied by arsenic in this sequence is determined not on the basis of the zonality index but, rather arbitrarily, by taking into account the above-mentioned mineral zonality. Arsenic$_1$ corresponds to the occurrence of tetrahedrite, whereas arsenic$_2$ corresponds to the occurrence of arsenopyrite.

Arkhon. This is a polymetallic metal deposit in the Northern Caucasus (on the right bank of the Arkhon-don River). The region is composed of metamorphic, intrusive, and igneous-sedimentary rocks. Numerous faults, which control the pattern of the polymetallic mineralization, are widespread in the region.

The polymetallic veins of the Arkhon deposit are located within steeply dipping fissures with a northeastern trend. The most common ore minerals are pyrite, sphalerite, galena, pyrrhotite, chalcopyrite, and arsenopyrite. Tetrahedrite, argentite, jamesonite, and chalcocite are less abundant.

Processing of the geochemical sampling data has shown that the ore veins, although they are not very thick, are surrounded by rather broad and extensive halos formed by lead, zinc, copper, silver, cobalt, and bismuth. The distribution patterns of arsenic and antimony could not be studied because the analytical technique was not sensitive enough to detect these elements.

Fig. 30 shows primary halos formed by silver, lead, zinc, copper, and cobalt around the "Slepaya" ("Blind") vein. The parts of the halos formed above the ore bodies have been plotted based on sampling data from the bedrock at the surface, and from data above the

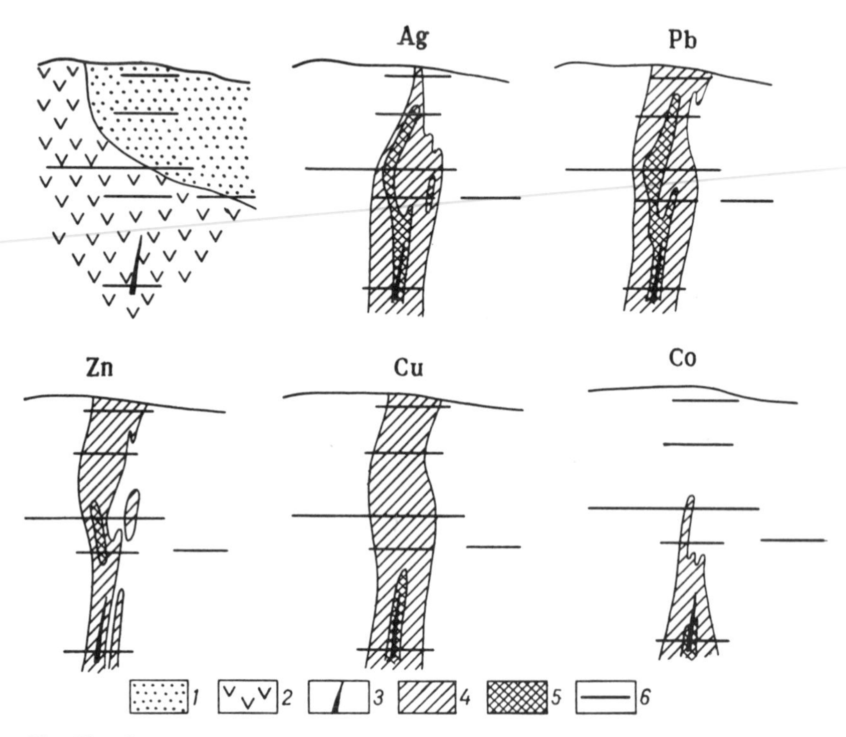

Fig. 30. Primary halos around the Arkhon ore body.
1 - sandstones; *2* - albitophyres; *3* - multielement vein; *4* and *5*- element contents, in background units (*4* - Ag: 1 to 3; Pb: 1 to 5; Zn: 1 to 3; Co: 1 to 2; *5* - Ag, Zn, and Cu: greater than 3; Pb: greater than 5; Co: greater than 2); *6* - cross sections sampled.

ore vein along various elevation levels (the relief above the vein is steeply dipping at an angle of between 35 and 40°). The halos have subsequently been plotted in a vertical plane. It can be readily seen that the halos formed by these elements are much larger than the ore body itself and, like the ore body, they are markedly extended vertically. The vertical extents of the halos over the ore body formed by silver, copper, and lead are considerable (more than 150 m).

Thus, a distinct vertical zonality exists in the structure of the primary halos. This zonality is due to the development of stronger silver and lead halos in the cross section above the ore bodies, whereas cobalt forms a strong halo in the lowermost level of the ore body (Fig. 30).

Uranium deposit (in granites). This particular deposit occurs at the southeastern termination of a regional fault which forms several branches in the vicinity of the ore field. Granites of Late Variscan age and their veined analogs serve as the enclosing rocks.

Two stages of mineralization have been recognized within the ore field: (1) the greisen stage, which is responsible for the formation of numerous greisenization zones; and (2) the hydrothermal stage, which resulted in the formation of the uranium ore bodies.

The greisenization zones vary in extent and average a few hundred meters in length (Fig. 31). Their thicknesses range from a few to several dozens of meters. The greisenization zone is a series of steeply dipping fissures fringed by material with a quartz-muscovite and a quartz-topaz composition.

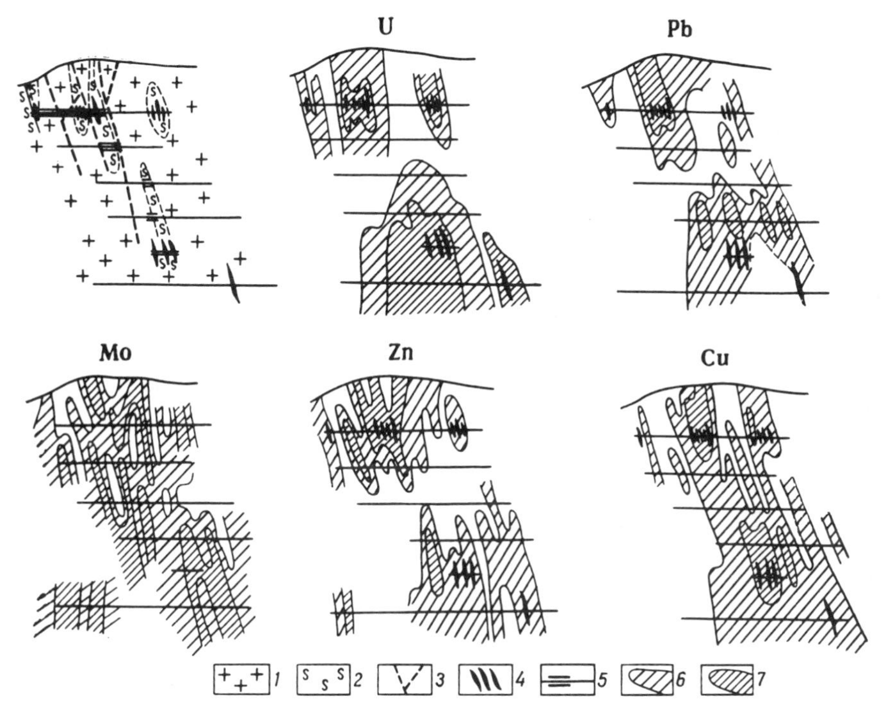

Fig. 31. Primary halos in the vicinity of hidden uranium ore bodies.
1 - leucocratic granites; *2* - greisenization zones; *3* - fractures; *4* - ore bodies; *5* - sampled mine workings and boreholes; *6* and *7* - primary halos (*6* - first anomalous field; *7* - second anomalous field).

The more recent hydrothermal stage is divided into two periods of mineralization. One is a quartz-carbonate-sulfide stage, and the other is a fluorite-uraninite stage. Veins and veinlets of the quartz-carbonate-sulfide stage are not numerous. They are composed basically of gangue minerals, such as quartz, calcite, fluorite, dickite, barite, albite, ankerite, chlorite, small amounts of galena, sphalerite, chalcopyrite, tetrahedrite, and hematite. These veins range in thickness from a few millimeters to 5 centimeters, and they seldom extend more than a few meters.

Veins and veinlets of the fluorite-uraninite stage intersect greisenization zones. They occur most frequently within greisenization zones, however, sometimes they are found outside these zones. The results of a detailed geochemical sampling of the greisenization zones outside the areas where the fluorite-uraninite mineralization stage is developed, have shown that the greisenization process was not accompanied by any appreciable introduction of elements typical of the uranium mineralization (Grigorian and Yanishevskii, 1968). This suggests that the formation of the uranium ore bodies and of the halos around them is due to the effect of the later uranium-bearing hydrothermal solutions on both the greisenized and unaltered enclosing rocks.

The ore bodies in the deposit are composed of numerous diversely oriented small veinlets which form a stockwork. They are lenticular in shape, and are elongated along the strike and dip. Their long axes coincide with the trend of the greisenization zones. The uranium ore bodies dip steeply, in conformity with the direction of the greisenization zones, and they are blind (hidden).

Geochemical sampling has revealed broad and distinct halos formed by uranium, lead, molybdenum, zinc, and copper around the ore bodies.

Fig. 31 shows primary halos developed in the vicinity of two groups of blind ore bodies: (1) the upper group (horizons II and III), and (2) the lower group (horizons VI and VII). The ore bodies in the lower horizons are the largest, and the uranium halos there have the maximum width and intensity. Lead, zinc, and copper, unlike uranium, form maximum concentrations above the ore bodies. Zonality is especially pronounced in the case of the lead and zinc halos. Therefore, these two elements are sensitive indicators of the blind ore bodies at depth. Each group of ore bodies is clearly indicated by specific halo "caps" formed by these elements. From Fig. 31 it can be seen that the halos formed by these elements in the vicinity of the upper group of the ore bodies are larger and stronger than the lower "wave". This occurs despite the fact that the lower ore bodies are larger. These data show that a certain zonality probably exists for the deposit as a whole, and not only in the halos developed around each individual ore body (or around a group of ore bodies). This zonality, which is related to the entire deposit, is expressed in the development around ore bodies located at higher levels (other conditions being equal), of stronger halos of the elements which usually accumulate above the ore bodies.

Mercury deposit. *Agyatag.* Interpretation of the results of geochemical sampling of the ore bodies and their enclosing rocks in mercury deposits is quite difficult. This is because of the complicated geological structure of such deposits and, primarily, because of the great spatial variability exhibited by the mineralization. Also significant is the fact that the elements which accompany mercury form very weak halos because of their low contents in ores. Therefore, the reliable detection and delineation of the halos formed by the elements accompanying mercury is a very complicated problem. This is due to (1) the appreciable effect of random analytical errors, and (2) particularly to the interferences associated with the redistribution of the background abundances of the chemical elements during the process of mineralization. It is obvious that similar redistributions also take place during the formation of halos in other types of deposits (for example, polymetallic). However, in these latter cases the effects of the redistribution are counterbalanced by an intensive introduction of elements during the processes of the halo formation.

Difficulties involved in studies of the mercury distribution patterns formed around ore bodies are also caused by other reasons. Most important is the extremely high mobility of mercury which leads to the broad dispersion of this element, and to the formation of numerous anomalies detached from the ore bodies. Furthermore, the reproducibility of the analytical results from the geochemical sampling for mercury is extremely low, not only because of the poor accuracy of the analytical procedures used, but also because of the large errors associated with the processing and storage of the samples. All these factors must be taken into account in studies of the geochemical halos associated with mercury mineralization.

Below we will describe primary halos detected in the Agyatag mercury deposit, located in the eastern part of the Azerbaidzhan SSR, within the Sevan-Karabakh mercury belt. The enclosing rocks in the deposit consist mainly of tectonic breccias, and partly of siliceous brown mudstone, dacite, serpentinite, and listwanite. The mercury mineralization is of the veinlet-dissemination type, is characterized by a very irregular distribution, and is confined to the mudstone tectonic breccias.

Primary halos from the Agyatag deposit were studied on the basis of the results from bedrock sampling at the surface, and of sampling in the underground mine workings (Fig. 32). Strong and broad halos formed by several elements were revealed in the cross section being described. Two groups of these halos can be identified in terms of the element distribution in the cross section: (1) the halos formed by mercury, arsenic, copper, lead, and, to a lesser extent, tin, decrease both in size and intensity with depth; (2) the halos formed by nickel, cobalt, molybdenum, beryllium, and tungsten, on the contrary, expand with depth. Zinc does not exhibit any distinct selective accumulation in the vertical direction.

The results of geochemical sampling of the ore bodies and of their enclosing rocks at the Agyatag deposit have made it possible to determine the characteristics of formation of primary halos, as well as to assess the ore reserves in the lower horizons of the cross section. As shown in Fig. 32, the halos of the indicator elements in the parts of the cross sections above the ore body thin out with increasing depth (mercury, arsenic, etc.), whereas the halos formed by cobalt, tungsten, and certain other elements characteristic of deeper parts of the cross sections develop with increasing depth. These data indicate a wedging out, rather than a "bending", of the ore body. The absence at depth of another "wave" of the halos formed by the elements normally occurring above the ore body, ruled out the possibility that there may be a blind ore body at still greater depths.

It is necessary to stress the difficulties involved in the prospecting for mercury deposits using monoelement halos, such difficulties being due mainly to the low intensity of the halos. Therefore, it seems worthwhile to use composite, in particular, multiplicative halos. Both total halos (composite of all elements, to detect and delineate the anomaly), and partial halos (two or three elements, to determine the erosion level of the anomaly), should be used.

Primary Halos in the Vicinity of Gently Dipping Ore Bodies

The characteristics of primary geochemical halos around gently dipping ore bodies are considered below based on examples from (1) the polymetallic skarn deposits at Tutly I (the

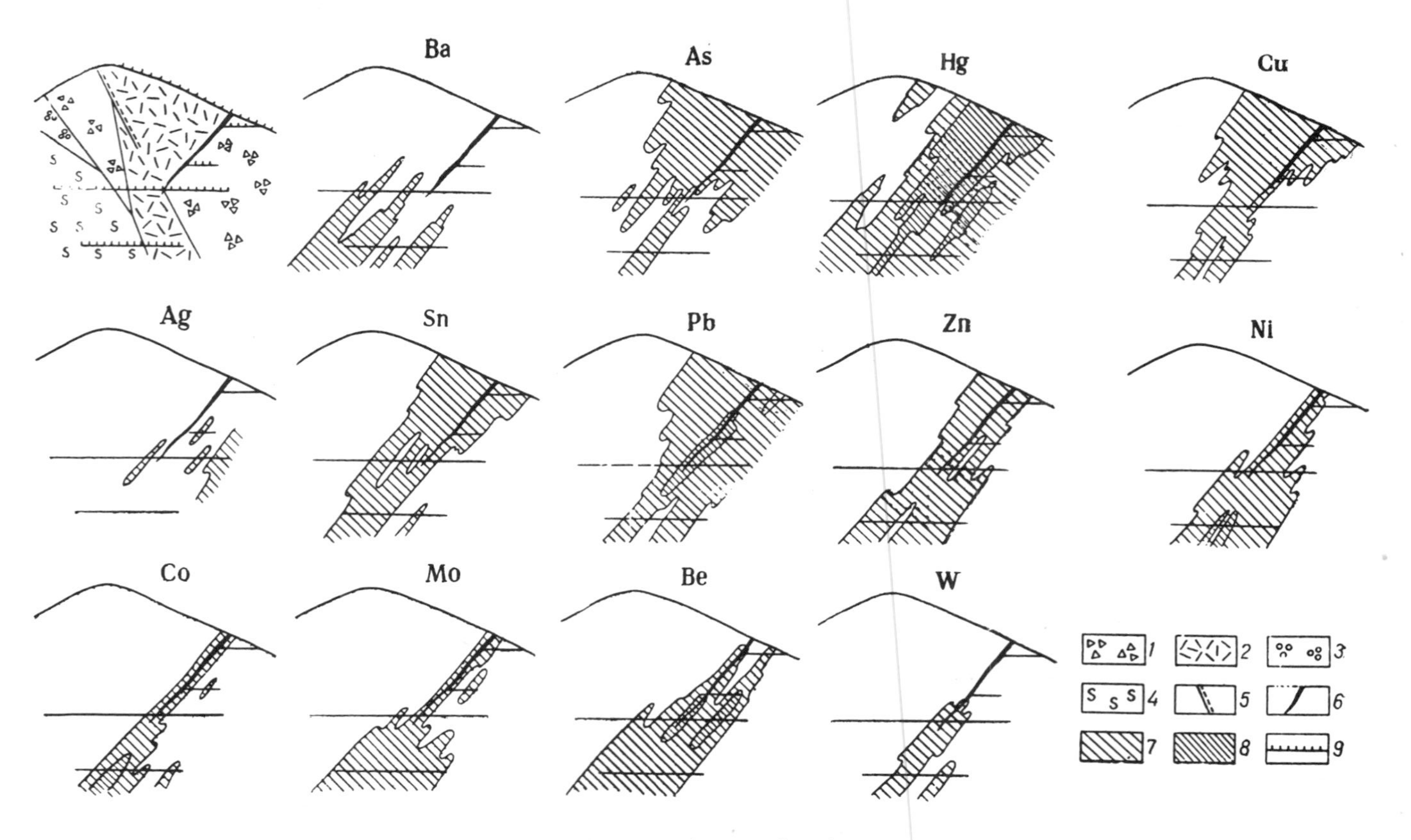

Fig. 32. Primary halos in the vicinity of the ore bodies at the Agyatag deposit.
1 - tectonic breccias; *2* - dacites; *3* - listwanites; *4* - gabbroic rocks; *5* - fault zone; *6* - ore body; *7* and *8* - element contents, % (*7* - Ba: 0.005 to 0.054; As: 0.003 to 0.04; Ag: 0.00005 to 0.00012; Cu: 0.003 to 0.02; Hg: 0.000003 to 0.00001; Sn: 0.0001 to 0.0003; Pb: 0.0005 to 0.001; Zn: 0.003 to 0.01; Ni: 0.1 to 0.1; Co: 0.005 to 0.01; Mo: 0.0001 to 0.0008; Be: 0.00005 to 0.0001; W: 0.00016 to 0.0006; *8* - Hg: 0.00001 to 0.003; Pb: 0.001 to 0.0015; Ni: 0.1 to 0.9; Co: 0.01 to 0.14; Be: 0.0001 to 0.0005); *9* - sampling intervals.

Kurusai ore field in Soviet Central Asia) in the Nikolaevskoye (Maritime Territory), and from (2) the Sarycheku porphyry copper deposit (in Soviet Central Asia).

Tutly I. The ore bodies at this deposit occur within the limestone-arkose conglomerate series of the Akchinsky suite (S_{2-3}) which overlie an eroded surface of andesite-dacite porphyries in the same suite (Fig. 33).

The region in which the deposit occurs has numerous steeply dipping fractures with northeastern and southern trends. Individual blocks in the conglomerate series are shifted along these dislocations.

The ore bodies are gently dipping (10 to 15°), lenticular, and in places nearly horizontal skarn deposits up to several meters in thickness are found. Garnet (andradite-grossular) and pyroxene are the principal skarn minerals.

The polymetallic mineralization occurs mainly in the pyroxene skarns, and the main minerals are galena, sphalerite, pyrite, tetrahedrite, and chalcopyrite.

Mineralization took place in the weakest zones, specifically in places where individual interlayers of the conglomerates became separated. This explains the conformable occurrence of the ore bodies and the enclosing stratum, as well as a regular increase in the thickness of the ore bodies in the vicinity of the fractures.

Lithological factors played an important role in the localization of the mineralization. Ore bodies have been found only in the conglomerate series. At the same time, metalliferous solutions percolating through the andesite-dacite porphyrite series, which are not favorable for ore deposition, only formed zones of dispersed mineralization, and no ore accumulations.

Primary halos around ore bodies of the deposit were studied by means of geochemical sampling of cores obtained from boreholes in several cross sections. The most intense and largest halos are formed by lead, zinc, silver, arsenic, and antimony (Fig. 33). These halos differ from those formed in steeply dipping bodies in that they form distinct bands in conformity with the shape of the ore bodies, and they exceed the latter in size. The ore bodies in the deposit have an average horizontal extent reaching dozens of meters; however, the halos extend for much more than 200 m.

A well defined zonality in the structure of the primary halos exists: arsenic and antimony accumulate selectively in frontal (leading) parts of the halos, whereas the maximum concentrations of lead and zinc are especially characteristic within the area of the ore-body. Silver occupies an intermediate position. This zonality is similar to that of the halos formed by steeply dipping ore bodies, and it also reflects the direction of movement of the ore-forming solutions.

Fig. 34 shows multiplicative halos plotted for two groups of indicator elements: antimony and arsenic (which are indicators of the frontal parts of the halos), on the one hand, and lead and zinc (the main components of the ore), on the other. Fig. 34 shows that the partial multiplicative halos formed by these elements exhibit a distinct zonality, which is expressed quantitatively through a plot of the ratio of the linear productivities of the halos

$$\frac{Sb \cdot As}{Pb \cdot Zn}.$$

Sarycheku. This porphyry-copper deposit is located in the southern part of the Saukbulak ore field, at the northern end of the Kuraminsky Range.

The area in which the deposit is located is composed of Late Caledonian alaskites, an-

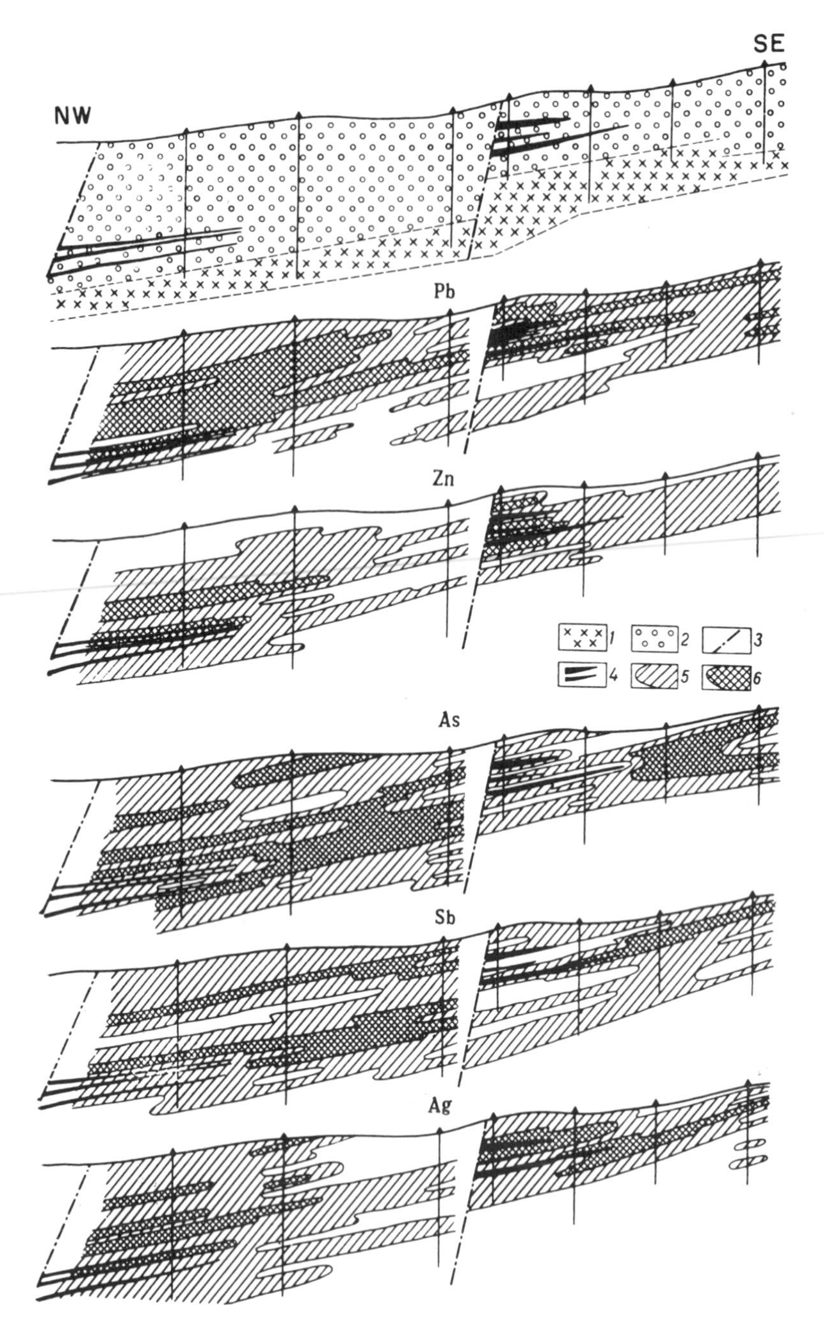

Fig. 33. Primary halos in the vicinity of the ore bodies in the Tutly I deposit.
1 - andesite porphyrites; *2* - conglomerates; *3* - faults; *4* - ore bodies; *5* and *6* - element contents, % (*5* - Pb: 0.003 to 0.03; Zn: 0.01 to 0.1; As: 0.005 to 0.01; Sb: 0.01 to 0.1; Ag: 0.00005 to 0.0001; *6* - Pb: 0.003 to 0.03; Zn: 0.01 to 0.1; As: 0.01 to 0.1; Sb: 0.1 to 0.3; Ag: 0.0001 to 0.03).

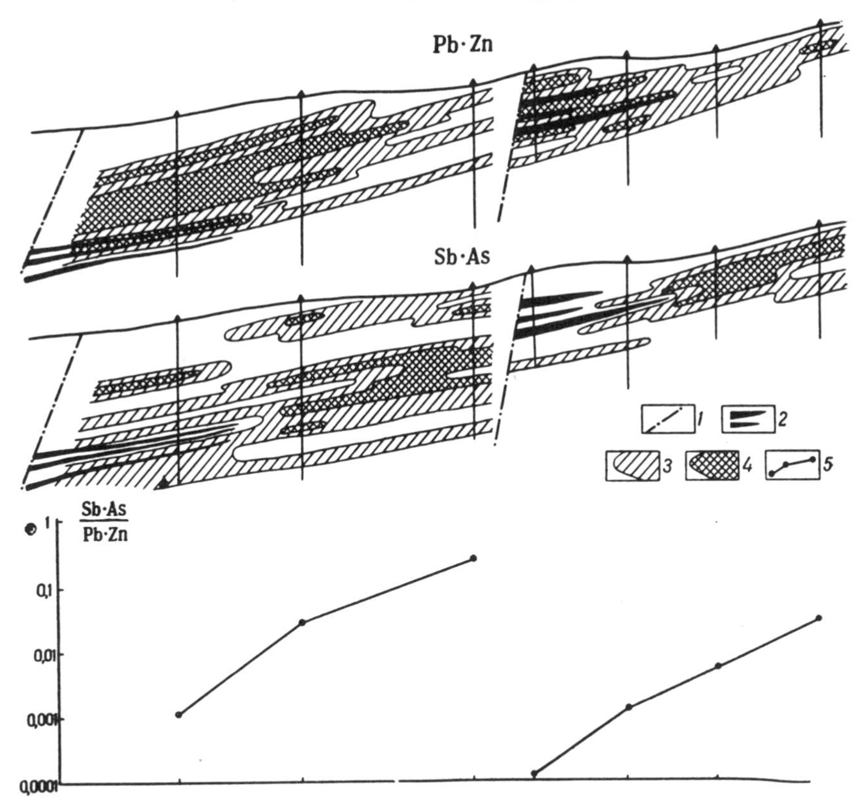

Fig. 34. Multiplicative halos in the vicinity of the Tutly I deposit.
1 - fault; *2* - ore bodies; *3* and *4* - values of products (*3* - Pb•Zn, 100 to $1000 \cdot 10^{-5}$; As•Sb, 1 to $10 \cdot 10^{-4}$; *4* - Pb•Zn, 1000 to $10000 \cdot 10^{-5}$; As•Sb, 10 to $200 \cdot 10^{-4}$); *5* - plots of the ratios of $\frac{Sb \cdot As}{Pb \cdot Zn}$

desite and andesite-dacite porphyries, quartz porphyries, "gray" granodiorites of Lower Devonian age, and arenaceous-carbonate rocks of Upper Devonian age (Fig. 35). The entire rock complex is pierced by intrusions of "rose-colored" granodiorites, syenite-diorite-porphyries of Permian age, and finally by young dikes ("black" granodiorite porphyries and lamprophyries). The main host rocks in the deposit are the quartz porphyries and the "rose-colored" granodiorites.

Morphologically, the primary porphyry-copper ores constitute a sheet-like deposit, gently pitching northeastwards. The mineralization developed in strongly jointed zones within quartz porphyries and "rose-colored" granodiorites. The mineral composition of the ore bodies is simple: pyrite, galena, sphalerite, barite, fluorite, and quartz occur together with chalcopyrite and molybdenite. Argentite and bismuthinite are not common.

Five stages of mineralization have been identified in the deposit: quartz-pyrite; quartz-molybdenite; quartz-chalcopyrite; quartz-polymetallic; and quartz-anhydrite-carbonate. Hydrothermal alterations in the rocks are represented by skarning, epidotization, silicification, sericitization, chloritization, and carbonatization.

Halos formed by several indicator elements have been revealed around the ore bodies in

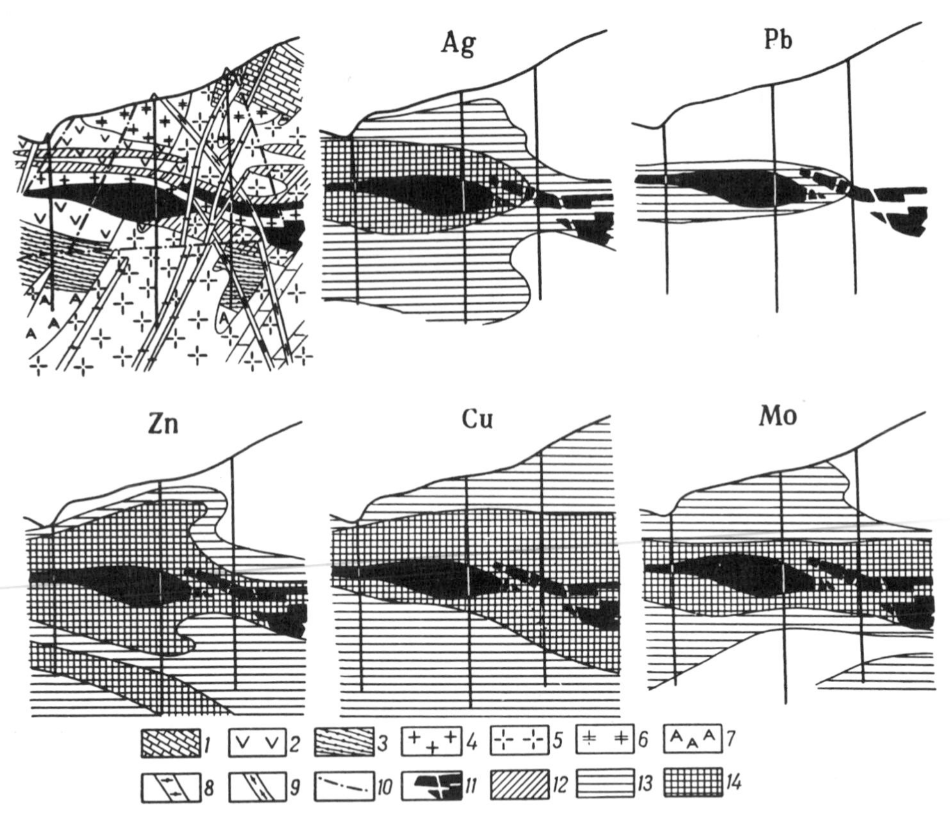

Fig. 35. Primary halos around ore bodies of the Sarycheku deposit.
1 - limestones; *2* - quartz porphyries; *3* - andesite porphyrites; *4* to *6* - granodiorites (*4* - Gushsai-type; *5* - rose-colored, *6* - gray); *7* - alaskites; *8* - granosyenite-porphyries; *9* - granodiorite-porphyries; *10* - fractures; *11* - ore body; *12* - sub-economic ores; *13* and *14* - element contents, % (*13* - Ag: 0.00003 to 0.0001; Pb: 0.005 to 0.24; Zn: 0.0037 to 0.0087; Cu: 0.01 to 0.11; Mo: 0.0003 to 0.001; *14* - Ag: 0.0001 to 0.0007; Zn: 0.0087 to 0.2; Cu: 0.11 to 1.5; Mo: 0.001 to 0.004).

the deposit. They are broad and extend along the trends of the ore bodies (Figs. 35 and 36). A zonality in the structure of the halos exists. It is due to the abrupt widening in those halos formed by silver, lead, and, to a lesser extent, zinc, along the rise of the ore bodies. In contrast, the halos formed by copper and molybdenum, expand in the opposite direction (see Figs. 35 and 36). The existence of this zonality is corroborated by the results of calculations of the ratios of the halo parameters, and is expressed as a sequence of indicator elements: silver - lead - zinc - copper - molybdenum (from the front of the halos toward the rear).

Similar to the Tutly I deposit, the zonality in the halos distinctly indicates the direction of movement of the metalliferous solutions, in all probability, from the ore-distributing Miskansky Fault (see Fig. 36).

In addition to the zonality discussed above, a vertical zonality also exists which is expressed as an appreciable upward shift of silver, which is the most mobile of the indicator elements.

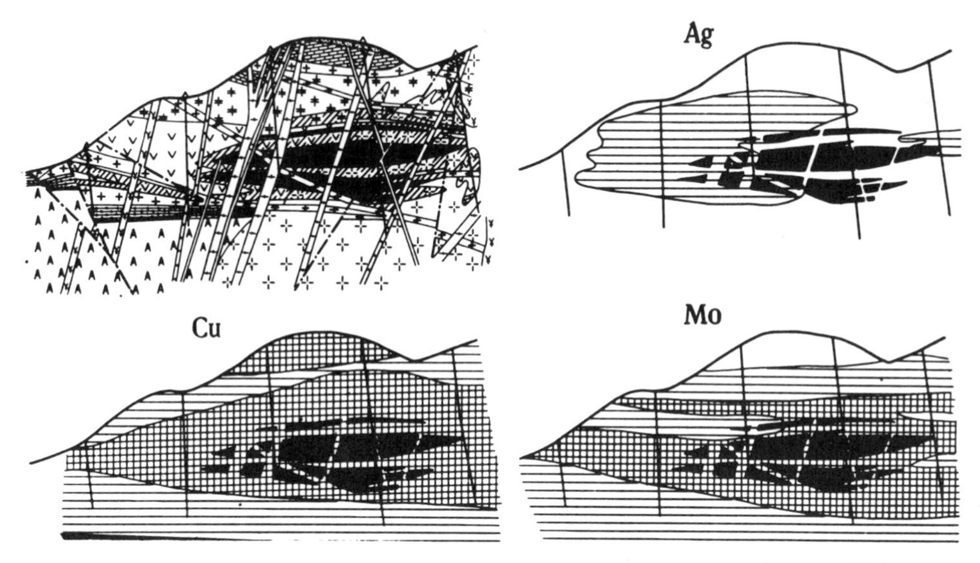

Fig. 36. Primary halos around ore bodies in the Sarycheku deposit (see Fig. 35 for conventional symbols).

Nikolaevskoye (Maritime Territory). This multi-element skarn deposit is made up of a sheet-like body occurring at the contact between Upper Triassic limestones and Upper Cretaceous igneous rocks (Fig. 37). The ore body, whose thickness reaches dozens of meters, plunges in northeastern and northwestern directions following the strike of the base of an igneous series, at angles of about 30°. This is a blind deposit, which occurs at depths of between 700 and 1100 m.

The igneous series overlying the deposit is composed largely of tuffaceous breccias and tuffs of quartz porphyries with thin interlayers and lenses of siltstones and sandstones.

The enclosing rocks are pierced by dikes of diabase porphyries, as well as by a small stock-like intrusion with a complex composition (gabbro-diorites, granite porphyries, and felsites) and are strongly altered. The sandstones and sedimentary breccias in the basement are chloritized and sericitized, and the limestones are altered to marble or metamorphosed to hedenbergite skarns at the contact with silicate rocks. The igneous and intrusive rocks above the ore body are strongly propylitized.

The mineralization occurs in the hedenbergite skarns. Sphalerite, galena, pyrite, and pyrrhotite are the principal ore minerals. The ore textures can be described as veined and disseminated.

Primary geochemical halos formed by several elements can be outlined at the deposit on the basis of the results obtained by sampling borehole cores. Fig. 37 shows the primary halos delineated in the cross section. The ore body occurs at a depth of 850 m.

Despite the gently pitching nature of the ore bodies, the Nikolaevskoye deposit differs appreciably from those considered above based on the characteristic development of the primary geochemical halos. The horizontal dimensions of the halos are not large, and they do not extend beyond the ore body. Another characteristic of these halos is their considerable vertical extent, which reaches 850 m. But even this distance is by no means a maximum

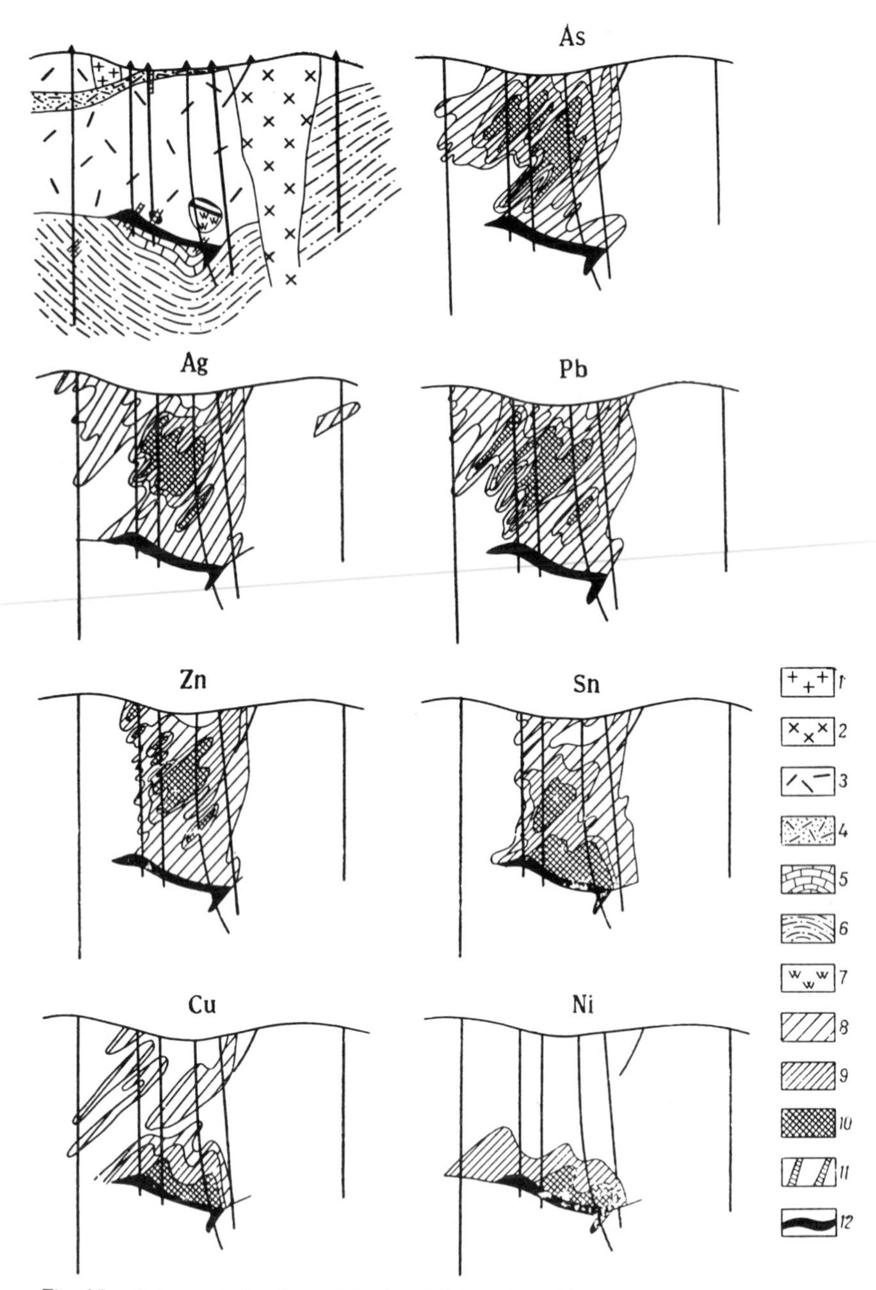

Fig. 37. Primary halos formed in the vicinity of the Nikolaevskoye ore body. *1* - quartz porphyries; *2* - gabbro-diorites; *3* - tuffs and tuff breccias of quartz porphyries; *4* - tuffs; *5* - limestones; *6* - siltstones and sandstones; *7* - siliceous rocks; *8* - to *10* - element contents, % (*8* - As: 0.003 to 0.005; Ag: 0.00003 to 0.0001; Pb: 0.004 to 0.01; Zn: 0.002 to 0.01; Sn: 0.0005 to 0.001; Cu: 0.003 to 0.01; *9* - As: 0.005 to 0.01; Ag: 0.0001 to 0.0005; Pb: 0.01 to 0.05; Zn: 0.01 to 0.1; Sn: 0.001 to 0.005; Cu: 0.01 to 0.05; Ni: 0.001 to 0.005; *10* - As: 0.01 to 0.1; Ag: 0.0005 to 0.001; Pb: 0.05 to 0.1; Zn: 0.1 to 0.3; Sn: 0.005 to 0.01; Cu: 0.05 to 0.1; Ni: 0.005 to 0.01); *11* - porphyrite dikes; *12* - ore body.

limit, because the halos are quite strong and broad at the surface.

Consequently, the Nikolaevskoye deposit, notwithstanding its gently pitching occurrence, can be considered an analog of a steeply dipping deposit based on the developmental features of its primary halos. This conclusion is corroborated by the distinct vertical zonality, which is similar to that observed around steeply dipping ore bodies with a polymetallic composition (the Kurusai, Harpenberg, and other previously described deposits). This zonality is characterized by the following mobility sequence of the indicator elements: arsenic - silver - lead - zinc - copper - tin - nickel (Fig. 37).

These data indicate that the primary halos at the Nikolaevskoye deposit were formed by ascending ore-bearing solutions. The economically important ores were deposited within skarned limestones which are particularly favorable as sites for the localization of mineralization. The paths for the movement of the ascending solutions must have been provided by steeply dipping fractures, some of which are now healed (sealed) by dikes formed by the diabase porphyrites (Fig. 37).

The above data suggest that the distribution of deposits with gently pitching ore bodies, as well as the characteristic features of the primary halos around them, are controlled primarily by the geologic-structural conditions at the site of mineralization. This observation must be taken into account both in the study and in the practical use of the primary halos.

Zonality in Primary Halos

Some specific types of zonality usually exist in the structure of primary halos. This zonality is due to the regularity in the spatial variations among the various characteristics of the halos.

Zonality in a halo is a vectorial concept. This is why the several types of zonalities, depending on direction, can be described. *Axial* zonality is expressed in the direction of movement of the ore-bearing solutions. In the case of steeply dipping metalliferous zones it coincides with the vertical (Fig. 38), whereas in the case of subhorizontal zones it coincides with horizontal zonalities.

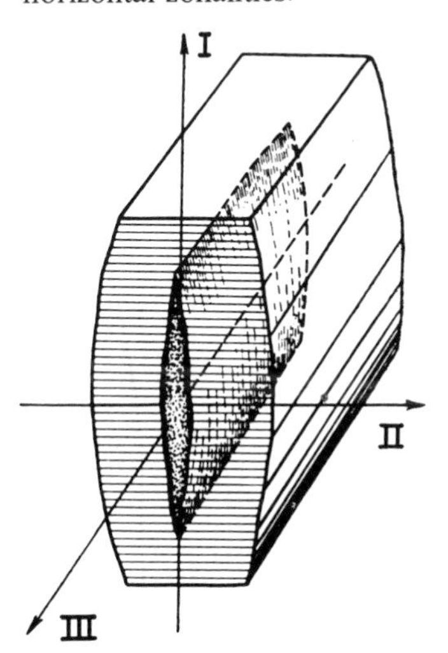

Fig. 38. Directions of axial (I), transverse (II), and longitudinal (III) zonalies in primary halos developed around a steeply dipping ore body.

Longitudinal zonality is a reflection of the halo structure along the strike, whereas *transverse* zonality, is a reflection of the halo structure across the strike. For subhorizontal ore bodies, the axial and longitudinal zonalities coincide with the horizontal, whereas transverse zonalities coincide with vertical zonalities (in the case of a concordant development of the halos). Axial zonality is the most important type from both theoretical and practical aspects. For the sake of brevity, axial zonality will be called simply zonality, in contrast to the longitudinal and transverse zonalities.

Table 30 lists sequences of the indicator elements which characterize axial zonality. The sequences are based on the procedures outlined above. This table compares zonality features of halos formed by hydrothermal deposits which differ in their composition and in their environments of formation.

Table 30. Zonality (axial) sequences of indicator elements.

	Type of deposit	Name of deposit	Zonality sequences
Steeply Dipping Bodies	Tungsten-molybdenum in skarns	Shurale	Ag, Pb, Zn, Mo, W, Ni, Co
	Bismuth in skarns	Chokadambulak	As, Pb, Ag, Zn, (Co, Cu, Bi), Ni
	Multielement in skarns	Nov. Kamarsai	As, Cd, Ag, Pb, Zn, Cu, Bi
		Aktash	(As, Sb), Ag, Pb, (Zn, Cu), Bi, Co, Sn, Mo
		Kurusai	Ba, (As, Sb), Ag, Pb, Zn, Cu, Bi, Co, (Mo, W), Sn
		Altyntopkan	Sb, Cd (Ag, Pb) (Sn_1, Zn), Cu, Bi, Ni, (Co, Mo, Sn_2, W, Be)
		Harpenberg	Sb, Ag, Pb, Zn, Cu, Sn, Bi (Ni, Mo, Co), W
	Gold ore	I	(Sb, As_1, Ag, Pb), Zn, Au, Mo, Cu, Bi (Co, Ni, As_2, W, Be)
		II	Sb, As, Ag, Pb, Zn, Au, Cu, Mo, Sn, Bi, Be, W, Co
		III	Ba, Sb, As, Ag, Pb, (Zn, Cu), Au, Mo (Sn, Bi, W)
	Tin ore (sulfide-cassiterite)	Zimneye, Ege-Khaya, Deputatskoye	Sb — (Ag, Pb, Zn) — (Sn, Cu, Bi) — (Co, Mo, W)
	Porphyry copper	Almalyk	Ba, As, Sb (Ag, Pb, Zn), Au, Bi (Cu, Mo) (Sn, Co, W, Be)
	Copper	Kafan	Ba, As, Pb, Zn (Ag, Sn), Cu, Bi, Co, Ni
	Multielement	Eastern Kanimansur	Ba, As_1, Ag, Pb, Zn, Cu, Bi, Co, As_2, W
		Arkhon	Ag, Pb, Cu, Mo, Co*
	Uranium	I (in felsites)	Ag, Pb, Zn, Cu, Mo, U*
		II (in granites)	Ag, Pb, Zn, Cu, Mo, U*
	Stratiform lead-zinc	Sumsar	Ba, As, Cu, Ag, Pb, Zn, Co, Ni, Be
	Mercury	Symap	Ba, Hg, Ag, Pb, Zn, Cu (Co, Ni, Sn), Mo
		Konchoch	Sb, As, Hg, Ag, Pb, Zn, Cu (Mo, Bi), (Co, Ni, W, Sn)
		Sakhalinskoye	As, Hg, Sb, Pb, Zn, Cu (Co, Ni)
		Agyatag	As, Hg, (Ag, Pb, Sn, Zn), Cu, Co, Ni (Be, Mo, W)
Gently Pitching	Multielement in skarns	Tutly I	Sb, As, Ag, Pb, Zn*
		Nikolaevskoye	As, Ag, Pb, Zn, Cu, Sn, Ni
	Porphyry copper	Sarycheku	Ag, Zn, Cu, Mo*
	Antimony-mercury	Tereksai	As, Sb, Hg, Cu, Ag, Pb, Zn, Be, Co, Ni
		Karakamar	As, Sb, Hg, Ag, Sn, Pb, Zn, Cu, Mo (W, Co, Ni)

* Only the main indicator elements are listed. Because of the lack of the samples no analyses were performed for a wider range of elements.

The data given in Table 30 show a striking similarity between the sequences of the halos formed in the vicinity of various deposits which differ in their composition and depositional environments. This suggests the following standard sequence for the principal indicator elements characteristic of the zonality of the deposits which have been investigated:

Ba—(Sb, As, Hg)** —Cd—Ag—Pb—Zn—Au—
(93)* (100) (87) (100) (80) (84) (87) (84)
—Cu—Bi—Ni—Co—Mo—U—Sn Be—W.
(86) (72) (50) (55) (48) (100) (66) (60) (72)

It follows from Table 30 that the zonality sequences are not absolutely identical and there are certain variations for individual elements. Therefore, the probability of occurrence of each element in the corresponding position in the sequence was calculated. The probabilities have been calculated with the help of single element zonality sequences, based on the results of studies of primary geochemical halos from many deposits.

The position of cadmium in the sequence was determined using the results of studies from a limited number of deposits and, consequently, its probability has not been calculated.

Certain elements show relatively low probabilities due to the shift in their positions in the zonality sequences for individual deposits. However, these shifts are insignificant (in most cases by one, and extremely rarely, by two or more positions). Low probabilities are characteristic mainly of the trace (admixed) elements which, in the majority of the deposits studied, form small and weak halos (nickel, cobalt, molybdenum, etc.). This suggests that a low probability is, to a certain extent, a consequence of fluctuations in the background contents of the elements or fluctuations in their redistribution, the relatively low accuracy of the analysis, etc., because halos with weak contrasts are more sensitive to the effect of this type of interference.

A conspicuous feature of the Sumsar and Tereksai deposits is the unusual position of copper in the zonality sequences. At these deposits, in contrast to the others, the copper halo has a higher position in the zonality sequence; that is, it is higher than that of lead and silver (Table 30). Copper in the ores and halos of these deposits is present mainly in the form of tetrahedrite. In those occurrences where copper is represented by chalcopyrite, it occupies a position on the right in the zonality sequence (after zinc). These data indicate the existence of mineral zonality in primary halos. This zonality is caused by the alternation in space of various minerals which are carriers of the same elements.

In all probability, similar mineral zonality is responsible for the characteristic behavior of arsenic in gold (Fig. 24) and multi-element deposits (Fig. 28). Strong halos formed by this element (arsenic) has been found in both the upper parts of the ore zone, and in the deep horizons where the ore bodies wedge out. This could, perhaps, be explained by the ap-

*The numbers in parentheses indicate the probability of occurrence of each element at the corresponding location (position) in the sequence (in %).

**Those elements whose relationships in the zonality sequence could not be unequivocally determined are given in parentheses.

pearance of arsenopyrite at greater depths, whereas in the upper horizons arsenic seems to be part of the tetrahedrite structure, as well as of the galena and sphalerite structures (the content of arsenic in these minerals reaches hundredths or even tenths of percent).

The validity of this statement is corroborated by the results of studies of the halos detected at the Sumsar deposit where arsenopyrite is totally absent, and where the copper ore containing arsenic (tennantite) is very abundant. Because of this, strong and broad halos formed by arsenic were detected at this deposit only in the upper (supraore) parts of the metalliferous zones.

These data show that the standard series of the zonality-indicating elements given above is valid only in the case of those deposits in which the forms of occurrence of the elements are stable (constant), or which are represented by minerals with identical spatial distributions. In those cases in which the elements are represented by different mineral forms within the halos, one and the same element in the zonality sequence may occupy different positions, as in the above-mentioned examples of copper and arsenic.

It is clear, however, that the existence of different mineral forms of the same element in halos will not always cause an appreciable change of this element in the zonality sequence. Changes in the zonality sequence may occur only in the case of similar abundances of the mineral forms whose distributions exhibit sharp spatial differentiation. For example, in some halos (as well as in ores) in a number of deposits, in addition to galena, other forms of the lead occurrences (various sulfates) were discovered. However, the amounts of these minerals, compared to those of galena, proved to be negligible; therefore, no shift of lead in the zonality sequence occurred. In the case of similar amounts of the "competing" mineralogical forms, the zonality sequence may exhibit considerable shifts, as illustrated in the above examples. This is why the quantitative, rather than the qualitative, nature of mineral zonality should be taken into account in the study of halo zonality and of the corresponding series upon which it is based.

Now that the above-mentioned mineral zonality has been taken into account, the standard zonality series of indicator elements becomes: (Sb, As_1, Hg) - Cu_1 - Cd - Ag - Pb - Zn - Sn_1 - Au - Cu_2 - Bi - Ni - Co - Mo - U - Sn_2 - As_2 - Be - W.

Naturally, examples of spatial differentiation of the elements in terms of the mineral forms in which they occur are not limited to those considered above. Further studies will make it possible to reveal additional features of mineral zonality, which will help to refine and expand the available standard element zonality series.

Halo zonality is a zonality of deposition, which is produced by changes in the internal equilibrium of an ore-forming solution (Ovchinnikov and Grigorian, 1970).

A relationship exists between the above generalized zonality in element distribution, and the stability of the common complexes they form in solution. The relative thermodynamic stabilities of the sulfide complexes formed by various metals have been calculated by Barnes and Czamanske (1970). These stabilities are expressed in arbitrary units (all of the complexes are of the same sulfide type):

Hg	Cd	Pb	Cu	Zn	Sn	Ni	Co
227	156	154	134	132	126	83	81

whereas the stabilities of chloride complexes at 25°C, according to Helgeson (1967), are: $Cu^{+2} < Zn^{+2} < Pb^{+2} < Ag^{+2} < Hg^{+2}$

The coincidence of the above series of elements with the standard zonality sequence, suggests that one of the major factors responsible for the geochemical zonality of primary halos is the stability of the compounds in which the chemical elements are transported by ore-bearing solutions.

The zonality in the primary halos also explains, in part, the reason for the stage-by-stage formation of hydrothermal deposits. All the deposits so far considered are of a multi-stage origin. Different stages may be, and sometimes are, characterized by specific indicator elements. However, their halos exhibit a single vertical zonality. This suggests that the individual stages in the formation of ore deposits discussed above are probably not separated by any appreciable amount of time. Thus, the alternation of the mineralization stages is a result of a spatial, rather than a time, differentiation from an ore-forming solution which is continually being supplied. The existing mutual intersections (overlap) of the products of the various stages may be due to movements within the ores which occur in different parts of the metalliferous region (Ovchinnikov and Grigorian, 1970).

In some cases, zonalities in the halos are disturbed (out of sequence). This can be explained by the spatial combination of two or more ore formations, which are different in composition. These are the so-called *multiformational* halos, and they are considered in detail later in this book.

Transverse zonalities in primary halos express their differences across the strike (width) of ore bodies. This type of zonality depends on the concentration of elements in ore bodies and halos and their mobilities, as well as on their background contents in the enclosing rocks. This is illustrated in the above (see Fig. 5) plots of element distributions around ore bodies (across their strikes), as well as by Table 31 which lists transverse zonality sequences in primary halos formed at deposits with different compositions. The elements in the sequences are given in the decreasing order of the widths exhibited by their halos at the upper levels of the ore bodies, or close to the surface.

Table 31. Transverse zonalities in primary halos.

Deposit	Element sequence
Lead-zinc in skarns	Barium, zinc, lead, arsenic, silver, copper, antimony
Veined lead-zinc	Lead, barium, zinc, silver, copper, arsenic, cobalt
Scheelite in skarns	Tungsten, molybdenum, copper, barium, zinc, lead
Gold ore in quartz	Gold, arsenic, bismuth, silver, lead, antimony, copper, beryllium, molybdenum, cobalt, zinc
Porphyry-copper	Gold, copper, molybdenum, silver, arsenic, antimony
Copper-bismuth	Copper, bismuth, lead, silver, arsenic, barium, zinc, cobalt
Uranium	Uranium, molybdenum, lead, copper, zinc, silver, mercury, arsenic, barium, copper, lead, zinc, nickel, silver, cobalt
Mercury	Mercury, arsenic, barium, copper, lead, zinc, nickel, silver, cobalt
Copper-molybdenum	Copper, molybdenum, zinc, lead, cobalt, nickel, tin, beryllium, tungsten, bismuth, arsenic, barium, silver
Sulfide-cassiterite	Tin, silver, zinc, lead, copper, molybdenum
Stratiform lead-zinc	Silver, lead, copper, arsenic, barium, cobalt, zinc, nickel

It follows from Table 31 that transverse zonalities, which depend on the ore composition, have specific features in the case of each deposit, unlike vertical (axial) zonalities which are the same for deposits differing in composition. The first place in the sequence is usually occupied by elements which are the major economically valuable ore components.

This dependence of the width of the halo on the ore composition makes it possible to use the transverse zonality as a criterion in the determinations of the composition of blind mineralization. Sequences of transverse zonalities, based on average contents expressed in units of the average background contents, can also be used for this purpose. These sequences do not differ appreciably from those given in Table 31. They are used in cases when, for some reason, the widths of the halos in the transverse direction could not be determined (e.g., the halos are not fully delineated because of limited sampling).

Longitudinal zonality is expressed as the regular variation in the parameters of a halo along the strike of the metalliferous zone, including halos formed in both the ore bodies and in their vicinity. The elements show a distinct differentiation in this direction, corresponding with the axial zonality. A longitudinal zonality also reflects the movement of solutions in the plane of the metalliferous zones.

Fig. 39 illustrates longitudinal geochemical zonality in the structure of primary halos. The contour lines show variations in linear productivities of the halos. The linear productivities were calculated using the data from boreholes and adits, which cross the metalliferous zone, as well as from bedrock sampling at the surface. The productivities were calculated at the points of intersection with the plane of the metalliferous zone (the example is from the Zimneye tin ore deposit).

Fig. 39 shows that as one goes away from the massif, the halos formed by antimony, lead, and silver become markedly stronger. The halos formed by copper (not shown), bismuth, and tungsten are stronger in the immediate vicinity of the massif. The sequence of longitudinal zonality (antimony, silver, lead, copper, bismuth, and tungsten) corresponds exactly to the vertical zonality sequence exhibited by primary halos of the tin ore deposits listed in Table 30.

Longitudinal zonality is especially pronounced if one considers changes in the ratios of the linear productivities in the multiplicative halos above (lead, silver, antimony) and below (copper, bismuth, tungsten) the ore body. This indicator ratio is $0.00n$ to $0.n$ (where n is 1 to 9) in the vicinity of the intrusion. It increases to $n \cdot 10^3 - n \cdot 10^4$ as one goes away from the intrusion (Fig. 39). Vertical zonality is also distinctly expressed in the plane of the ore zone. It can be seen in the shift in the fields of the maximum values of the indicator ratio in an upward direction in the cross section. This indicates that the composite vector of the movement of the metalliferous solution in the zone below the ore body has two components: a horizontal component (longitudinal zonality) and a vertical component (vertical zonality).

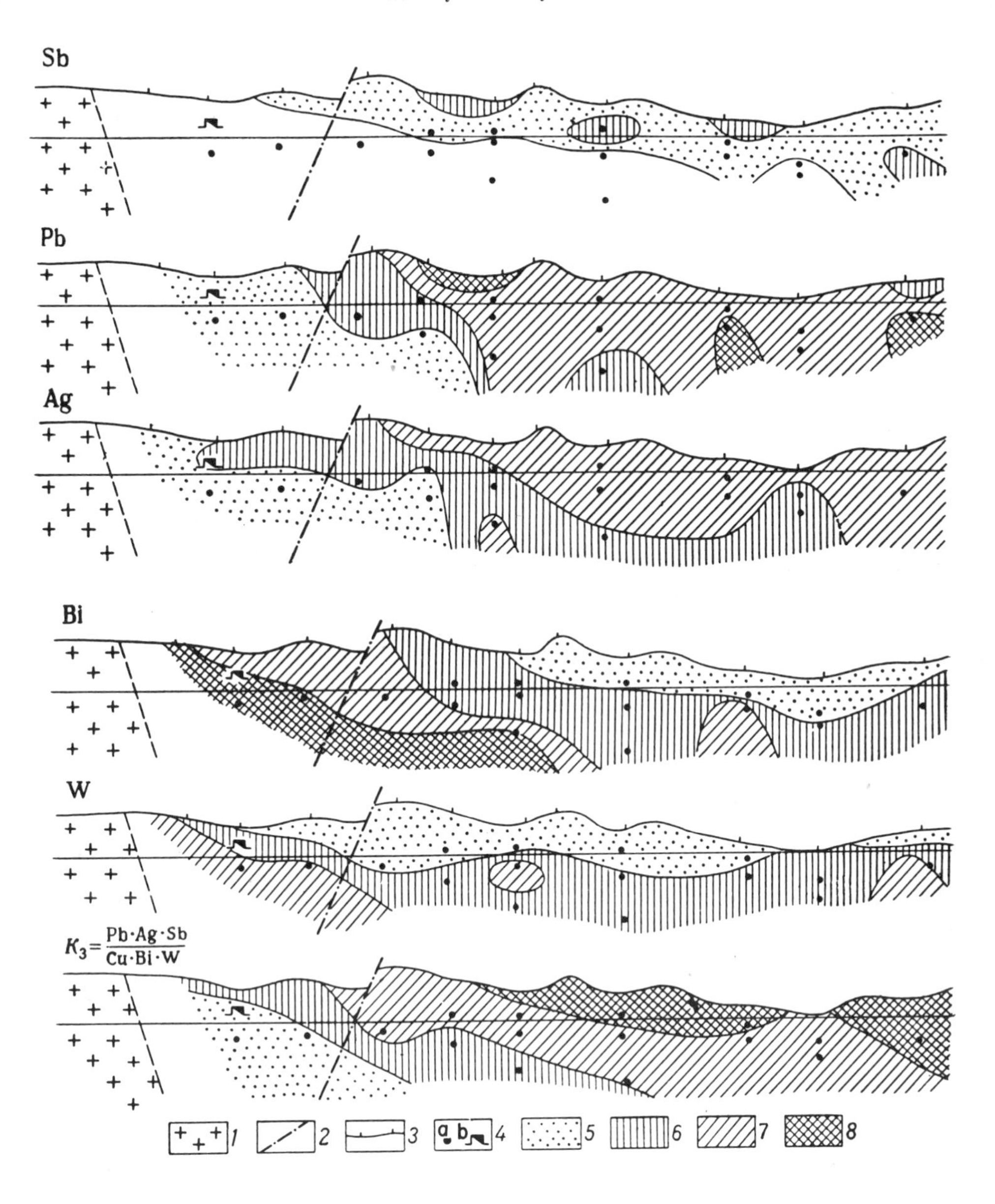

Fig. 39. Variation in the productivity values in the halos formed by antimony, lead, and silver in the longitudinal direction (vertical projection).
1 - biotite granites; *2* - line of contact between projections of the ore zones; *3* - geochemical profile; *4* - points of intersection of ore bodies (*a* - by boreholes, *b* - by adits); *5* to *8* - element productivities, m % (*5* - Sb < 0.01; Pb < 1; Ag < 0.01; Bi < 0.003; W < 0.01; K_3: < 0.01; *6* - Sb: 0.01; Pb: 1 to 2.5; Ag: 0.01 to 0.03; Bi: 0.003 to 0.01; W: 0.01 to 0.03; K_3: 0.01 to 1; *7* - Pb: 2.5 to 5; Ag: 0.03; Bi: 0.01 to 0.03; W: 0.03; K_3: 1.0 to 1000; *8* Pb: 5; Bi: 0.03; K_3 < 1000).

Multiformational Halos

We have discussed primary halos which accompany ore bodies and deposits formed during a single stage of mineralization. They characteristically show a close coincidence in space of the halos formed by all the indicator elements; smaller halos are inscribed within the outlines of larger halos.

Studies have shown that, in some cases, there is a coincidence in space of ore formations, which differ in composition and environments of ore accumulation, and these may produce complex halos which are called *multiformational.**

Fig. 40 shows primary geochemical halos formed in the vicinity of a copper-bismuth mineralization. Uranium halos, associated with a more recent stage of uranium mineralization, were found in the same cross section. As can be seen from the figure, the uranium halos are detached from the copper-bismuth ore body (they occur above the ore body, as well as in its hanging wall). The elements accompanying uranium (lead, zinc, copper, and molybdenum) are also characteristic of the copper-bismuth mineralization. This is why they indicate both the ore body as well as the fields of anomalous uranium contents. Molybdenum is the most representative in this respect.

A similar "shift" in the location of the epicenters of the halos produced by the copper-bismuth and uranium mineralization is indicated in Fig. 41. From this figure it can be seen that (in plan), the uranium and molybdenum halos are clearly shifted to the north relative to the halos formed by the elements which are typomorphous for the copper-bismuth mineralization.

Multiformational halos are characterized by specific correlations between contents of certain indicator elements. The correlation is usually negative between the contents of the elements which are typomorphous for the different ore formations. It is essentially positive for the halos associated with a single ore formation. Table 32 lists the values of the coefficient of rank correlation between the element contents in multiformational halos formed as a result of the superimposition of the uranium on the multi-element mineralization.

The negative correlation between the contents of uranium and of the elements indicating multi-element mineralization is probably due to the different ways in which the ore-bearing solutions are introduced during the formation of the respective ore bodies and of their halos. This is illustrated in Fig. 42 where plots of the distribution of lead (an example of polymetallic mineralization) and uranium (an example of uranium mineralization) are inclined away from each other, indicating different paths for the circulation of the ore-bearing solutions during the formation of the halos.

---

*The word "multiformational" is an exact translation from the Russian. It is important to keep in mind that the *formational* portion of the word is used in the sense of *composition* (possibly origin), and *not* in a stratigraphic sense. Hence, a multiformational halo is not necessarily one which crosses two or more stratigraphic formations, but rather it is a halo formed by the coincidence in space of two or more ore formations with different compositions (and possibly origins of formation). Perhaps, the terms "multistage" or "multimineralization" halos would be better in English but, nevertheless, the authors' term has been retained. Editor.

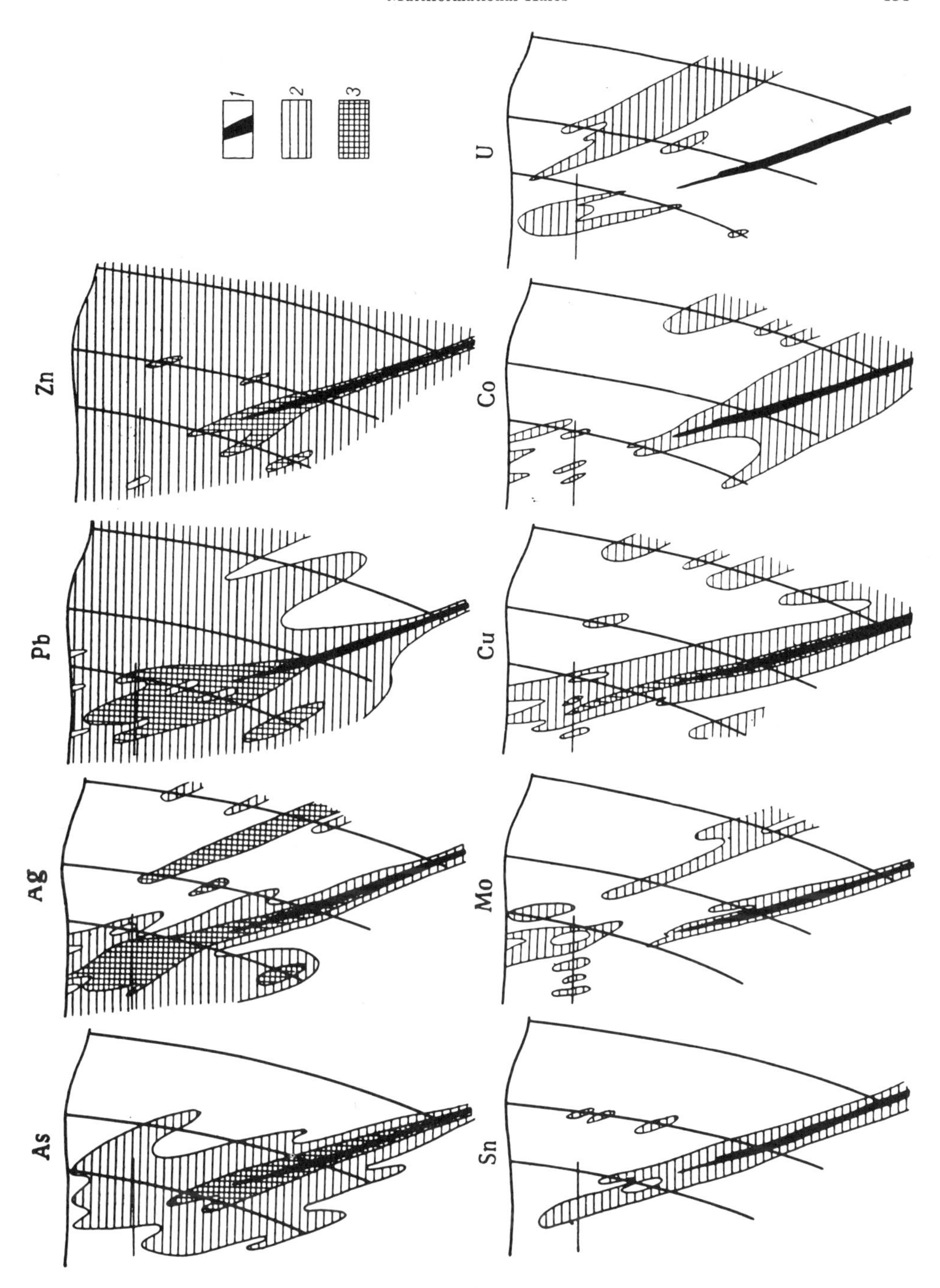

Fig. 40. Primary halos around ore bodies of a deposit.
1 - ore body; *2* - first anomalous field; *3* - second anomalous field.

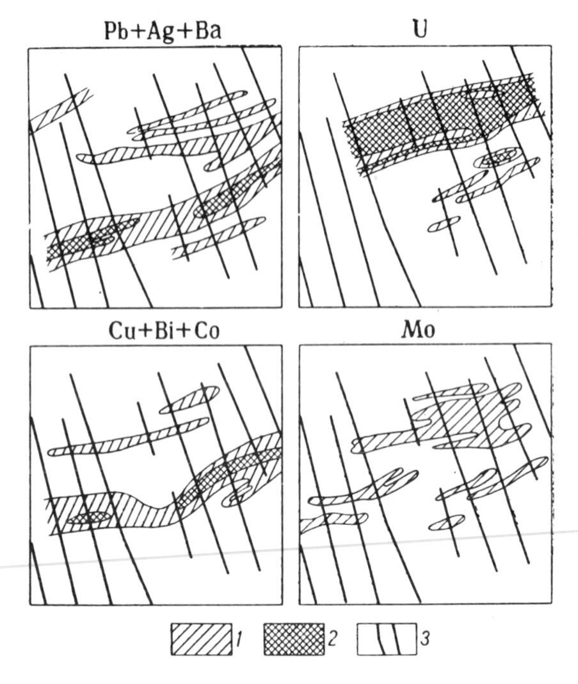

Fig. 41. Primary halos formed by indicator elements at the surface (based on Fig. 40). *1* - sum of elements equal to 6 - 20 background units; 0.00005 to 0.0001 % uranium; 0.0005 to 0.001 % molybdenum; *2* - sum of elements equal to 20 or more background units; 0.0001 and 0.001 % uranium; *3* - sampling profiles.

In some cases, multiformational halos form as a result of the vertical alternation of ore bodies and deposits which belong to different ore formations. The Shurale deposit (Tadzhikistan) is an example of this effect. The rare-metal mineralization there (scheelite-molybdenite) is replaced by a polymetallic mineralization at depth.

Table 32. Coefficients of sequential correlation of chemical element contents in multiformational halos.

Pair of chemical elements	Halos in polymetallic mineralization		Pair of chemical elements	Halos in uranium mineralization	
	borehole 318	borehole 309		borehole 318	borehole 309
Lead-zinc	+ 0,54	+ 0,069	Uranium-zinc	−0,74	−0,9
Lead-silver	+ 0,64	+ 0,72	Uranium-silver	−0,64	−0,11**
Lead-copper	+ 0,65	+ 0,95	Uranium-copper*	—	0,14**

* The correlation coefficient of the copper and uranium contents could not be determined for borehole 318 because the halos formed by these elements did not coincide in space.
** The correlation is insignificant (5 % significance level).

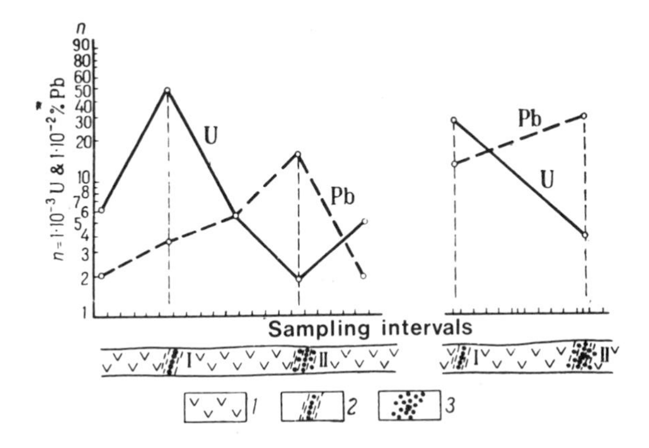

Fig. 42. Plots of the distribution of uranium and lead around ore-bearing zones (I - uranium mineralization; II - polymetallic mineralization).
1 - tuffaceous lavas; *2* - ore-bearing zones; *3* - sulfide dissemination.

Fig. 43 shows primary halos formed by chemical elements around the Shurale I ore-bearing zone. The ore body is represented by a zone of albitized rocks. A zone of disintegrated rock is located in the middle of the albitized rocks, and is composed of quartz veins of different thicknesses. The zone has a northeastern trend (35 to 45°) and dips steeply southeastwards. Zones of albitization are superseded by pyroxene skarns, and scheelite mineralization gives way to molybdenite mineralization.

The ore minerals in the central part of the zone are represented by scheelite, chalcopyrite, and less commonly by molybdenite.

The chemical elements are classified on the basis of their vertical distribution features. The molybdenum halo is the widest and the strongest in the upper portion of the cross section; it gradually narrows down the pitch of the ore bodies (Fig. 43). Tungsten and cobalt form halos which expand with depth. The nickel halo does not show any appreciable changes with depth. The halos formed by silver, lead, and partly by zinc, narrow with depth, and then expand abruptly at the level of the deepest borehole (Fig. 43). The element distribution in the latter group is anomalous for the occurrence of rare-metals and is due to another ore formation located at depth. When the halos formed by the elements of this group were compared with the primary halos in other ore deposits, it became clear that the halos in the lower part of the cross section were formed above lead-zinc skarn bodies. The complete wedging out with depth of the halos formed by tungsten, molybdenum, nickel, and cobalt unequivocally point to the absence, at depth, of blind ore bodies with rare-metal mineralization. The deep horizons in the deposit are promising for polymetallic ore bodies. This is indicated by the broad and strong halos at depth which are characteristic of this type of mineralization (Fig. 43).

Criteria for the interpretation of multiformational geochemical halos are discussed below.

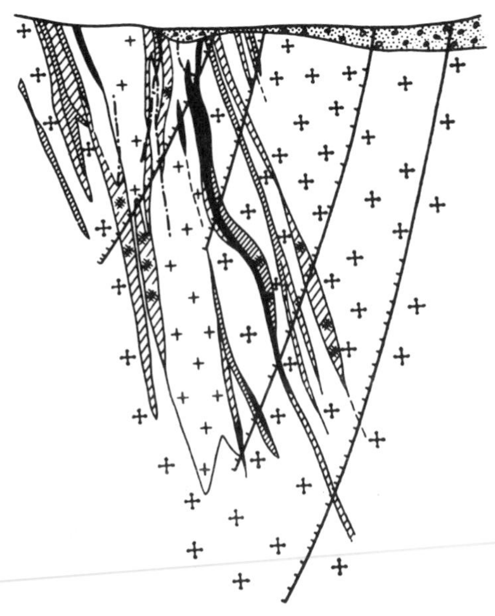

Fig. 43. Primary halos formed in the vicinity of the Shurale ore bodies.
1 - alluvium; *2* - granodiorites; *3* - albitized granodiorites; *4* - strongly albitized granodiorites; *5* - albites; *6* - fractures; *7* - skarning; *8* - sampling interval; *9* - ore body; *10* and *11* - element contents in halos, % (*10* - 0.000005 to 0.00001 Ag; 0.001 to 0.01 Pb; 0.005 to 0.01 Zn; 0.0001 to 0.0003 Bi; 0.0003 to 0.0005 Be; 0.0001 to 0.001 Sn; 0.001 to 0.25 Cu; 0.0001 to 0.001 Mo; 0.003 to 0.001 Ni; 0.0003 to 0.001 Co; 0.0003 to 0.02 W; *11* - 0.00001 to 0.015 Ag; 0.01 to 0.1 Pb; 0.01 to 0.2 Zn; 0.005 to 0.001 Be; 0.001 to 0.01 Sn; 0.01 to 0.009 Mo; 0.01 to 0.002 Co).

Certain Geochemical Features In Zones With Dispersed Ore Mineralization

It is well known that each ore-producing region includes numerous mineral occurrences and geochemical anomalies which represent zones of dispersed mineralization. Zones of dispersed mineralization are areas within which increased concentrations of indicator elements have formed as a result of the effect of ore-forming fluids on the enclosing rocks. These concentrations are higher than the background, however, they do not form economically important mineralization. Nevertheless, in many cases, these manifestations of ore occur in environments favorable for economic mineralization. This is sometimes used as justification for detailed prospecting. At present, there is a lack of reliable criteria for interpreting zones of dispersed ore mineralization, and thus it is not possible to reject the nonpromising zones from further investigation. This results in considerable expenses during subsequent detailed studies, including expensive pitting and drilling.

Zones of dispersed ore mineralization are especially "risky" in geochemical exploration for blind mineralizations. The search for blind ore bodies and deposits usually involves detection of geochemical halos formed above ore bodies. These halos, like zones of disseminated ore mineralization, are characterized by low indicator-element concentrations which

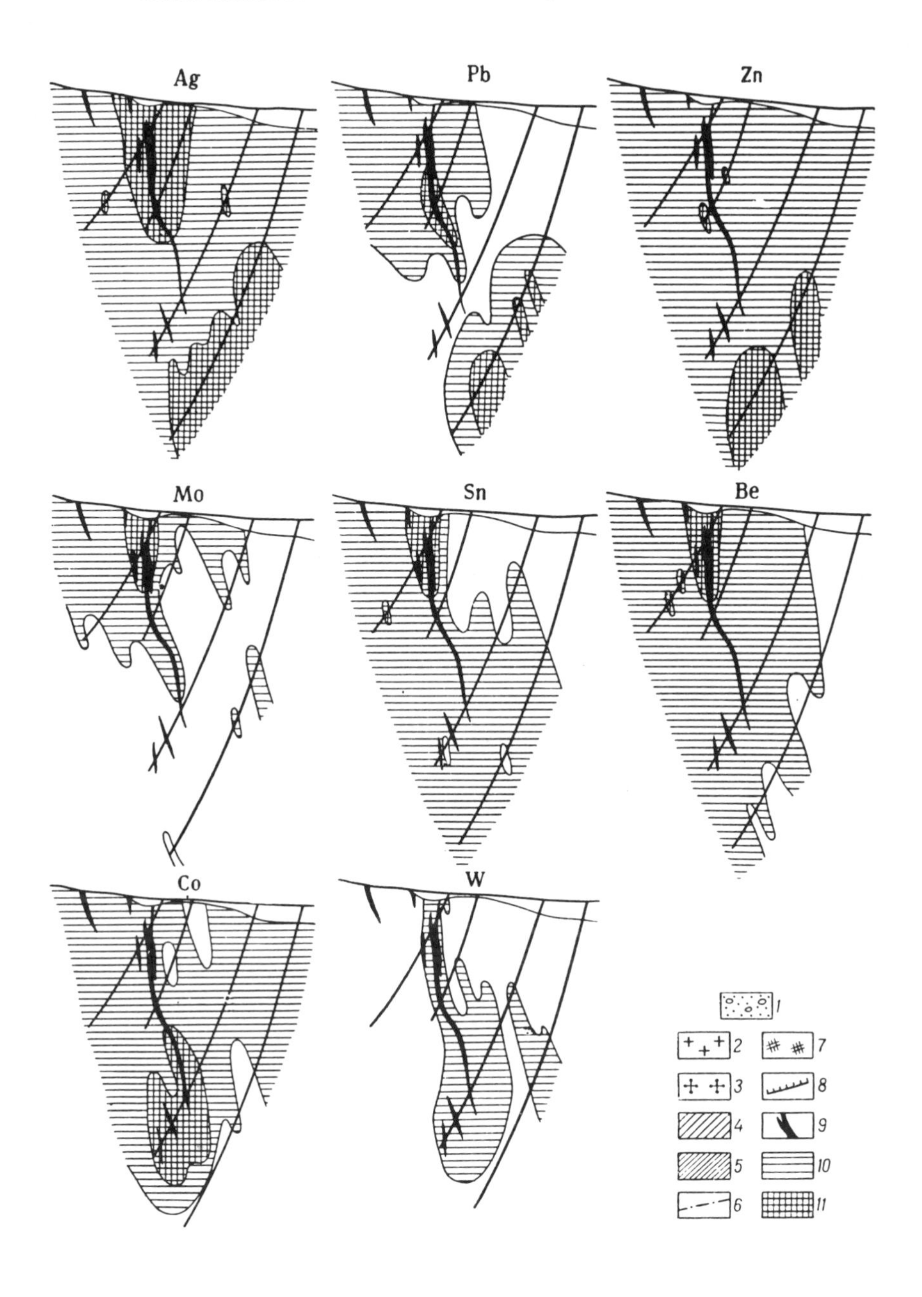

usually are close to the Clarke values. In addition, some other characteristics are also similar. This may result in the erroneous interpretation of geochemical sampling data. A comparative study of the characteristics of primary halos in economically important hydrothermal mineralizations, and in zones of dispersed ore mineralizations, have made it possible to recognize certain geochemical dissimilarities in these two types of formations. These findings may be used for the assessment of the potential reserves in regions with geochemical anomalies.

The geochemical characteristics of zones of dispersed mineralization are considered below based on examples from polymetallic deposits in the Rudny Karamazar (Tadzhikistan). In this case, typomorphous elements which occur in polymetallic vein-type

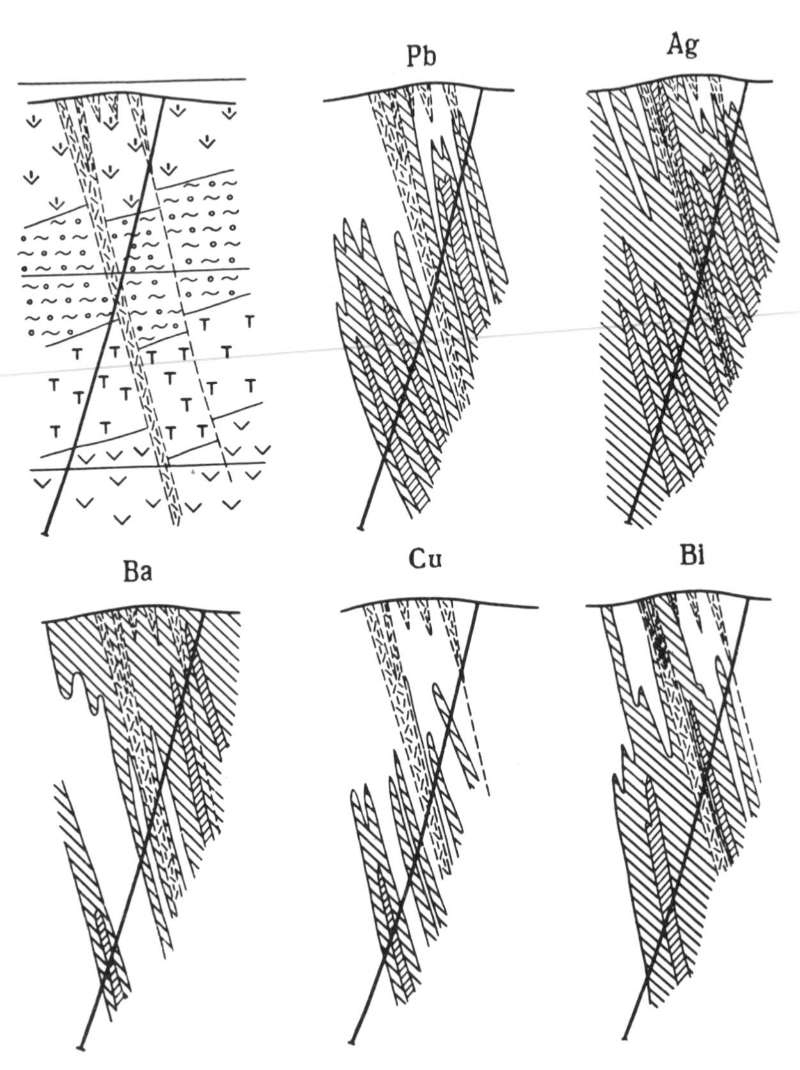

Fig. 44. Geochemical anomalies in the Kyzyltash-Kokchegirtke ore showing.
1 - andesite porphyrites; *2* - felsite flows; spherulite-porphyries; *3* - tuffs and tuffaceous lavas; *4* - andesite and andesite-dacite porphyrites; *5* - zones of rock disintegration (*a*) and fractures (*b*); *6* - element contents, %: 0.005 to 0.015 Pb and Cu; 0.00002 to 0.0001 Ag; 0.02 to 0.1 Ba; 0.0001 to 0.001 Bi; 0.0003 to 0.001 Co, Mo, and Sn; 0.005 to 0.01 As; 0.008 to 0.03 Zn; 0.001 to 0.003 W; *7* - above the upper limit of the contents listed.

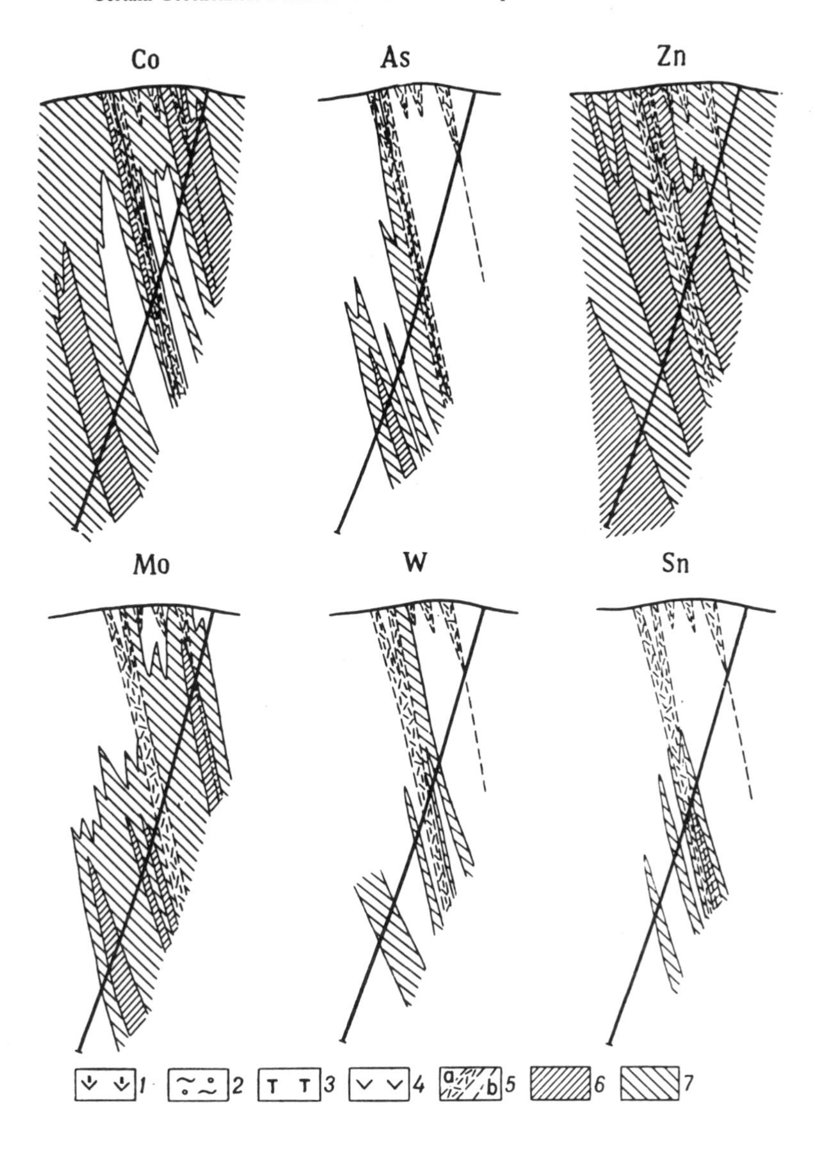

deposits in acidic igneous rocks, are used as the bases for comparison. The geological structure of these deposits and the characteristics of their primary geochemical halos were discussed earlier in this book. It may be added that all the polymetallic deposits discovered in Central Karamazar have halos formed by numerous chemical elements, and all have a distinct vertical zonality. This zonality is expressed through a regular differentiation of the elements in the vicinity of the ore bodies. The widest and the strongest halos are formed by barium and silver, and they are found in the upper parts of the metalliferous zones. The halos of these elements narrow with depth. At the same time, the halos formed by some other elements such as cobalt, copper, bismuth, molybdenum, and tungsten, expand with depth.

The results of geochemical bedrock sampling in regions with dispersed mineralization were compared with corresponding studies of the halos found in zones with economic mineralization. The comparison showed that these types of anomalies are characterized by identical elemental compositions, as well as by similar concentrations of the indicator elements. No appreciable differences in plan in the morphology of the anomalies could be found.

Cross sections through the geochemical anomalies represented by zones of dispersed mineralization and by halos from economic deposits, enabled the certain differences to be distinguished between them (Grigorian et al., 1973). Fig. 44 shows anomalous fields of the principal indicator elements typical of the polymetallic deposits in Central Karamazar which were found in the section across the Kyzyltash-Kokchegirtke ore showing. Detailed studies at the surface, in combination with a survey of the deep horizons, showed that this ore showing does not contain any economically important mineralization, but is a zone of non-economic, dispersed ore mineralization. The enclosing rocks are represented by andesite porphyrites. Several parallel fracture zones were revealed at the surface of the ore showing. These zones are composed of limonitized, chloritized, sericitized, and silicified rocks impregnated with pyrite, galena, sphalerite, and rarely chalcopyrite.

It can be seen in Fig. 44 that geochemical anomalies were found in the cross section for all the indicator elements which are typomorphous for polymetallic deposits in the region covered by the studies. The main metalliferous zone, in contrast to the halos from economic mineralization, is not marked by fields with maximum concentrations of the indicator elements (Fig. 44). This feature is especially distinct in the structure of the composite additive anomalies (Fig. 45).

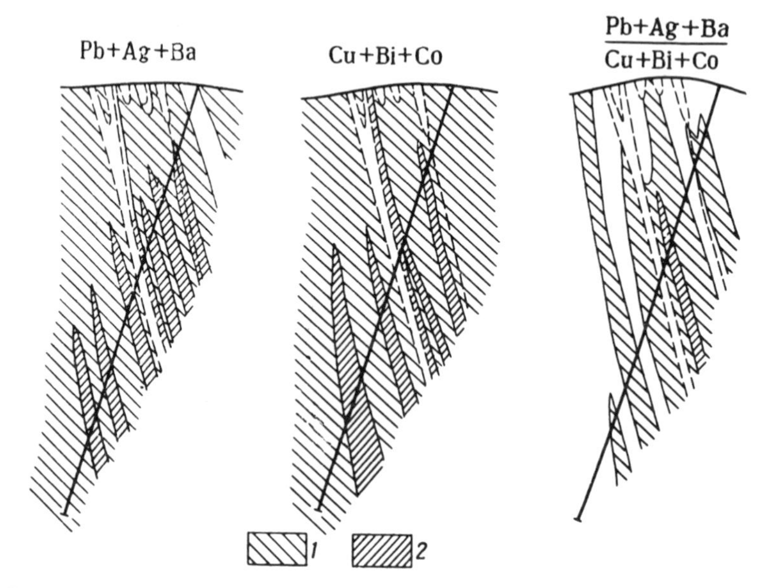

Fig. 45. Additive anomalies in the dispersed mineralization zone of the Kyzyltash-Kokchegirtke ore showing.
1 and *2* - total contents of elements and their ratios (*1* - 6 to 20 background units; 1 to 5; *2* - more than 20 background units; more than 5).

Another characteristic feature of the anomalies in the cross section being studied is the virtual absence of vertical zonality. There is no noticeable differentiation in the distribution of the elements above and below the ore bodies.

In addition, there is no vertical zonality in the structure of the additive halos (Fig. 45). Also, no vertical zonality waş found in the structure of the halos delineated on the basis of the value of their additive indexes (ratios for each sample) in background units of the elements below and above the ore bodies.

Unlike the zones in the areas of dispersed (non-economic) mineralization, distinct vertical zonality is always present in halos formed in the vicinity of concentrated economic mineralization (see Figs. 28 and 29).

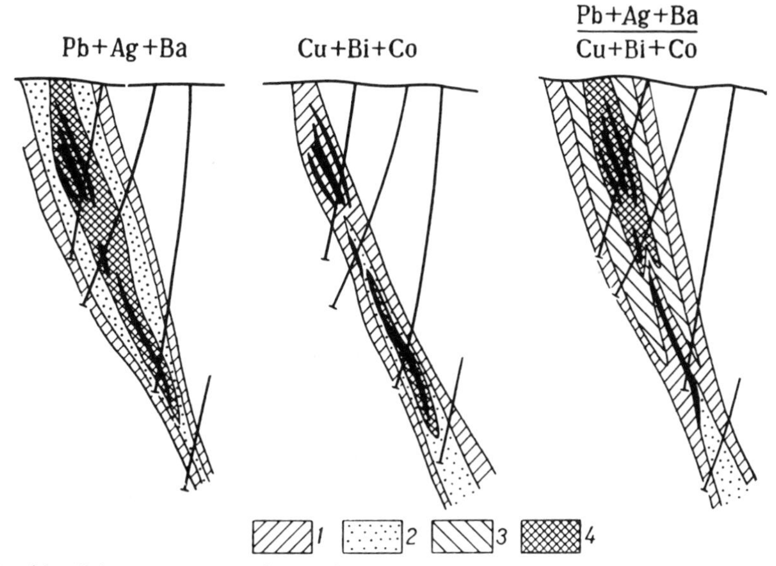

Fig. 46. Primary halos and fields of values of the additive index around ore bodies (the Eastern Kanimansur deposit).
Content, in background units: *1* - Pb + Ag + Ba, 6 to 20; Cu + Bi + Co, 6 to 20; their ratio, 0.1 to 1.0; *2* - Pb + Ag + Ba, 20 to 100; Cu + Bi + Co, 20 to 100; their ratio, 1 to 3; *3* - element ratio, 3 to 10; *4* - Pb + Ag + Ba > 100; Cu + Bi + Co > 100; their ratio, 10 to 50.

Fig. 46 shows additive geochemical halos around ore bodies in the Eastern Kanimansur polymetallic deposit (Central Karamazar). It follows from Fig. 46 that there is a distinct zonality in the structure of individual additive halos. This zonality can also be detected with the aid of an additive index.

The differences between the zones of dispersed (non-economic) polymetallic mineralization and halos of economic mineralizations, in terms of their zonality features, are corroborated by calculated results. Table 33 lists values of the ratios of the linear productivities of the individual additive anomalies of the indicator elements from an area in the Kyzyltash-Kokchegirtke region. This parameter was calculated for three levels: 1490 m (surface), 1400 m (the upper part of the borehole), and 1330 m (the lower part of the borehole). This table also gives data for an area of the "Operyayushchaya" zone (Fig. 47)

Table 33. Ratios of linear productivities of additive anomalies.

Type of anomaly	Region	Location of anomaly	Ratio of linear productivities $\frac{Ba + Ag + Pb}{Cu + Bi + Co}$
Zones of disseminated mineralization	*ore showings*		
	Kokchegirtke	Surface (horizon 1490 m)	1,2
		Borehole 873 (horizon 1400 m)	1,3
		Borehole 873 (horizon 1330 m)	0,9
	Operyayushchaya		
		Surface (horizon 1450 m)	1,7
		Borehole 885 (horizon 1350 m)	2,2
Primary halos at the level of the lower part of economic mineralization	*Deposits*		
	Kaptarkhona	Surface (horizon 1550 m)	1,4
	Kanimansur	Borehole 659	1,1
	Taryekan	Borehole 79	0,81

which represents disseminated mineralization. For the second area (the "Operyayushchaya" zone), the calculations were performed at two levels (the surface and the borehole). It can be seen from Table 33 that the variations in this parameter with depth is not great (it is within the accuracy of the calculations). This indicates that there is no distinct vertical zonality in the distribution of the dispersed ore mineralization. It was also established that any one of the cross sections in the area of dispersed mineralization studied corresponds to primary geochemical halos formed at a level in the lower parts of the economic mineralization. This conclusion is based on the absolute values of the indicator ratios (see Table 33).

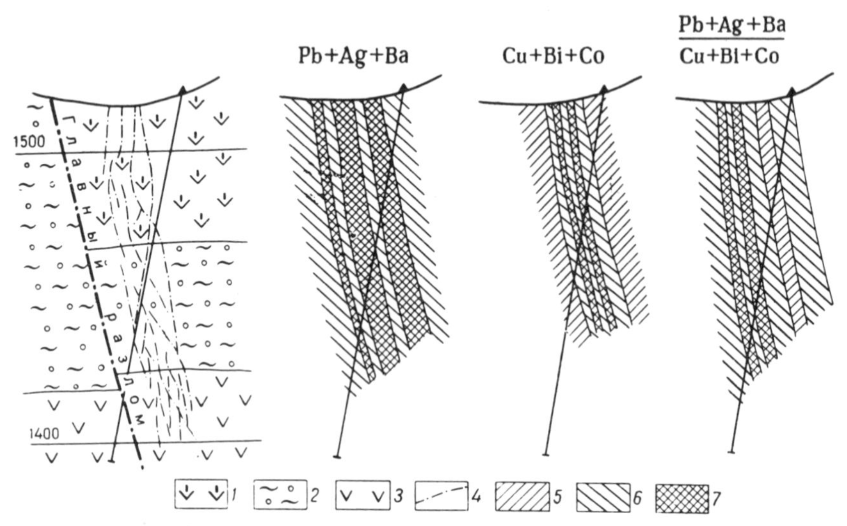

Fig. 47. Additive anomalies in cross section in the zone of dispersed mineralization (the "Operyayuschaya" zone).
1 - quartz porphyries; *2* - spherulite-porphyries; *3* - andesite porphyrites; *4* - zones of fractures; *5* to *7* - additive contents of elements (in background units) and their ratios (*5* - 6 to 20; 0.1 to 0.5; *6* - 20 to 30; 0.5 to 1.5; *7* - 30 to 50; 1.5 to 5).

Significant differences between zones of dispersed mineralization, on the one hand, and halos of economic mineralization, on the other, may be revealed if one constructs anomalies based on the values of the *additive index*. This index is the ratio of additive contents (in geochemical background units) of the indicator elements, calculated for each sample. Elements which have a tendency to accumulate in halos above the ore are selected for the numerator. Elements which tend to concentrate below the ore are selected for the denominator. Fig. 46 shows such anomalies delineated by contour lines in the section across an economic mineralization in the Eastern Kanimansur deposit. Maximum values of the additive index indicate ore bodies. As one goes away from these bodies toward the hanging and the footwall sides of the cross section, the value of the index decreases regularly, which leads to a contrasting transverse zonality in the structure of the halos in the vicinity of ore bodies. In addition to the transverse zonality, vertical zonality is also found. It is expressed in Fig. 46 as a distinct shift in the clusters' maximum values of the additive index from the bottom to the top in the cross section.

The pattern is different in zones of dispersed mineralization. As can be seen from Figs. 45 and 47, they show no vertical zonality. Their transverse zonality differs from that exhibited by halos representative of economic mineralization; specifically, there is no distinct increase in the additive index toward the center of the anomaly.

Similar features have been found in zones of dispersed mineralization which accompany

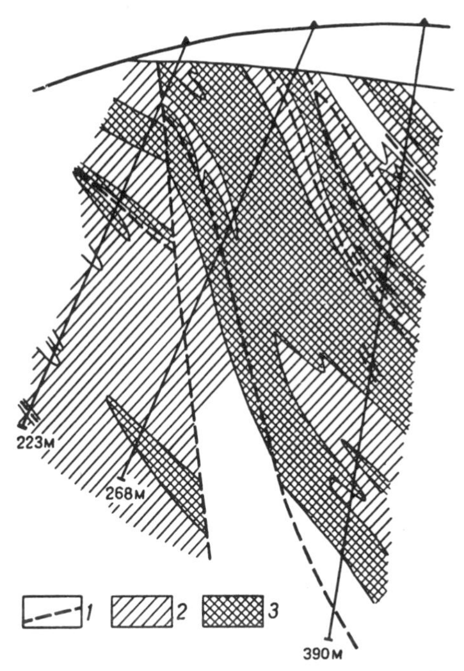

Fig. 48. Variations in the values of the multiplicative ratio of elements above the ore body (Sb, Hg, As, and Ag), to those below the ore body (Cu, Bi, W, Zn), which indicate a mercury deposit (Chalboi).
1 - faults; *2* and *3* - absolute values of multiplicative ratios (*2* - 0.0001 to 0.1; *3* - 0.1 or higher).

ore deposits differing in composition and in environments of formation (tin, gold, mercury, etc.).

Fig. 48 shows a variation in the multiplicative index calculated for groups of elements above an ore body (antimony, mercury, arsenic, and silver) and below an ore body (copper, bismuth, tungsten, and zinc) in a section through a deposit in the Chalboi area (a zone of dispersed mercury mineralization). The *multiplicative index* is the ratio of products of the contents of elements below and above the ore body. It is calculated for each sample. Special investigations have shown that similar results are obtained in studying geochemical anomalies with the help of additive and multiplicative indexes. Nevertheless, the multiplicative index is preferred because its determination does not require normalization of the element contents in terms of the background values.

It can be seen in Fig. 48 that there is no distinct vertical zonality in the cross section (similar data were obtained with the help of the additive index). In contrast to this, halos accompanying economic mercury mineralization of the same type exhibit a distinct vertical zonality. This is illustrated in Fig. 49 where the multiplicative index varies in a cross section through economic ore bodies in the Konchoch deposit.

Also shown in Fig. 49 are the plots of the variations with depth of the ratios of linear productivities of the individual multiplicative halos related to economic mineralization (along a cross section through the Konchoch deposit) and in a zone of dispersed mineraliza-

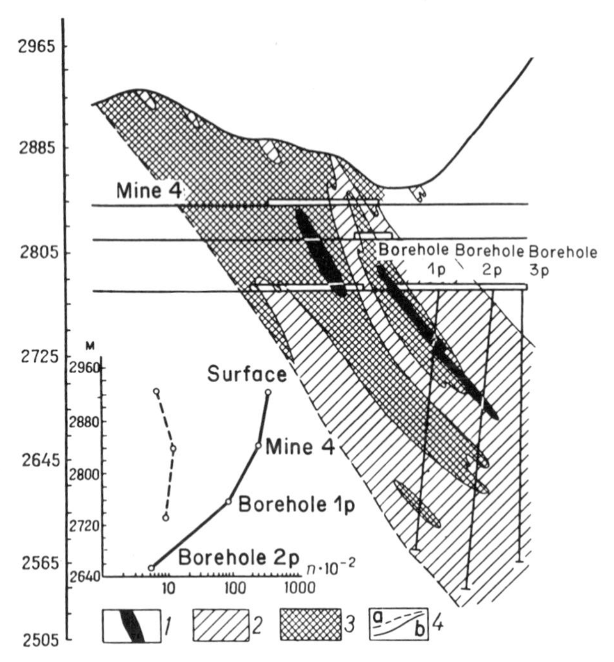

Fig. 49. Variations in the multiplicative index in cross section, and plots of the ratios of linear productivities of individual multiplicative ratios (from the Konchoch deposit). *1* - ore body; *2* and *3* - values of the multiplicative index (*2* - 1 to 100; *3* - more than 100); *4* - plots of variations in the linear productivities of the multiplicative anomalies (*a* - Chalboi area, *b* - Konchoch deposit).

tion (the Chalboi area). These plots completely confirm the patterns established with the examples based on the polymetallic deposits. There is a distinct vertical zonality in the structure of the halos formed in the vicinity of the economic mineralization (the graph of the indicator ratio values decreases monotonically with depth). This type of zonality is practically absent in the zone with dispersed mineralization. In this case, too, the zone of dispersed mineralization corresponds to the lower levels of the economic mercury mineralization, in terms of the indicator ratios.

The above specific features in the distribution of the indicator elements within zones of dispersed ore mineralization may be used for interpreting the results of lithogeochemical (rock geochemistry) sampling in order to differentiate zones of dispersed mineralization from anomalies which represent halos of economic mineralization. The absence of distinct zonalities may serve as such a criterion, as a first approximation. Of course, for this purpose it is necessary to have enough data to characterize the distribution of the elements from at least two horizons, i.e., at those levels where there are boreholes or underground mine workings.

If the data are available from only one level, then the anomalies represented by zones of dispersed mineralization may be located on the basis of the nature of the horizontal zonality revealed with the help of the additive or of the multiplicative indexes. Zones of dispersed mineralization differ from halos representative of economic mineralizations in that the former do not have any distinct horizontal zonality.

In the interpretation of those anomalies found as a result of geochemical sampling, it would be an extremely serious error to classify zones of dispersed mineralization as geochemical halos formed above an ore body and to conclude that they are promising for blind, or weakly eroded, economic mineralization occurrences which should be studied further. If dispersed mineralization is erroneously classified as a halo below an ore body, this will not lead to erroneous assessment of its potential reserves, because both these indications are evaluated as not promising, and are not included in any subsequent prospecting. There is yet another possibility of an erroneous evaluation of dispersed mineralization zones: they may be considered as halos formed at some level in the middle of economic mineralization. This type of error is less probable because in such cases there must be outcrops of economic ores on the surface, and such outcrops are absent in zones of dispersed mineralization.

It follows from the above that a successful interpretation of zones of dispersed mineralization requires, first of all, an elaboration of the effective criteria which will help in their differentiation from genuine halos formed above the ore bodies with economic mineralization.

The following criteria for recognizing zones of dispersed mineralization can be outlined. They are based on the above geochemical distinctions between the halos from economic mineralization and zones of dispersed (non-economic) ore mineralization.

1. *Vertical zonality.* Vertical zonality is practically absent in zones of dispersed mineralization but, in contrast, it is present in halos of economic mineralization.

2. *Ratios of parameters, and in particular, ratios of the productivities of the halos formed above and below the ore body.* These ratios in zones of dispersed mineralization differ markedly from such ratios for halos from promising blind, or slightly eroded, mineralization. (The zones of dispersed mineralization are characterized by constant, and essentially low, values of the indicator ratios).

3. *Transverse zonality revealed with the help of multiplicative or additive indexes.* Within halos from economic mineralization, unlike those in zones of dispersed mineralization, the ore-bearing regions are indicated by clusters of maximum values for these indexes.

4. *Elemental composition of the anomalies.* The halos formed above the ore body in zones of economic mineralization usually do not contain halos of the elements characteristic of the levels below the ore body. This implies that zones of dispersed mineralization are characterized by constant elemental compositions.

Criteria and Methods for the Interpretation of Endogenic Geochemical Halos

An interpretation of any geochemical anomaly is aimed at an assessment of its ore potential. It is hardly necessary to prove that this problem is extremely complex and that its successful solution requires the application of a combination of many different methods. However, this book deals with geochemical methods and, therefore, particular attention is placed on the use of geochemical criteria.

Below techniques for interpreting geochemical anomalies revealed and delineated in the course of detailed geochemical sampling of bedrock (at a scale of 1:25,000 to 1:10,000, or larger) are discussed. It is desirable to consider the methods for the interpretation of geochemical anomalies separately for the following cases: (a) the search for blind mineralizations; (b) the assessment of ore potential at depth in those places where ore mineralization (showings) appear at the surface.

As noted above in the description of the geochemical characteristics of zones with dispersed ore mineralization, halos from dispersed mineralization are quite similar to those formed by economic mineralizations above ore bodies, and may be mistaken for the supra-ore halos indicative of promising blind ore bodies. In order to avoid such errors in the interpretation of the geochemical sampling results, it is first necessary to isolate and to reject those geochemical anomalies which represent zones of dispersed ore mineralization.

Interpretation of Zones of Dispersed Mineralization

The criteria used for the interpretation of geochemical anomalies represented by zones of dispersed mineralization, are based on the above-mentioned geochemical differences between zones of dispersed mineralization and primary halos due to economic mineralization.

Zones of dispersed mineralization have halos which differ from halos from blind ore bodies (formed above these ore bodies) in that the former have much lower vertical geochemical zonality index values. For this purpose, the ratio of the productivities of the individual* composite (additive or multiplicative) anomalies formed by the indicator elements indicative for above ore body and below ore body levels for the particular type of deposit, are used. These ratios within the zones of dispersed mineralization correspond to the primary halos formed at the level of the middle and lower parts of the ore bodies. Since there is no distinct vertical zonality in the structure of the zones from the dispersed

*The word "individual" as used here and previously (e.g., in connection with Fig. 49), means that additive or multiplicative ratios should be calculated for each single (individual) sample, and not for averages.

mineralization, the indicator ratio does not exhibit any significant vertical variation. In contrast to this, the primary halos from concentrated (economic) mineralization are characterized by distinct and monotonical vertical variations in the indicator ratio.

To illustrate a technique for the interpretation of geochemical anomalies representative of zones of dispersed mineralization, we shall consider the following two examples from geochemical surveys carried out in Rudny Karamazar (Grigorian et al., 1973).

1. **The Akpet ore showing.** This ore showing is located northeast of the Tyryekan deposit (Central Karamazar). The area is composed of felsite porphyries, tuffs, tuffaceous breccias, and spherulite porphyries from the Oyasaiskaya Suite.

A geochemical survey was undertaken in order to assess the ore potential at depth. Anomalies due to lead, zinc, barium, silver, molybdenum, copper, bismuth, cobalt, and tungsten were traced along the zone of the Akpet Fault (Fig. 50). Below are the parameters

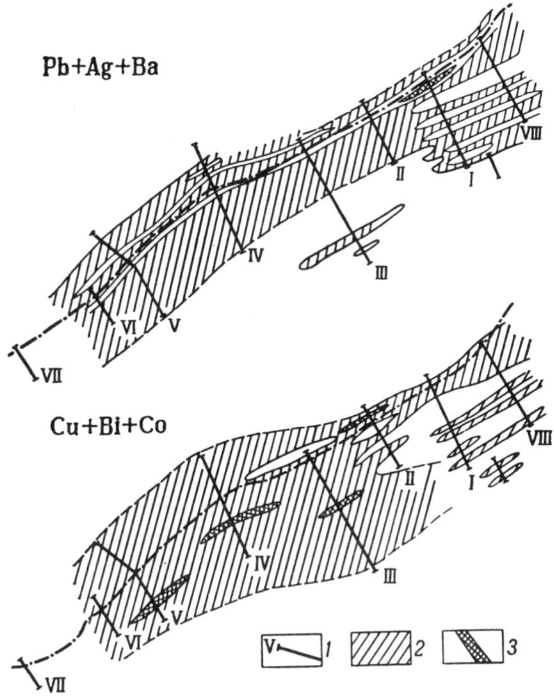

Fig. 50. Primary halos of elements at the surface in the Akpet area.
1 - geochemical sampling traverse number; *2* and *3* - contents, in background units (*2* - from 6 to 20; *3* - more than 20).

of the anomalies revealed in the region along the central sampling traverse. For comparison, parameters found for the zone of dispersed mineralization in the Kyzyltash area are also listed.

Area	*Ratio of linear productivities* $\frac{Ba + Ag + Pb}{Cu + Bi + Co}$
Akpet ore showing	0.4
Kyzyltash (standard cross section)	1.1

It follows from the above data that the anomaly in question has parameters similar to those of dispersed zones of mineralization. This is indicated by the similarity of the parameters exhibited by the geochemical anomalies in the region studied, and in the dispersed mineralization zone in the Kyzyltash region. Based on these data, a negative assessment of the possible ore potential in the region was made, and it was recommended that the area be excluded from further detailed exploration. Nevertheless, three boreholes were later drilled in the region at sites of the most pronounced mineralization. All the boreholes have intersected, at different depths, a metalliferous zone without economic mineralization. This fully confirmed the assessment based on the results of the interpretation of the geochemical sampling data. Geochemical sampling of the cores recovered from the boreholes did not reveal any changes in the parameters of the anomaly with depth (Fig. 51) and thus corroborated the dispersed character of the multi-element mineralization in the region.

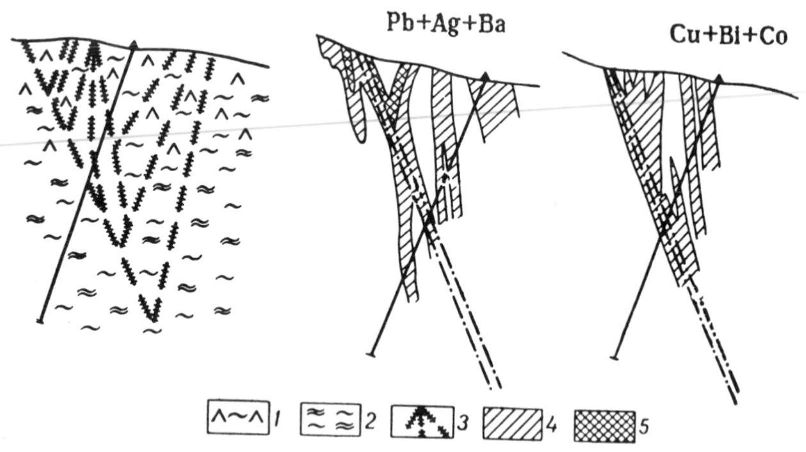

Fig. 51. Additive geochemical anomalies in cross section.
1 - felsite porphyries; *2* - tuffs and tuffaceous breccias; *3* - fractures; *4* and *5* - total content of elements, in background units (*4* - from 6 to 20; *5* - more than 20).

2. **Southern Taryekan locality.** This area is situated south of the Taryekan deposit. Tuffs and tuffaceous lavas of the Oyasaisky Suite and felsite porphyries are abundant in the area and they are intruded by dikes composed of diabase porphyrites.

The "Yuzhnaya" (Southern) tectonic zone, extending in a northeastern direction for about 1.5 km, is the major ore-bearing zone. It is composed of silicified, chloritized, and sericitized rocks with quartz veins which are impregnated with chalcopyrite, pyrite, bismuthinite, and hematite. Locally, increased contents of lead and silver have also been detected.

Geochemical sampling of the bedrock was carried out in order to assess the ore potential in the zone at depth. Anomalies due to indicator elements which are typomorphous for polymetallic mineralization were found. In terms of the indicator ratios, the level of the erosion surface of the anomalies corresponded to a level below the middle portion of any economic mineralization. However, the absence of any economic ore outcrops in the area suggested that the anomalies, in all probability, represent a zone of dispersed mineralization and are not promising at depth. Despite the negative evaluation, a borehole (borehole No. 9) was drilled in the area. It intersected the "Yuzhnaya" zone at a depth of 370 meters, but did not reveal any mineralization and thus confirmed the interpretation of the geochemical sampling results. Below are the parameters of the anomalies in the Southern Taryekan area:

Profile and borehole numbers	*Ratio of linear productivities* $\frac{Ba + Ag + Pb}{Cu + Bi + Co}$
II - II	0.26
III - III	0.67
IV - IV	0.61
Borehole No. 9	1.3

The Search For Blind Mineralization

Techniques for the interpretation of geochemical anomalies are essentially dependent on the mode of occurrence of the ore deposits. This is why the problem of the interpretation of anomalies are considered below for two extreme cases. These are when the ore bodies: (a) dip steeply; (b) occur almost horizontally.

In order to establish the modes of occurrence of hypothetical ore deposits at depth, it is most important to know the geologic-structural setting of the area being studied. It is especially important to have a knowledge of geologic-structural position of the ore deposits in the region. Results from geophysical studies are also essential. Geochemical criteria, and in particular the characteristic features of primary halos, may also be helpful in some cases.

An idealized model of the primary halos (Fig. 52) shows, in the case of the steeply dipping ore bodies, that halos closely overlapping in plan and distinctly extending along the ore-enclosing structure, should be expected at the surface (along faults, fissures, and zones of heavy jointing). These halos attenuate rapidly as one goes away from the ore-enclosing structure. In the case of gently pitching ore bodies (see Figs. 33, 34, and 52), the primary halos at the surface are definitely different. These halos are much larger, less distinct, and are detached from the faults (presumably channels for the ore solutions). Further, these halos often exhibit horizontal zonality, which is expressed as a larger separation of halos composed of indicator elements which are typical for the supraore cross sections from the feeding structures (e.g., faults), in accordance with the generalized sequence of the halo zonality.

The search for blind mineralization in steeply dipping structures.

The interpretation of geochemical anomalies in the search for blind mineralizations consists of two major stages:

(a) The determination of the type of ore mineralization responsible for the given anomaly, and;

(b) The determination of the level of the anomaly relative to the mineralization (specifically, when an anomaly has been found at the surface, the level of the erosion surface with respect to the mineralization should be estimated).

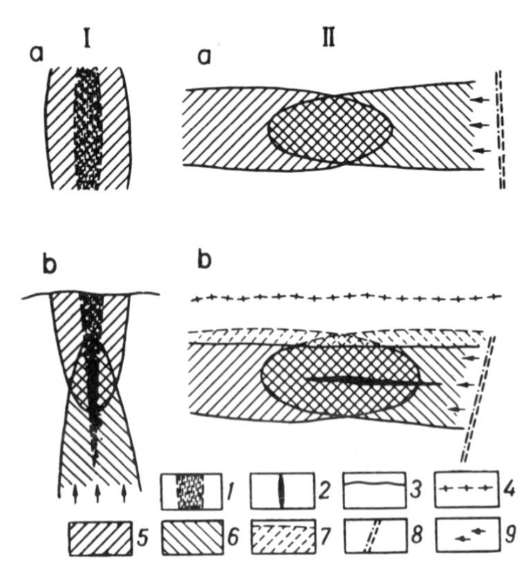

Fig. 52. Generalized patterns of primary halos formed around steeply dipping ore bodies (I), and around gently dipping ore bodies (II), in plan (a) and in the cross section (b).

1 - metalliferous zone; *2* - ore body; *3* - outlines of present-day surface; *4* - position of the surface at the time of formation of the deposit; *5* - halo above the ore body; *6* - halo below the ore body; *7* - halos removed by erosion; *8* - fractures; *9* - direction of movement of the ore-bearing solutions.

The type of the ore mineralization is, in most cases, determined: (1) on the basis of the geological conditions in the area being studied (by analogy with known deposits in the region); (2) by using data on the nature of the rock alterations in the vicinity of the ore bodies; and (3) on the typomorphous association of minerals established by detailed investigations of the local anomalies. Geochemical criteria, such as transverse zonality, can also be used for this purpose (see Table 31).

In some cases, absolute values of the average contents of the indicator elements in geochemical anomalies may be sufficiently reliable criteria for the determination of the type of ore mineralization. For example, an average content of lead (which is an indicator element in nearly all types of endogenic ore deposits), does not exceed 0.01 % in any cross section from primary halos of mercury deposits. A corresponding estimate for lead in halos formed by polymetallic sulfide deposits reaches hundredths, or tenths, of a percent. Mercury is another such element. An average content of this element in the halos of polymetallic deposits does not exceed $1 \cdot 10^{-4}$%, whereas in mercury and antimony-mercury deposits it amounts to thousandths, or hundredths, of a percent. The above examples show that those elements which are the main components of ores are the best indicators of the type of mineralization.

Determination of the level of a geochemical anomaly relative to the hypothetical mineralization. In the search for blind mineralization the problem is essentially one of identifying the most promising of numerous geochemical anomalies represented by supraore geochemical halos. In certain cases, if a geochemical anomaly has been delineated at the surface, the problem becomes one of determining the level of the erosion surface.

The solution to this problem is especially important in the search for blind mineralization, because in ore regions which are particularly highly eroded, the number of anomalies represented by halos below the ore bodies is great. Therefore, the accurate selection of anomalies is essential to the successful search for blind mineralizations. Fig. 53 shows an

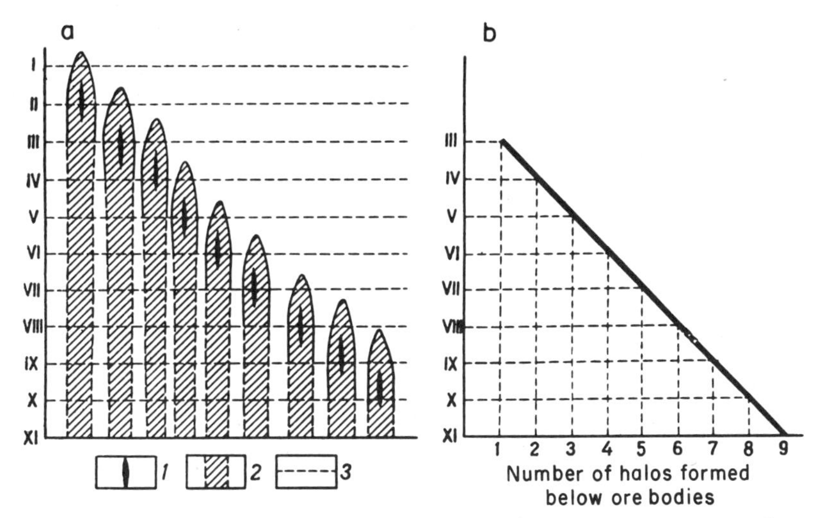

Fig. 53. An idealized diagrammatic representation of the possible occurrence of ore bodies within a metalliferous interval (after Smirnov, 1955) showing (a) primary halos, and (b) the plot of the increase in the number of the halos formed below the ore bodies with depth.
1 - ore body; *2* - halo; *3* - level of erosion surface.

echelon-like arrangement of ore bodies within a general ore-bearing interval (cross section) (Smirnov, 1955), as well as the number of primary halos fringing each ore body. The general halo in the vicinity of an ore body includes all the halos formed by all the indicator elements for the given types of mineralization. In determining the vertical extent of the halos above and below the ore bodies, the results obtained from studies of primary halos from ore deposits which differ in the composition and in the environments of formation have been taken into account. It has been established that, in some cases, it is possible to delineate the supraore halos to their complete wedging out up the pitch of the ore bodies. With respect to halos below the ore bodies, no data on their vertical extent are presently available, because they could not be traced to the point where they completely wedge out. In some cases, however, it has been possible to trace the complete wedging out of the halos formed by individual elements below the ore bodies (these elements are generally those typical of the supraore region). However, due to the alternation of the halos formed by different elements, no wedging out of the composite halos at accessible depths has been established.

This suggests that those halos formed below ore bodies are appreciably more extended vertically (by comparison with those formed above the ore bodies). The halos formed below ore bodies are known to indicate the paths for the movement of the ore-bearing solutions and can, therefore, be traced to considerable depths.

The great vertical extent of the halos formed below ore bodies in comparison with supraore halos (in the case of the echelon-like arrangement of the ore bodies), leads to a progressive increase in the number of the halos below the ore as the depth of the erosion surface increases. This is illustrated by the plot in Fig. 53. If the erosion surface is shallow, the supraore halos will represent only those ore bodies which occur at the highest levels. In the case of the total erosion of an entire metalliferous interval, all eroded ore bodies will be represented by halos formed below the ore body. The number of geochemical anomalies in the latter case may increase by several times, however, these anomalies are not of practical interest. Consequently, the development and the implementation in exploration practice of effective criteria for the differentiation between those halos formed below and above the ore body is an essential condition for the successful location of hidden mineralization. This problem of differentiation can be solved with the help of the vertical zonality of the primary halos which has been considered above.

The supraore geochemical anomalies are usually identified by using ratios of the average contents and productivities of the halos of the pairs of elements, as well as by using ratios of the parameters of the individual composite (additive and multiplicative) halos of the supraore elements to the corresponding parameters characteristic of the halos formed below ore bodies. In some cases, the elements of the main economically valuable components of ores are also used in the denominator. The elements above and below the ore bodies are chosen on the basis of the results of the study of zonalities of the primary halos of known (standard) deposits. Particular attention is paid to the elements which are indicators of vertical zonality.

The level of the erosion surface of the anomalies is evaluated by comparison between the values of the above-mentioned parameters with the corresponding parameters of halos formed around known ore bodies of the same type. As noted above, it is more reliable to use parameters of the composite halos for the evaluation of the level of the erosion surface of the geochemical anomalies; however, this does not mean that paired ratios should not be considered. The latter may successfully be used for those deposits whose primary halos are strong, extensive, and which are characterized by distinct vertical zonalities.

The results of investigations in an area of the Kansky ore field (Central Asia) are given below as an example of the method for predicting blind mineralization on the basis of vertical zonality of primary halos.

Several geochemical anomalies were detected using results from geochemical bedrock sampling at the surface at the peripheries of the Kansky lead-zinc deposit. The ratios of the productivities of the individual multiplicative halos formed by the indicator elements of the multi-element mineralization both above (barium, silver, and lead) and below (zinc, copper, and cobalt) the ore body were calculated. A comparison of these values with the data obtained by studying halos around known ore bodies made it possible to conclude that some of the anomalies were supraore halos with promising potential for blind mineralization. Fig. 54 shows the distribution of lead within a supraore anomaly. The indicator ratio for the given anomaly is 5 (the indicator ratio is more than 0.5 based on standard cross sections for supraore halos, and less than 0.002 for halos below the ore). Subsequent drilling proved economically important blind mineralization at depth and thus confirmed the assessment based on the results of geochemical sampling of the bedrock (Fig. 55).

Quantitative interpretation of geochemical anomalies formed above ore bodies is the most

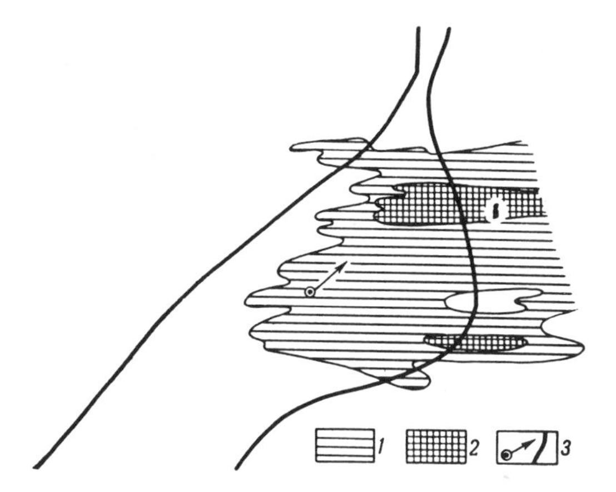

Fig. 54. Primary halos formed by lead (at the surface).
1 and *2* - content of lead, $n \cdot 10^{-3}\%$ (*1* - 3 to 10; *2* - more than 10); *3* - borehole and sampling profiles.

complicated problem. Quantitative interpretation would be expected to evaluate the possible extent of blind mineralization. However, no reliable methods for solving this problem (quantitative interpretation) are presently available. Development of such methods will be the objective of future studies. In individual favorable cases it is possible to judge, more or less reliably, the extent of the blind mineralization using correlations between the parameters of the ore bodies and of halos formed in their vicinity.

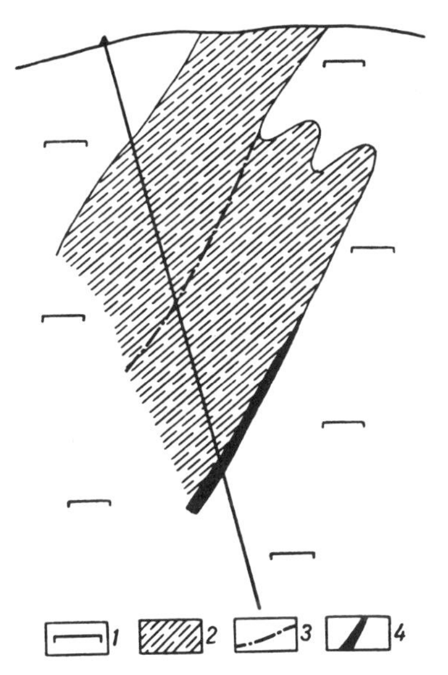

Fig. 55. Section across blind mineralization revealed as a result of the study of a geochemical anomaly.
1 - serpentinites; *2* - ore-bearing xenolite (calcareous-shale rocks); *3* fractures; *4* - blind ore body.

Studies have shown that wider, and more extensive, halos usually form around thick ore bodies. A definite relationship between the extent of halos and the thickness of the ore bodies is usually found for ore bodies from identical environments of mineralization.

By way of example, Fig. 56 shows plots of the variation with depth in the values of the linear productivity of halos formed around ore bodies from a uranium deposit in granites.

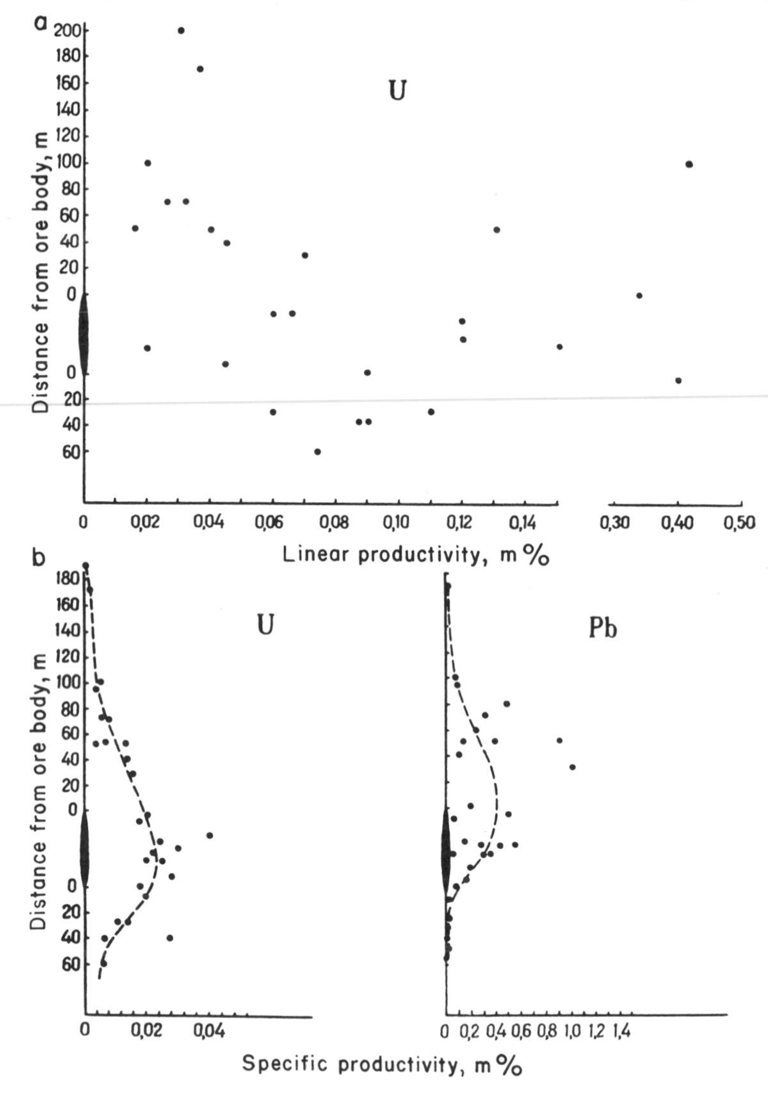

Fig. 56. Plots of vertical variation in the linear (a) and specific (b) productivities of the halos.

This deposit was described earlier. Each point in the plots corresponds to the value of the linear productivity calculated for a horizontal cross section of the halo. Plotted on the X-axis are linear productivities of the halos formed by uranium and lead which are the principal indicators of uranium mineralization. On the Y-axis, distances (in meters) from the ore body in the upward and downward directions are given. The upper plot is a cluster of randomly scattered points. The pattern becomes essentially different if, on the X-axis, one plots the relative values of linear productivity, rather than absolute values. The relative values are ratios, which are calculated by dividing the absolute productivities by the maximum thicknesses of the corresponding ore bodies (this is called *specific productivity*).

In contrast to the case of uranium, the points on the plot for lead are characterized by a greater scatter and by a sharp upward shift, which reflects vertical zonality (strong halos of lead above the ore bodies).

The above criterion may be used for the quantitative interpretation of anomalies within areas adjacent to a standard (known) deposit. Despite the well documented dependence of the halo productivity on the dimensions of the ore body, it is extremely difficult to evaluate the extent of blind mineralization based on the parameters of the halos. This is because, in addition to the extent of the blind mineralization, other factors, such as the geologic-structural environment of the mineralization, physicomechanical properties of the enclosing rocks, etc., have an appreciable effect on primary halos. Notwithstanding this, when one chooses the most promising anomalies of the same type within regions where the above correlation between the halo parameters and the ore bodies has been established based on examples from known deposits, preference should be given to those anomalies which are characterized by maximum productivity. Anomalies are considered as being of the same type if they: (a) are represented by the same type of ore mineralization; (b) are characterized by identical geologic-structural environments and similar erosion levels; and (c) are developed within the same rocks. The latter conditions is especially important, because the intensity and the size of a halo depends, to a large extent, on the type of the enclosing rocks, primarily on their physicomechanical properties; stronger and larger halos develop in rocks with increased porosity and jointing. This must be taken into account during the quantitative interpretations of geochemical anomalies.

This criterion of quantitative interpretation of geochemical anomalies, based on the correlation between parameters of ore bodies and of halos, is relative rather than absolute. Therefore, it may be of value in interpreting groups of similar anomalies rather than isolated anomalies. When used in this manner, the most promising anomalies could be selected from the group of anomalies of the same type, which are considered as having been formed above the ore body.

The "depth range" to which primary halos may be used in the search for blind mineralization. To this point we have considered the problems involved in the search for blind mineralization in those cases in which the primary halos surrounding blind mineralization are detectable at the present-day erosion surface and are recognizable by geochemical sampling. Naturally, this is not the only possible case. Sometimes, when ore bodies occur at great depths, their halos may not be detected at the surface. This means that in those cases in which primary halos are not revealed at the surface, a negative evaluation of the area may be applicable only to a certain depth. We are referring here to the "depth range" in the search for blind mineralizations using primary halos. In the case of steeply dipping ore bodies this depth range is determined by the vertical extent of the halos formed over the ore bodies.

Table 34. Vertical extent of halos formed above ore bodies (as observed).

Deposit	Extent, meters
Tungsten-molybdenum (Chorukh-Dairon)	200
Bismuth (Chokadambulak)	200
Copper-gold (Parkovoye)	340
Lead-zinc in skarns (Kurusai)	450
Lead-zinc in skarns (Nikolaevskoye)	850
Tin ore (Verkhneye)	270
Polymetallic vein (Arkhon)	130
Uranium (in granites)	200
Fluorite (Myshikkol)	100

Studies have shown that the vertical extent of primary halos may be quite large. In all the cases listed in Table 34, sufficiently strong and broad halos were revealed at great distances from the upper boundaries of the blind ore bodies. This implies that the figures reported are by no means the limit and, in fact, halos formed above ore bodies may be much larger.

Table 34 gives the sizes of monoelement halos. If necessary, the depth range which can be considered in the search for blind mineralizations, based on primary halos, may be considerably increased by using the composite halo methods (additive and multiplicative).

In this connection, the problem of ore occurrences in regions where no geochemical anomalies have been revealed, should be given special consideration. Final evaluation of the existence of possible blind mineralization in such regions must be based on the "depth range" of the method, i.e., it is necessary to ascertain whether or not the "depth range" is sufficient for the given type of mineralization.

One possibility is that such regions may be evaluated as being nonpromising. However, in the case of an insufficient depth range, this evaluation may be applicable only to a limited depth, which is equal to the vertical extent of the supraore halo. This problem can be solved only with respect to a specific type of ore deposit within the environments of a particular region.

Prospecting for subhorizontal blind ore bodies. The methods involving the use of primary halos in the search for subhorizontally-occurring ore bodies (in contrast to steeply dipping bodies) have not been studied sufficiently. Consequently, the techniques considered below require further refinement.

The geochemical criteria of prospecting for horizontally-occurring ore bodies are based on the developmental features of the primary halos around such ore bodies. This, in turn, is controlled by the geologic-structural environment of the mineralization and, in particular, by the morphological characteristics of ore bodies. This is why successful utilization of primary halos in the search for this type of deposit is impossible unless one takes into account all the geologic-structural characteristics of the region and the specific area being studied.

The characteristics of primary geochemical halos around ore bodies which occur almost horizontally, were discussed above using examples based on the skarn-polymetallic deposits of Tutly I (Kurusai ore field), Nikolaevskoye (Maritime region), and the Sarycheku porphyry copper deposit (in Central Asia).

It was pointed out that primary halos are extended in the plane of the ore body, however, their vertical extents are limited. This implies that the depth ranges which can be used in the search for nearly horizontal blind ore bodies (if one uses primary halos), will be much smaller than is the case with steeply dipping bodies. Consequently, there may be cases when the halos are also blind due to the relatively deep occurrence of the ore bodies. This possibility must be taken into consideration in the interpretation of the results of geochemical sampling in regions which are potentially promising for flat-lying ore bodies.

When geochemical anomalies due to subhorizontal ore bodies are being interpreted, the problem of differentiating between the halos formed above and below the ore bodies can be solved quite simply, provided there is a stratigraphic control at the mineralization site. The stratigraphic control may occur when ore bodies lie within a certain favorable lithologic horizon (e.g., the Tutly deposit within skarned conglomerates), or at the contact of different rocks (e.g., the Nikolaevskoye deposit in the Maritime region, which occurs at the contact of carbonate and igneous rocks).

In cases where there is no such stratigraphic control in the distribution of mineralization, vertical zonality of the halos formed in the vicinity of subhorizontal ore bodies may be used for differentiating between anomalies formed above and below the ore bodies. This zonality is expressed as the relative enrichment of the above-ore halos of certain elements. These elements are antimony and arsenic in the Tutly I deposit, and silver and lead at the Sarycheku deposits. Vertical zonality in the halos around flat-lying bodies is, however, quite indistinct in most cases. Consequently, at this time, it can only be stated that, in principle, this zonality possibly can be used. However, the efficiency of its practical implementation requires further study, including proof in field situations.

From Figs. 33 and 35, it was seen that primary halos formed by the principal components of ores are especially closely correlated in space with the ore bodies. The major ore components include lead and zinc in polymetallic deposits (copper and molybdenum in porphyry copper deposits). In particular, these elements form the strongest halos within the interval of the ore bodies. It is, therefore, recommended that drilling be conducted at the epicenters of anomalies of the above-mentioned elements during prospecting in promising areas. This is because other halos may be shifted with respect to the ore bodies corresponding with the zonality in their distribution. Such a distinct shift was established for silver at the Sarycheku deposit (Fig. 36), as well as for arsenic and antimony at the Tutly I deposit (Fig. 34).

The restricted depth range in the search for subhorizontal ore bodies based on primary halos, makes it especially appropriate to use a wide variety of special procedures to extend the depth range in the search for blind bodies (e.g., analyses of heavy fractions, construction of multiplicative and additive halos). In these cases, valuable information can be obtained from studies of geochemical specialization of steeply dipping faults which not infrequently function as "channels" connecting the surface with the deep ore bodies. This may be illustrated by the Nikolaevskoye deposit, where blind mineralization at a depth of nearly 900 meters was detected with the help of the data from geochemical sampling of the bedrock at the surface (Fig. 37).

The steeply dipping faults at this deposit, along which ore-bearing solutions probably have circulated, extend directly through the ore body. This permitted accurate localization of the deposit using the results from the geochemical sampling. In many cases, however, the steeply-dipping ore-conducting fractures occur outside the ore bodies (Fig. 36), and this must also be taken into account in the interpretation of the results from geochemical sampling.

The use of the primary halos will appreciably contribute to the effectiveness of drilling operations conducted during prospecting for subhorizontal ore bodies. The rather large extent of the halos in plan (in the plane of the ore body) will create favorable conditions for the identification of areas promising for blind mineralization, particularly with help from drilling a widely-spaced network of boreholes. In this case, not only areas where ore bodies are revealed should be considered as promising, but also those areas where geochemical anomalies from indicator elements which are characteristic of the given type of mineralization are detected in the course of core sampling. The promising areas thus revealed should be subsequently drilled along a more closely-spaced network in order to detect and delineate ore bodies. In this stage, as well as in the previous one, geochemical sampling of core samples, and the interpretation of the results from these samples are essential, because these data may be very helpful in the proper location of the boreholes, as well as for changes in the direction of drilling. In the latter case, it is recommended that halo zonality be used in the determination of the direction of movement of the ore-bearing solutions. To illustrate this point we will consider a section across the Sarycheku deposit (Fig. 57).

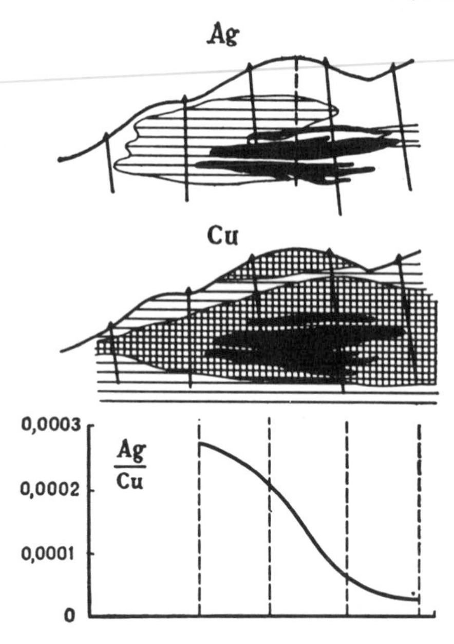

Fig. 57. Primary halos and the plot of the variation in the ratio of linear productivities of the halos along the cross section (the Sarycheku deposit; see Fig. 35 for conventional symbols).

Fig. 57 shows an ore body revealed by boreholes drilled down to 200 meters. In the first stage of exploration, the network of boreholes could have been wider-spaced, in view of the considerable dimensions of the primary halos in the plane of the ore body. From Fig. 57 it is possible to see where two boreholes spaced at a distance of 600 m in the cross section would not reveal any ore body (this is the least favorable situation). Primary halos may be invaluable in such cases (which are frequently encountered in exploratory drilling). In such a situation, the very fact that strong primary halos, formed by indicator elements which are typomorphous for the given type of mineralization are present in the region and can be detected by geochemical sampling of cores from boreholes, will make it possible to avoid an erroneously negative evaluation of the ore potential in the area. Moreover, the most likely sites for the occurrence of the ore body may be determined by using the results from studying the distribution patterns of the indicator elements within the halos. In the particular case being considered, the area between the boreholes is just such a promising location, because the boreholes revealed the head and root portions of the halos, which is indicated by a sharp decrease in the ratio of the linear productivities of the halos formed by the pair silver-copper (Fig. 57).

Assessment of Ore Showings

Studies of primary geochemical halos have been directed mainly at developing techniques for the search for blind ore bodies. It is, therefore, not fortuitous that primary halos are practically always considered only in connection with the search for blind mineralization. Studies have shown, however, that a distinct vertical zonality in the distribution of the indicator elements may be revealed not only when comparing halos above and below ore bodies, but when comparing cross sections of the halos formed at different levels of ore bodies. This suggests that vertical zonality of primary halos may be used both for differentiating between the halos formed below and above ore bodies (the search for blind mineralization), and for the evaluation of the depth of the erosion level of the outcropping ore bodies. The assessment of the latter is a major component in the evaluation of the ore reserves at depth.

Separate consideration of the techniques for the interpretation of geochemical anomalies in search for blind mineralizations and for the evaluation of ore reserves, does not mean that these studies are carried out independently. It is only for reasons of convenience that they are considered separately. In practice, these studies are generally conducted together because each ore showing may indicate blind bodies either at depth or on the flanks. This implies that geochemical sampling for the evaluation of ore showings must also be done outside the limits of ore mineralization in areas which are promising for blind mineralization. Experience has not infrequently shown cases where known ore showings appear to be unpromising because of the deep erosion level, whereas supraore halos from blind ore bodies may be detected in the immediate vicinity.

The great practical significance of using primary halos for the evaluation of the ore potential at depth becomes evident if one takes into account the fact that in practically all metalliferous regions numerous ore showings exist which must be evaluated in order to select the most promising ones.

Recent experiences have confirmed the effectiveness of using the criterion of zonality of primary geochemical halos in both the evaluation of the level of the erosion surface represented by the ore showings, and in the rejection of nonpromising (deeply eroded) ore showings from further consideration.

In order to illustrate this statement we will consider the results of geochemical studies carried out at the Kamazakbulak and Marazbulak ore showings (Grigorian et al., 1970).

These copper-bismuth ore showings are situated in Central Karamazar and are confined to the Karamazar Fault and feather joints. The enclosing rocks in the region are represented by Upper Paleozoic extrusives (liparite porphyries of the Shurabsai Suite).

The Kamazak-Marazbulak Fault is the major ore-enclosing structure in the ore field. This fault can be traced for 5.5 km in a northeastern direction, and it crosses the Kamazakbulak and Marazbulak ore showings. In the east, it is bounded by the Zhelezny Iron Fault. Ore bodies with economic amounts of copper and bismuth have been found at the surface in the regions being described.

In order to evaluate the prospective ore reserves at deep horizons represented by these ore showings, a geochemical sampling of the bedrock was conducted along profiles across

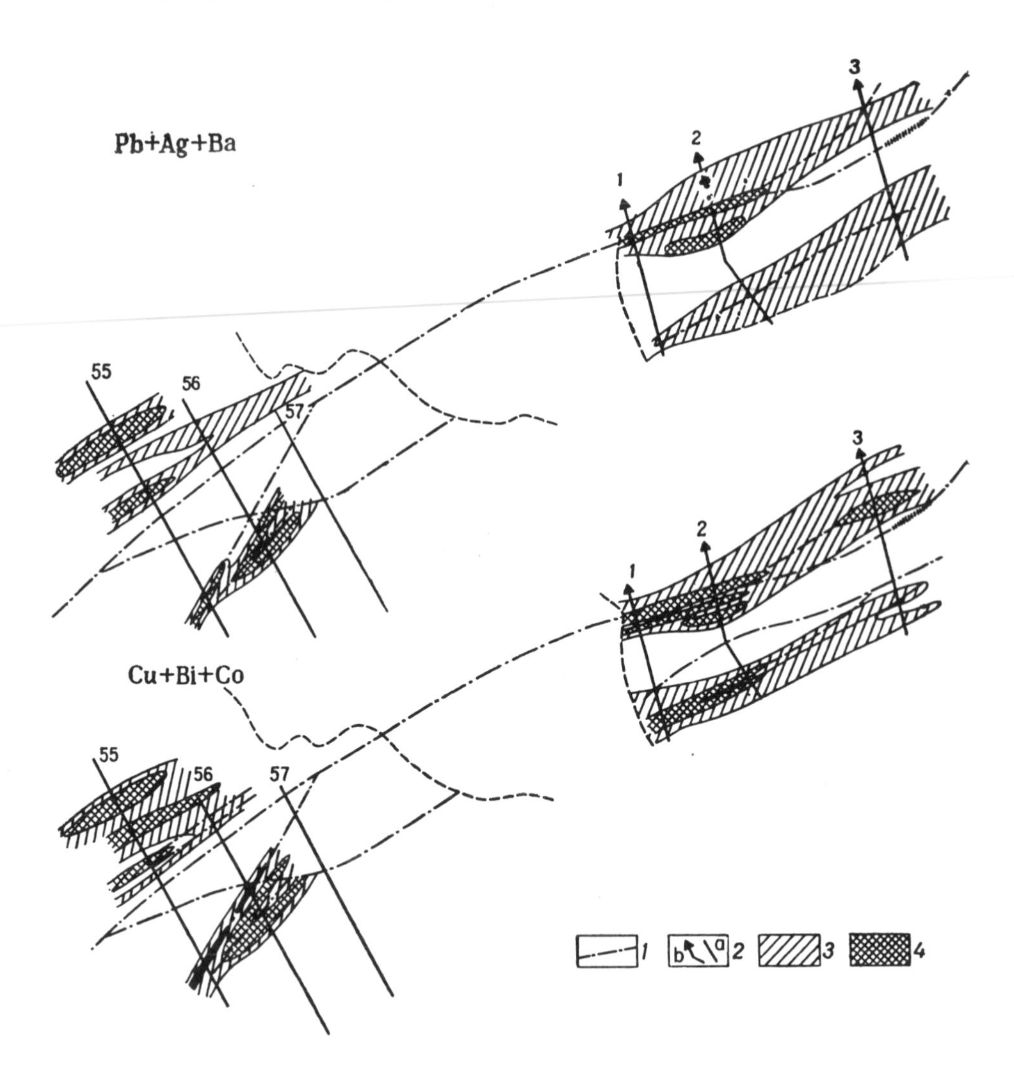

Fig. 58. Primary additive halos at the surface in the Kamazakbulak and Marazbulak regions.
1 - tectonic zones; *2* - geochemical profiles (*a*) and exploratory boreholes (*b*); *3* and *4* - element contents, in background units (*3* - 6 to 20; *4* - more than 20).

the strike of the major fault zone (Fig. 58). As a result, anomalies formed by several elements were revealed. Their parameters are listed in Table 35 (with the exception of barium whose halos are practically absent).

Table 35. Ratios of linear productivities of halos.

Region	Anomaly No.	Profile No.	Ag/Cu	Pb/Cu
Kamazakbulak	I	56-56	0,001	0,04
		57-57	0,001	0,04
	II	55-55	0,001	0,02
		56-56	0,001	0,01
Marazbulak	III	1-1	0,001	0,03
		2-2	0,003	0,1
	IV	2-2	0,002	0,016
Standard	Upper parts of the mineralization		0,02-0,4	0,3-2,5
cross-section	Roots of the mineralization		0,002	0,09

A comparison between the parameters of the anomalies detected and the data obtained from studying halos formed in the vicinity of known deposits, has shown that the ratios of linear productivities of the halos produced by the principal indicator elements along all profiles, correspond to the root portions of the copper-bismuth mineralization (Table 35). These data provided sufficient grounds for cancelling drilling operations in these areas, although they had been planned on a large scale. However, despite the negative evaluation, the exploration team drilled five boreholes in the region because the ore showings were characterized by a favorable geologic-structural environment and, in addition, because economic outcrops were found at the surface. The boreholes did encounter an ore-enclos-

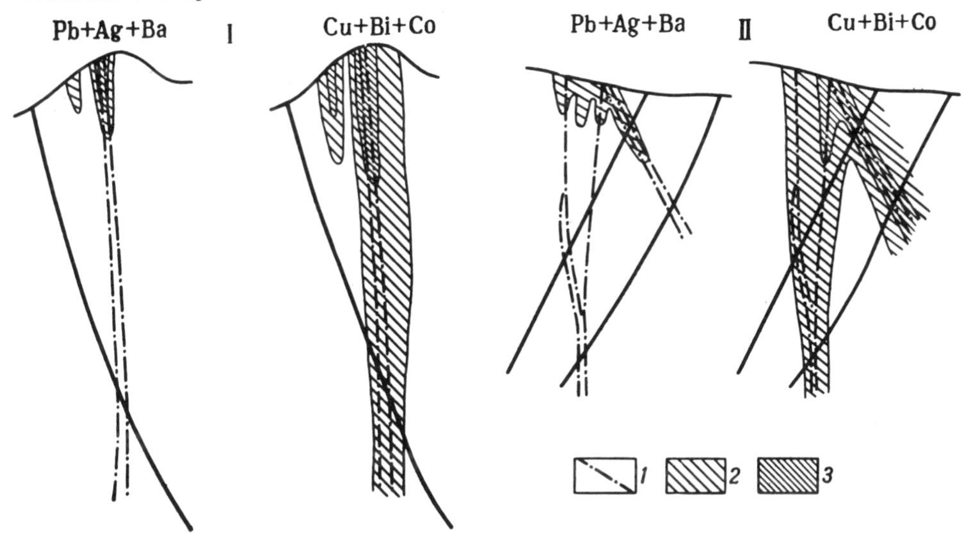

Fig. 59. Primary additive halos in cross section at the Marazbulak (I) and Kamazakbulak (II) ore showings.
1 - Fault zones; *2* - Additive halos; *3* - Fields of maximum values: more than 20 background units.

ing fault zone at depths ranging from 100 to 300 meters. However, no sulfide mineralization was found, and thus the negative evaluation based on the results of the geochemical sampling was confirmed. Subsequent sampling of the cores from the boreholes fully confirmed the assessment. As shown in Fig. 59, both the mineralization and the halos wedge out completely with depth.

The Maidantal region is situated on the east edge of the Taryekan deposit. It is composed of Upper Paleozoic extrusive rocks of acid and intermediate composition. These rocks are represented by felsite- and spherulite-porphyries, tuffaceous sandstones and andesite porphyrites intruded by isolated dikes of diabase porphyrites and felsite-porphyries with north and northeastern trends.

The region is broken by a series of faults, mainly with north and northeastern strikes. Fracturing and hydrothermal alteration of the rocks can be observed. A geochemical sampling of the bedrock was carried out along profiles across the strike of the major faults with the purpose of evaluating the ore potential in the region. Processing of the sampling data in the region revealed several geochemical anomalies, three of which turned out to be promising for blind and weakly eroded mineralizations (Fig. 60).

In order to determine the level of the erosion surface of the geochemical anomalies relative to the ore-bearing interval, the ratios of linear productivities for individual additive halos formed by the elements indicating the copper-bismuth mineralization above and below the ore body were calculated (Table 36).

A comparison between these anomalies and the standard cross section from known mineralization has shown that supraore anomalies exist in the area. So far, only one of the anomalies which was considered promising on the basis of geochemical sampling results has been tested by drilling (Table 36, Anomaly II). The drilling revealed ore reserves of economic significance at depth.

When the Maidantal area was compared with the above-described ore showing from Kamazakbulak and Marazbulak, it became clear that if their ore potentials were compared, without taking into account the level of the erosion surface (based only on studies at the surface), then preference erroneously would undoubtedly be given to Kamazakbulak and Marazbulak, because these areas are characterized by outcrops of economic ore.

In most cases, the differences in the level of the erosion surface in different areas can be revealed by a simple visual comparison between geochemical anomalies. This applies, in particular, to individual composite anomalies constructed for groups of supraore elements. The corresponding calculations further confirm the general estimates. Fig. 60 shows that the Maidantal area differs sharply from the Kamazak-Marazbulak ore showings by having stronger additive halos of supraore elements (lead, silver and barium).

The above, as well as experiences of this type from studies in several other regions in the country, suggest that the determination of the level of the erosion surface of geochemical anomalies, based on the vertical zonality of primary halos of the corresponding type of mineralization, must always precede the evaluation of the ore potential in any regions which are considered promising for endogenic mineralization.

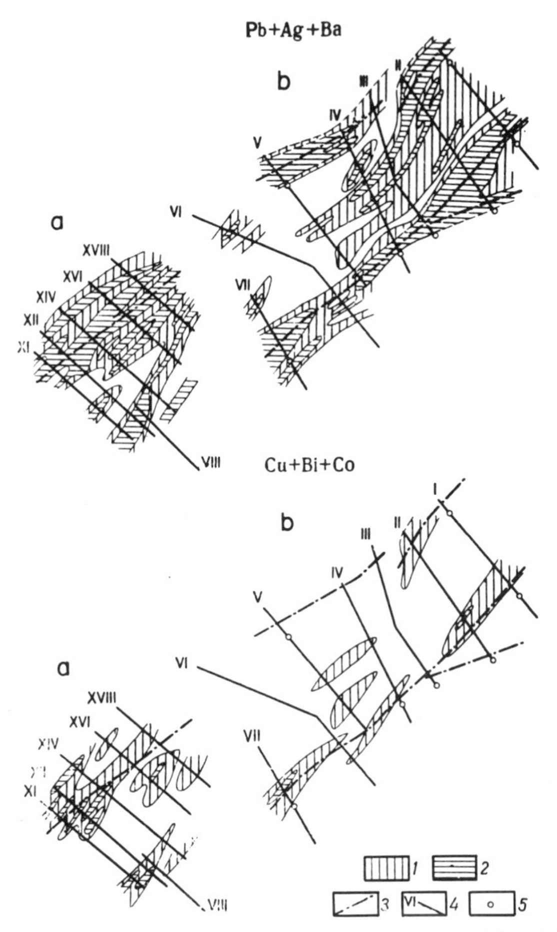

Fig. 60. Primary additive halos at the surface of the Taryekan (a) and Maidantal (b) deposits.
1 and *2* - element contents, in background units (*1* - 6 to 20; *2* - more than 20); *3* - fault zones; *4* - geochemical profiles; *5* - exploratory boreholes.

Table 36. Indicator ratios (Maidantal area).

Anomaly	Profile	Ratio
I	I-I	3,8
	II-II	8,9
	VII-VII	4,3
II	III-III	10,2
	IV-IV	8,6
III	V-V	12,6
Standard cross section	Upper parts of the mineralization	1,0

Interpretation of Multiformational Anomalies

To this point, we have considered only the criteria used for the interpretation of the anomalies resulting from a single ore formation. However, it is known that, in some cases, superimposition in space of the manifestations of various ore formations may result in a mixed halo — the so-called *multiformational halos* (Ovchinnikov and Grigorian, 1970) described earlier. This must be taken into account in the interpretation of the results of geochemical studies.

In many cases, the possibility of encountering geochemical anomalies of the multiformational type during prospecting depends primarily on the metallogenetic features of the regions under study. Such anomalies are most likely to be found in those regions with different ore formations, especially at the contact of ore fields formed by deposits which have different compositions.

In some cases, the multiformational character of the geochemical anomalies may be recognized as a result of detailed structural-mineralogical observations, which make it possible to reveal, at the surface, manifestations of various ore formations, in particular their typomorphous mineral associations. However, in cases where ore bodies in one of these formations are blind, a reliable interpretation of the anomalies requires additional criteria, because mineralogical indications of the blind mineralization may not be expressed at the surface. Such criteria may also be geochemical, based on the characteristics of the multiformational halos considered above.

As mentioned previously, the halos around the ore bodies of each ore formation, corresponding with the composition of the ores, are characterized by specific combinations of indicator elements. This means that the appearance of anomalies due to certain elements, which are not typical of the known ore formation, must be considered as an indication of a possible expression of another ore formation. For example, anomalies of uranium at polymetallic, gold, and other deposits, may indicate an expression of a new (uranium) ore formation, because uranium is generally not typical of halos formed at these deposits.

The "through" indicator elements, that is, those which are typomorphous (characteristic) for several ore formations, may also be used for the determination of the multiformational nature of the anomalies. In this case, the ratios of the dimensions (widths) of the halos formed by these elements are used; these dimensions vary with the composition of the deposit. For example, uranium halos are the largest at uranium deposits (the uranium is mobile). The halos formed by the other elements are contained within the uranium halos. An inverse relationship between the dimensions of the halos formed by these indicator elements must be considered as a possible sign of multiformational anomalies. In this case, the ratio of the linear productivities of the halos calculated in the section across the trend of the halo is especially informative. In cases where it is impossible to determine the width of the halos, because the profiles are too short, one can use the ratio of the average anomalous contents of the elements because, as already pointed out, the series of the indicator elements based on the ratios of linear productivities and on the ratios of the average anomalous contents, are practically identical. It is natural that in the determination of the formational affiliation of the anomalies with the help of the series of the transverse halo zonalities, only results with identical analytical sensitivities should be used.

In some cases, it is possible to identify, within a single multiformational geochemical anomaly, epicenters of maximum concentrations of the indicator elements of each formation. As shown by experience, the most informative method in such cases is that of constructing composite halos. This is illustrated by the above (see Fig. 41) distribution maps of the indicator elements within multiformational anomalies formed as a result of the coincidence in space of the halos associated with the copper-bismuth and uranium mineralizations. The epicenter of the additive halos clearly indicates the zone of the copper-bismuth mineralization, whereas the halos formed by uranium and molybdenum coincide with the fracture which branches out diagonally from the major zone (Fig. 41).

A correlation analysis may also be used in the interpretation of multiformational anomalies. As previously mentioned, an appreciable positive correlation usually exists between the indicator element contents in the primary halos associated with the expression of a single anomaly. The correlation is generally negative in the case of multiformational anomalies.

Procedures for the interpretation of multiformational geochemical anomalies are considered below with an example from the Central Orlinaya Gorka area, located at the eastern margin of the Kurusai ore field, within which only deposits and ore showings of polymetallic composition were known prior to geochemical studies. Geochemical investigations (Grigorian et al, 1969) revealed anomalies which differ appreciably from the halos formed around ore bodies of the polymetallic formation. In addition to the principal indicator elements of the polymetallic ores, these anomalies are also characterized by gold. Further, they exhibit another feature — increased contents of molybdenum, as well as of arsenic and antimony, relative to the halos formed by polymetallic deposits. This is illustrated in Table 37, which lists the average contents of the chemical elements in the anomaly revealed in the Central Orlinaya Gorka region. For comparison, the element contents in the halos formed around ore bodies of the skarn-polymetallic formation are given.

Table 37. Average contents of indicator elements in primary halos.

Indicator Elements	Deposit			
	Northern Kurusai II			
	Supraore halos	Halos at upper levels of ore bodies	Central Orlinaya Gorka	Akcheku
Cu (%)	0,01	0,02	0,1	0,07
Mo (%)	0,0002	0,0003	0,004	0,007
As (%)	not detected		—	—
Ag (%)	0,00005	0,0008	0,0004	0,00013
As/Cu	—	—	0,06	0,03
As/Mo	—	—	1	0,3
Cu/Mo	50	67	25	10

Interpretation of geochemical anomalies generally involves the solution of the following problems: (1) determination of the type of an ore formation (of ore formations in the case of multiformational anomalies) which caused the anomaly; and (2) evaluation of the ore potential of the particular type of mineralization.

The combination of the geochemical data suggested the multiformational nature of the anomalies in the region being studied. These anomalies formed as a result of the coincidence, in space, of the halos formed by a polymetallic (ore bodies of polymetallic composition are known to exist in the region) and by some other unidentified type of ore formation.

The multiformational character of the halos being described is especially distinctly expressed in cross section. Fig. 61 shows halos of several elements, delineated on the basis of the results of sampling of bedrock at the surface, as well as from adits and cores collected

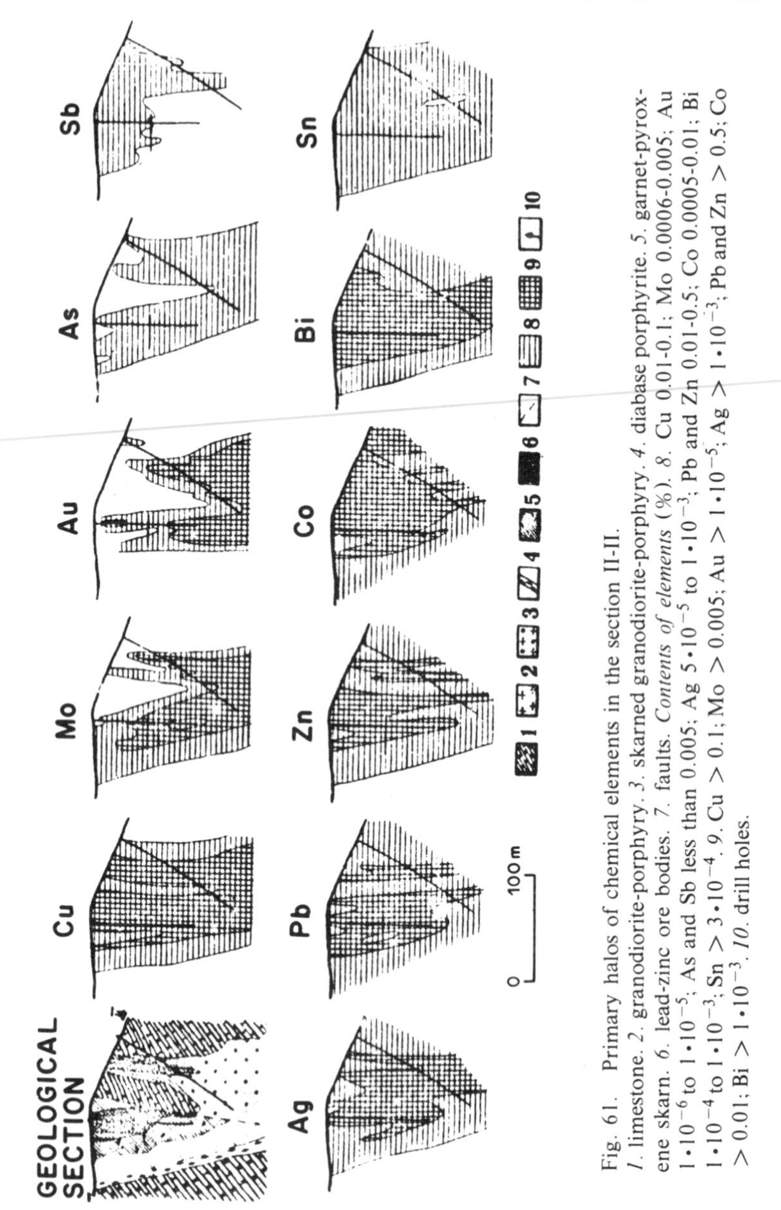

Fig. 61. Primary halos of chemical elements in the section II-II.
1. limestone. *2*. granodiorite-porphyry. *3*. skarned granodiorite-porphyry. *4*. diabase porphyrite. *5*. garnet-pyroxene skarn. *6*. lead-zinc ore bodies. *7*. faults. *Contents of elements* (%). *8*. Cu 0.01-0.1; Mo 0.0006-0.005; Au $1 \cdot 10^{-6}$ to $1 \cdot 10^{-5}$; As and Sb less than 0.005; Ag $5 \cdot 10^{-5}$ to $1 \cdot 10^{-3}$; Pb and Zn 0.01-0.5; Co 0.0005-0.01; Bi $1 \cdot 10^{-4}$ to $1 \cdot 10^{-3}$; Sn $> 3 \cdot 10^{-4}$. *9*. Cu > 0.1; Mo > 0.005; Au $> 1 \cdot 10^{-5}$; Ag $> 1 \cdot 10^{-3}$; Pb and Zn > 0.5; Co > 0.01; Bi $> 1 \cdot 10^{-3}$. *10*. drill holes.

from boreholes which were drilled for the purpose of surveying the polymetallic mineralization. The limestones wedge out at depth. This seems to account for the limited possibility of finding a skarn-polymetallic mineralization in the region. This conclusion is confirmed by a distinct wedging out with depth of the halos formed by the indicator elements which are typomorphous for this type of mineralization: lead, zinc, and silver (Fig. 61).

In the determination of the type of a recently discovered ore formation in the region, the elemental composition of the anomalies was taken into account, as well as the patterns of the distribution of the elements both in plan and in the cross section. Anomalies due to gold have been revealed in the region and these are of significant interest. Their intensities and widths sharply increase with depth (Fig. 61). The anomalies produced by this element are undoubtedly caused by a new ore formation, because gold is not characteristic of lead-zinc mineralization of the Kurusai type. The presence of rather broad and distinct anomalies formed by antimony are another indication of a new ore formation. The halos formed by this element around the lead-zinc ore bodies are extremely narrow and, in some cases, it is difficult to detect them. Arsenic behaves similarly.

Fig. 61 shows that the anomaly patterns formed by gold and molybdenum expand with depth rather noticeably. Molybdenum, like gold, cannot be considered as typical indicators of lead-zinc mineralization, because at such deposits it forms rather narrow and weak halos with contents of less than 0.001 % (Table 37). The width of the molybdenum anomaly with a content higher than 0.005% in the region being described, exceeds 40 to 50 meters.

Copper is another element whose behavior also seems anomalous in comparison with the halos characteristic of lead-zinc mineralization. Copper forms the widest and the strongest anomalies in the region. Its content in all samples collected in the cross sections is higher than the minimum anomalous (threshold). The anomaly field with the content of copper exceeding 0.1% is more than 80 m wide. The copper halos become stronger at depth.

Two groups of indicator elements can be distinguished clearly in terms of their patterns of distribution in vertical cross section: (1) elements whose halos show a distinct tendency to wedge out at depth (antimony, lead, zinc, silver, and cobalt); (2) elements whose halos expand with depth and become stronger (gold, copper, and molybdenum).

A comparison between the anomalies detected, and the halos from deposits of other ore formations, made it possible to consider them as analogs of primary halos of porphyry copper (with gold) deposits of the Almalyk type (Fig. 26).

This conclusion is corroborated by similar average contents of molybdenum and copper (principal indicator elements of the porphyry copper mineralization) in the anomaly from the Central Orlinaya Gorka region and in the supraore halos of the Akcheku porphyry copper deposit (Table 37).

After the results of the geochemical sampling first established the evidence for porphyry copper mineralization in the region, the old mine workings were inspected again. This included their channel samples. The studies fully confirmed the conclusion that the previously overlooked mineralized intervals, in fact, contained mineral parageneses typical of porphyry copper ores with economic amounts of copper.

After the multiformational nature of the anomalies was ascertained, it was found necessary to evaluate the ore potential for each formation. The previous exploration indicated that there is no polymetallic mineralization of economic importance in the area because the ore bodies known at the surface wedged out with depth. The results of geochemical sampling confirmed this conclusion, i.e., the halos formed by the indicator elements typomorphous for the polymetallic mineralization (lead, zinc, and silver) do wedge out with depth (Fig. 61). If there had been blind ore bodies of polymetallic composition at depth, there would have been no such wedging out of the halos. The wedging out of the polymetallic mineralization halos with depth would have been sharper, if there had been no superimposition of the halos from the porphyry copper mineralization, because the latter is also characterized by halos formed by lead and silver. However, due to the higher concentrations of these elements in the halos of the polymetallic mineralization, the latter "suppress" the halos associated with the formation of the porphyry copper, indicating a distinct wedging out of the polymetallic mineralization with depth.

In view of the abundance of the strong halos in the region formed by lead, zinc, and silver, these elements cannot be used for evaluating the prospective porphyry copper mineralization in the region. Therefore, in order to solve this problem, the characteristics of the distribution of the elements mainly associated with the porphyry copper mineralization were taken into account. These elements included copper, molybdenum, gold, and, in part, arsenic. A comparison between the ratios of the average contents of these elements with the parameters of the supraore halos of the standard (known) porphyry copper deposit (Table 37), suggests a conclusion based on the supraore nature of the anomalies detected. This conclusion agrees well with the behavior of the halos formed by the principal indicator elements of the porphyry copper mineralization. It can be seen from Fig. 61 that the copper and molybdenum halos expand with depth. Especially representative in this respect is the behavior of gold whose halos appear essentially only at depth. This undoubtedly suggests a supraore nature of the anomalies, because the strongest halos of this metal at the known porphyry copper deposits develop at the ore body level.

The above data, and in particular the large dimensions of the anomalies and their sharp expansion at depth, made it possible to conclude that the Central Orlinaya Gorka area is promising for blind mineralization. Exploration work was recommended. It was also recommended that special studies, including geochemical investigations, be carried out in order to assess potential porphyry copper mineralization throughout the region and to identify areas most favorable for deposits of this type.

Interpretation of multiformational geochemical anomalies is an extremely difficult problem whose solution is possible only if a comprehensive approach is used. This takes into account all the geological data, and primarily the geologic-structural features of the specific areas.

Further investigation of multiformational geochemical halos and the development of more reliable criteria for their interpretation is an extremely pressing problem because such halos, as shown by recent experiences, are much more widespread than was previously believed. Studies to this effect are also important because they provide more definite information on primary halo zonality, which is a major criterion in the interpretation of geochemical anomalies. In many cases, changes in the halo zonality at hydrothermal deposits, and the appearance of an "inverse" zonality, may be the consequence of multiformational halos.

Such an "inverse" geochemical zonality may be illustrated by the Shurale deposit described earlier. The indicator elements of the high-temperature rare-metal mineralization occurring there (tungsten, molybdenum, etc.), are replaced at depth by halos formed by lead, zinc, and silver, which are typical indicators of polymetallic mineralization (Fig. 43).

Detailed geochemical studies at this deposit, as well as at several others, have shown that in this case we are dealing with multiformational halos formed as a result of the coincidence in space of polymetallic and skarn-scheelite mineralizations. These halos are the products of two totally independent stages of mineralization, which may be expressed both separately ("single formational" halos with direct zonality) and in combination (multiformational halos). In the latter case, the spatial relationships between the products of these stages of mineralization may vary widely. The character of the halo zonality will definitely depend on this variation. An "inverse" zonality is observed at the Shurale deposit, whereas the pattern is different at the Tutly III deposit (Kurusai ore field), i.e., the skarn-polymetallic mineralization is replaced by a skarn-scheelite mineralization at depth, so that a "direct" zonality is revealed in the halos.

The Tyrnyauz tungsten-molybdenum deposit is another characteristic example of an extremely complicated geochemical zonality. A study of the geochemical zonality of the ore bodies and of the primary halos formed in their vicinity (Abramson and Grigorian, 1972) showed that multiformational halos are also widespread at this deposit. The multiformational halos form as a result of the coincidence in space of primary halos formed by skarn-scheelite and quartz-molybdenum mineralizations.

A detailed study of the characteristics of the development of primary geochemical halos around more or less separated ore deposits of skarn-scheelite and quartz-molybdenite compositions showed that the zonality of their halos is in perfect agreement with the above typomorphous series of element zonalities. In particular, the skarn-scheelite mineralization is characterized by halos of molybdenum, tin, and barium shifted upwards relative to the tungsten halos. For molybdenite bodies, on the other hand, tungsten is an indicator of that part of the halo below the ore bodies. This pattern is sharply disturbed in multiformational halos formed in regions of closely spaced ore bodies with both types of ore formations. In such cases, the character of the zonality depends mainly on the spatial relationships between ore bodies of different compositions.

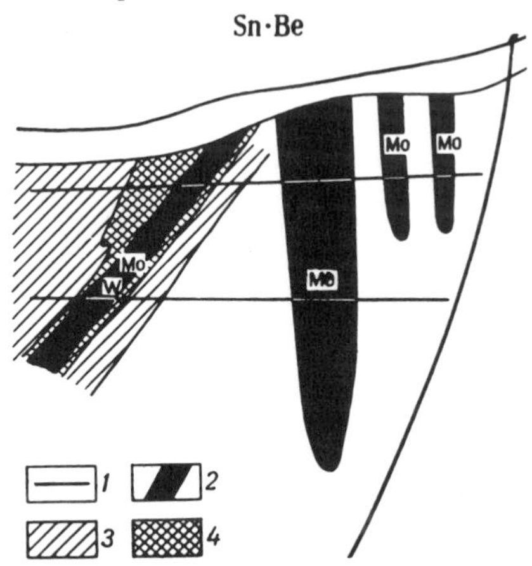

Fig. 62. Multiplicative halos of beryllium and tin in cross section.
1 - mine workings; *2* - ore bodies: skarn-scheelite (W, Mo) and quartz-molybdenite (Mo); *3* and *4* - products of element contents, % (*3* - 1 to $10 \cdot 10^{-7}$; *4* - more than $10 \cdot 10^{-7}$).

Studies carried out at the Tyrnyauz deposit showed that, in the interpretation of multiformational geochemical anomalies, the use of those elements which are indicators of only

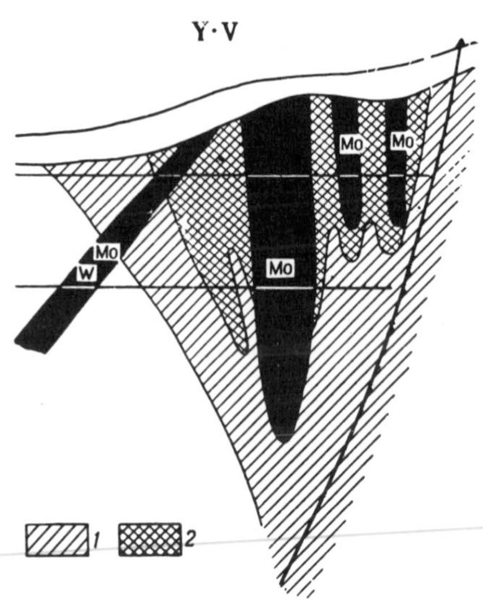

Fig. 63. Multiplicative halos of yttrium and vanadium in cross section.
1 and *2* - products of element contents, % (*1* - 1 to $10 \cdot 10^{-6}$; *2* - more than $10 \cdot 10^{-5}$).
Other symbols as in Fig. 62.

one type of mineralization is most reliable. A comparative study of the features exhibited by the primary halos of the deposit made it possible, in particular, to establish that beryllium and tin are such specific indicators of the skarn-scheelite mineralization. The halos of these elements always accompany the skarn-scheelite mineralization exclusively, and they cross the quartz-molybdenite ore bodies (Fig. 62). The latter are also characterized by specific indicator elements which are not characteristic of the skarn-scheelite mineralization. These are yttrium and vanadium, which form distinct supraore halos when associated with quartz-molybdenite ore bodies (Fig. 63). This selective development makes the halos of these elements reliable indicators of the corresponding type of mineralization.

The above suggests that the causes of the variations in the zonality of primary geochemical halos, in most cases, should be sought in: (1) the development of multiformational halos and; (2) the spatial coincidence of ore bodies and deposits, which differ in the composition and environments, and which formed during different stages of mineralization.

6

Secondary (Supergene) Lithogeochemical Halos and Dispersion Trains From Mineral Deposits

Introduction

Geochemical prospecting for ore bodies involving the use of secondary halos and dispersion trains are, at present, among the most important exploration techniques. They are based on secondary dispersion halos which form in surface materials (soils, eluvium, drift, etc.) during the supergene weathering of ore deposits. Consequently, a *secondary dispersion halo* is a local zone of abnormally high concentrations of certain elements which are indicators of mineralization. This zone forms within unconsolidated rock materials which overlay and surround outcrops of the ore bodies and their primary halos at the presently existing level of the erosion surface.

Dispersion trains are specific types of secondary halos. They usually have the form of linearly extending secondary halos (1) in the alluvial deposits of river valleys, and (2) from springs. They owe their existence to the supergene destruction of ore deposits, as well as to the destruction of their primary and secondary halos, in a given drainage basin.

The techniques based on secondary dispersion halos have been used for nearly 40 years. They were first applied in the Soviet Union in the early 1930's when N. I. Safronov, A. P. Solovov, and others, formulated the main concepts and outlined the range of geological problems which could be solved using the patterns of the secondary dispersion halos which develop from mineral deposits.

The following figures may give some idea of the importance of secondary halos. According to Bekzhanov et al. (1972), in Kazakhstan alone more than 60 million geochemical samples were collected and processed during the period between 1947 and 1969. Twenty new deposits and 150 ore showings were found. Similar data are available for some other regions in the Soviet Union. Outside the USSR, according to the data obtained by The Association of Exploration Geochemists, about 5 million geochemical samples [in the Western World] were collected and analyzed during the period from June, 1970 to June, 1972.

As experiences obtained from large-scale geochemical surveys accumulated, the techniques of sampling and analyses were continually refined, and the theoretical aspects of the problems involved were investigated.

The studies by Solovov were a major contribution to the development of mineral prospecting techniques using secondary halos and dispersion trains of the indicator elements. This author classified secondary dispersion halos of the chemical elements, and formulated the main concepts for the interpretation and the quantitative evaluation of supergene geochemical anomalies.

The problems concerned with the theory and practice of geochemical prospecting for mineral deposits based on secondary halos and dispersion trains, were further developed and improved on the basis of studies by A. N. Eremeev, E. M. Kvyatkovskii, V. V. Polikarpochkin, and others. Geochemists abroad, such as J. S. Webb, H. E. Hawkes, J. S. Tooms, and others, also made valuable contributions.

A. I. Perel'man, M. A. Glazovskaya, V. V. Dobrovol'skii, and others, did much to theoretically substantiate the methods. They studied the influence of particular landscape-geochemical environments on the migration of the chemical elements and on the formation of secondary dispersion halos.

The procedural problems of lithogeochemical exploration for mineral deposits using their secondary halos and dispersion trains, are discussed in sufficient detail in numerous publications by the above-mentioned authors and their associates. Therefore, these problems will be treated below only to the extent which is necessary for the practical utilization of the method, as well as for the use of secondary halos and dispersion trains in combination with other prospecting techniques.

As previously mentioned, secondary dispersion halos which result from the erosion of ore deposits, form within a broad range of natural environments. The main factors operative here are the processes of physical and chemical weathering, which rework initial rocks and ores, and form unconsolidated rock materials.

The products of the supergene alteration of mineral deposits under the effect of the mechanical, physical, physicochemical, and biochemical processes, migrate to the unconsolidated materials where they become fixed (immobilized) either in the form of their own mineral aggregates, or as various compounds which are combined chemically or by sorption with the unconsolidated rock materials surrounding the secondary halo.

Classification of Secondary Halos

In terms of their physical state, secondary dispersion halos are usually classed as *mechanical* (fragmental) and *hydromorphic* (salt). Mechanical halos are those secondary halos in which the ore components are present as primary or secondary minerals which are stable in the supergene zone, and which migrate mechanically. Hydromorphic halos are those which formed as a result of element migration in the form of compounds soluble in natural waters [these are frequently called *"chemical"* in English. Editor]. Secondary dispersion halos are often composed of both types, i.e., *mechanical-hydromorphic.*

Depending on the character of the enclosing rocks and on the depositional environment, secondary dispersion halos are divided into *residual* (formed by the products of weathering on top of the previously existing ore body or its primary halos), and *superimposed* (within the outlines of which there was no primary ore mineralization prior to the development of the halo).

With respect to the present surface, the halos are classified as *exposed* (open), i.e., they emerge onto the present-day surface, and *buried* (closed), i.e., they form [or are presently found] at some depth below the surface. Exposed halos may be found and studied by means of sampling the upper soil horizons, whereas buried halos are found by sampling at depth.

In the practice of geochemical exploration, the level of the secondary halo which is most representative for the collection of samples, has been called the *representative sampling horizon*. This is usually a certain genetic horizon in a soil or in an unconsolidated deposit, and its position determines the depth of the geochemical sampling.

Based on geochemical prospecting experiences involving secondary dispersion halos in various natural zones, Solovov has been able to classify the entire range of such halos into the most typical groups which are likely to be encountered in actual practice. There are seven such groups (called types), and they are described below.

Type 1. **Residual exposed halos which form in actively eroding regions within presently occurring eluvial-diluvial deposits (Fig. 64).**

Residual dispersion halos always contain clastic disintegration products of the mineral deposit and its primary geochemical halos. Under arid climatic conditions characteristic of deserts, where physical weathering predominates considerably over chemical weathering, the mechanical component of the residual halo may predominate appreciably. Residual dispersion halos of wind-blown origin also exist, and they consist almost exclusively of clastic weathering products of ore deposits. In temperate humid regions, the residual halos are

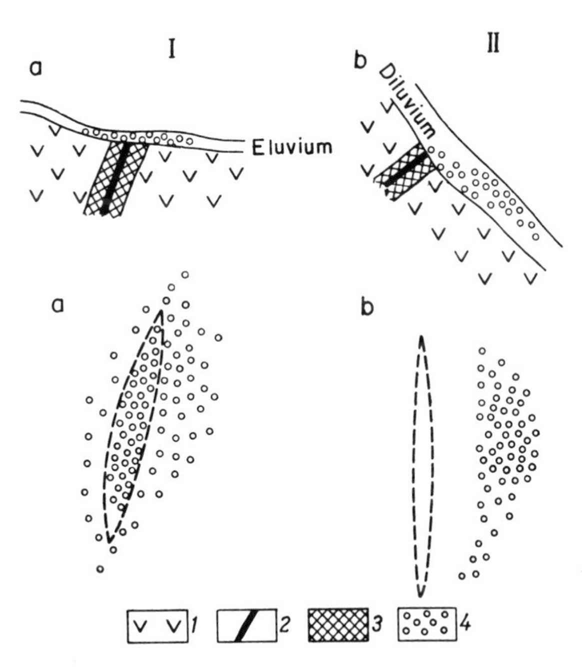

Fig. 64. Residual exposed secondary dispersion halos. I - nondisplaced. II - displaced: (a) in cross section; (b) in plan.
1 - enclosing rocks; *2* - ore body; *3* - endogenic geochemical halo; *4* - residual dispersion halo.

usually the combined mechanical-hydromorphic type. The mechanical fraction is represented by particles of the ore minerals which are resistant to weathering, as well as by sparingly soluble alteration products of the ore and accessory minerals which contain indicator elements typical of the given mineralization.

The hydromorphic fraction of these halos is associated, to a large extent, with the adsorption of soluble metal ions by clay particles, as well as by iron and manganese oxides, which may be present in the halo. Based on geochemical data obtained from exploration carried out in different climatic zones under projects sponsored by the UN Development Program, the coefficients of concentration in residual halos, as one goes from the 20 - 35 mesh fraction to the clay fraction, range from 20 to 40 for copper, and from 5 to 10 for molybdenum, zinc, and lead. This relationship is not characteristic of tin, tantalum, niobium, beryllium, or tungsten because these elements are present in halos largely in the mechanical component.

The phenomena of mechanical migration predominate in the process of the formation of residual dispersion halos. This applies also to the easily mobile elements which are sorbed by the finely dispersed particles within the anomaly. An inversely proportional relationship usually exists between the mobility of an indicator element and its accumulation in halos of this type. The more mobile the element, the more readily it can be moved beyond the halo. This results in a relative depletion in residual dispersion halos of those elements which migrate readily in various given environments.

Solovov devised a theoretical model for residual exposed halos. This model makes it possible to quantitatively describe the processes which take place within these halos. Solovov also proposed several formulas for the quantitative comparison and interpretation of such halos.

Thus, in the case of an ideal secondary halo over an eroded ore body, if it is assumed that the major metal reserves are concentrated in the ore body itself (disregarding the primary halo), a plot of the contents of the ore elements (C_x) along the profile perpendicular to the strike of the ore body, will appear as shown in Fig. 65, and will obey the equation:

$$C_x = \frac{M}{\sigma\sqrt{2\pi}} \cdot e^{-\frac{x^2}{2\sigma^2}} + C_\Phi,$$

where C_x is the concentration of a metal at a point in the halo at a distance x from the ore vein;

M is the quantity of the metal in the halo;

σ is the dispersion coefficient which takes into account the complexities of the local environments;

C_Φ is the background content of the elements.

The parameters M and C_{max} of the secondary halo are related as follows:

$$M = C_{max}\sqrt{2\pi},$$

where C_{max} is the maximum content of an element at the central point of the halo.

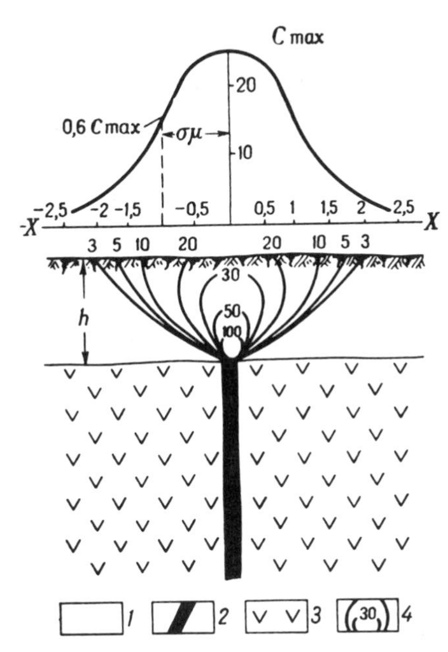

Fig. 65. Vertical cross section through an idealized exposed residual dispersion halo also showing a plot of the distribution of the ore elements (after Solovov).
1 - eluvium; *2* - ore body; *3* - enclosing rocks; *4* - lines connecting points with equal contents of metal.

The amount of a metal in a halo is usually determined using the following conventional formulas (Solovov, 1959):

$$P = \Delta x\left(\sum_{n}^{1} C_x - nC_\Phi\right)$$

$$P = \Delta x l\left(\sum_{n}^{1} C_x - nC_\Phi\right)$$

where Δx is the distance between the sampling points along the profile;
l is the distance between the profiles;
$\sum_{n}^{\prime}$ is the arithmetic sum of all anomalous concentrations of the element;
C_Φ is the value of the local geochemical background;
n is the number of points involved in the calculation.

The shape and the size of the residual dispersion halos depend on the topographic features of the locality and on the climatic conditions which determine the character of the supergene destruction of the mineral deposit. In the general case, the faster the weathering process, the larger the area of the residual halo. This area is usually larger than the ore body being destroyed, including the primary halo around it. When a mineral deposit situated on a slope is destroyed, this causes a distortion in the shape of the halo (Fig. 64). Detached residual halos may form on steep slopes.

Type II. **Superimposed exposed halos of the diffusion type, which form in the accumulation-denudation environments in arid, moderately humid, and humid regions (Fig. 66).**

Superimposed dispersion halos of this type form in association with the phenomena of diffusion of solutions into alluvial (allochthonous) unconsolidated rocks and soils overlying eroded ore deposits or their primary geochemical halos. In some cases, superimposed hydromorphic (salt) diffusion halos grade into residual mechanical-hydromorphic halos at depth. The unconsolidated deposits and soils, which may be penetrated by a diffusion column, are usually up to 2 - 4 m thick. However, in the case of large mineral deposits, the diffusion halos may penetrate a distance of up to 8 to 10 meters. If distinct layering of the soil exists, biogenic accumulation processes may also participate in the formation of exposed superimposed dispersion halos (Fig. 66).

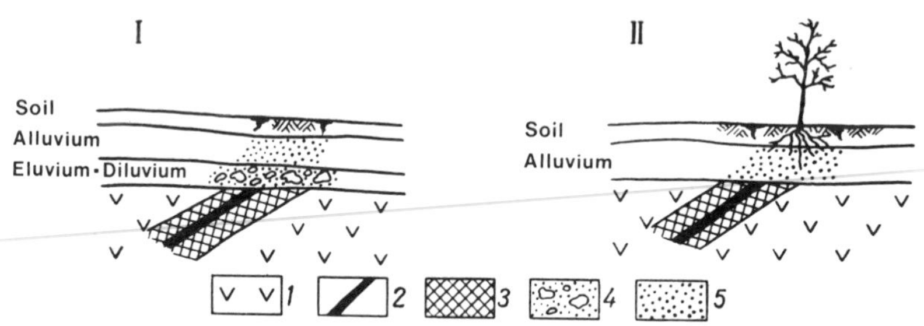

Fig. 66. Diffusion (superimposed) exposed secondary dispersion halos (I - formed over residual halo; II - with a biogenic component in the upper part).
1 - enclosing rocks; *2* - ore body; *3* - endogenic geochemical halo; *4* and *5* - dispersion halos (*4* - residual; *5* - diffusion).

The distribution of the indicator elements in superimposed diffusion halos in soils depends, to a large extent, on the character of the soil profile (Fig. 67). In poorly drained humid soils, the accumulation of metals occurs in the A horizon which is rich in organic matter. In the more common differentiated and well drained soils, the maximum accumulation of the indicator elements occurs in the B horizon, into which some components from the upper A horizon seem to be introduced. The concentration of metals in the B horizon is characteristic of a wide range of soil types, from loamy and chernozem soils in regions with temperate climates, to typical laterite soils in humid tropical regions (Tables 38 and 39). Thus, with the exception of poorly drained soils, as well as some types of soils from arid and semiarid regions where there is no appreciable difference in the distribution of indicator elements along the soil profile, the B horizon is the most representative horizon for sampling exposed superimposed dispersion halos with a diffusion origin.

For an estimation of the rate of accumulation of an indicator element in a soil profile relative to the rock on which the given soil formed, it is useful, in some cases, to use *soil-accumulation coefficients*, C_{ac}. These are the ratios of C_{soil} to C_{pr}, where C_{soil} is the content of an indicator element in the soil, and C_{pr} is its content in the parent rock. Soil-accumulation coefficients exceeding 1.0 indicate accumulation of an element in the soil horizon. On the other hand, soil-accumulation coefficients of less than unity are typical of elements which are being removed from the soil profile. Table 40 lists soil-accumulation coefficients for the genetic horizons of mountainous taiga permafrost soils which occur in Eastern Transbaikalia.

Table 38. Contents of some indicator elements in sulfide deposits classified on the basis of the different genetic soil horizons (in ppm); based on the −80 mesh fraction (after Hawkes and Webb, 1962).

Podzolic soil (North Carolina)									Laterite soil (Central Africa)			Poorly drained soil (Zambia)			
Soil profile			Background			Anomaly			Soil profile	Anomaly		Soil profile		Back-ground	Anomaly
Horizon	Depth, cm	Type of soil	Pb	Cu	Zn	Pb	Cu	Zn	Type of Soil	Cu	Pb	Horizon	Type of soil	Cu	Cu
A_1	0—2	Humus	100	20	160	440	150	260	Humus	130	350	A_1	Humus	90	360
A_2	2—5	Silty loam	190	24	140	840	300	300	Sandy horizon	160	140		Gleyed parent rock	80	125
B_1	5—40	Silty clay	230	34	140	1000	380	280	Compact laterite with iron concretions	400	880				
B_2	40—73	Silty clay	370	57	160	1300	750	410							
C	73	Eluvium	180	59	110	1700	1100	440	Eluvium	200	170				

Table 39. Average contents of some indicator elements in rare-metal deposits classified on the basis of the different genetic soil horizons in mountainous taiga and steppe landscapes from Eastern Transbaikalia, in ppm, (after Berengilova).

Mountainous taiga soil											Steppe chernozem soil									
Soil profile			Background				Anomaly				Soil profile		Background				Anomaly			
Horizon	Depth, cm	Type of soil	Sn	Be	Li	Pb	Sn	Be	Li	Pb	Depth, cm	Type of soil	Sn	Be	Li	Pb	Sn	Be	Li	Pb
A_1	0—3	Humus	6	10	40	26	50	29	250	340	0—50	Humic loam	3	3	32	19	20	22	170	37
A_2	3—15	Silty loam																		
B_1	15—30	Medium loam	8	13	50	31	120	63	500	410	50—70	Humic loam	3	3	33	20	22	24	190	39
B_2	30—60	Medium loam	10	19	65	37	130	83	700	450	70—105	Clayey loamy sand	3	4	30	20	29	22	200	42
B_K	—	—	—	—	—	—	—	—	—	—	105—120	Carbonate loamy sand	2	3	35	18	15	24	250	40
BC	60—120	Sandy loam	12	30	75	52	260	100	1000	570										
C	120—150	Shale eluvium			90					570	120—170	Shale eluvium	2	3	27	10	14	8	200	35

Table 40. Soil-accumulation coefficients for different genetic horizons from mountainous taiga permafrost soils within geochemical anomalies (after Berengilova).

Group	Element	Soil horizon A	B	BC-C	Behavioural characteristic of element
I	Pb	1,0	1,2	1,4-1,6	Element accumulates along the soil profile
II	Ta	0,7	1,0	1,4-1,5	Average content of the element in
	Nb	0,7	1,0	1,2-1,4	the soil profile is identical to
	Sn	0,6	1,1	1,0-1,2	that in the parent rock
III	Zn	0,4	0,8	1,0-1,2	Element is removed from the soil
	Rb	0,6	0,8	1,1-1,2	profile
	Be	0,4	0,8	0,9-1,1	

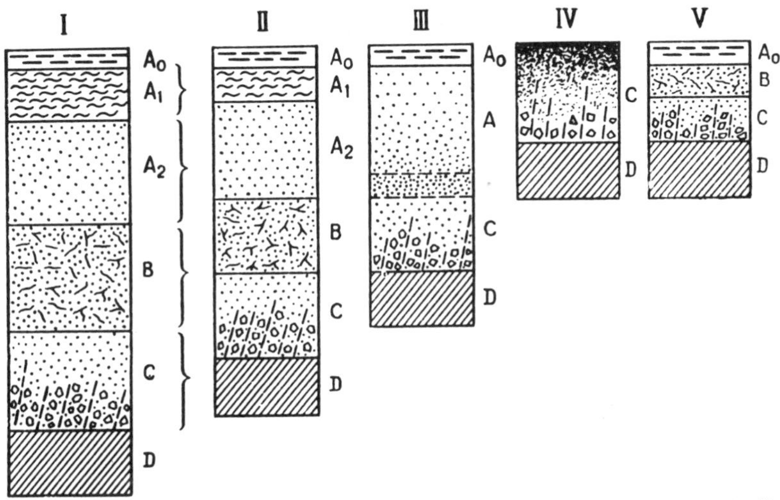

Fig. 67. Typical cross sections through soil profiles (after Andrews-Jones, 1968).
I - general cross section; II - podzolic soil; III - chernozem (semi-desert); IV - desert soil; V - recently formed mountain soil.
A_0A_1 - partly disintegrated organic remains; humus (dark-colored zone of organic accumulation); A_2 - light-colored zone with the preferential removal of trace elements; B - illuvial horizon (usually characterized by trace-element accumulation); C - horizon of disintegrated bedrock; D - bedrock.

If one understands the behavior of the indicator elements within a soil profile, it is much easier to evaluate the dispersion halos which may be detected. Sorption phenomena play an important role in the formation of this type of halo. Therefore, the ratio of the quantity of metal which can be leached out by weak reagents at low temperatures, to the total metal content in superimposed halos, is usually much higher than in residual halos. Sometimes this ratio will reach 40 to 50%, or even 80%, of the total metal content (Hawkes and Webb, 1962). Superimposed dispersion halos which originate by diffusion are more typical of those chemical elements which are characterized by a greater ability to migrate. However, chemical elements which are usually relatively immobile in supergene solutions (tin, tungsten, beryllium, etc.), may also accumulate in halos of this type if they are formed with the participation of biogenic processes.

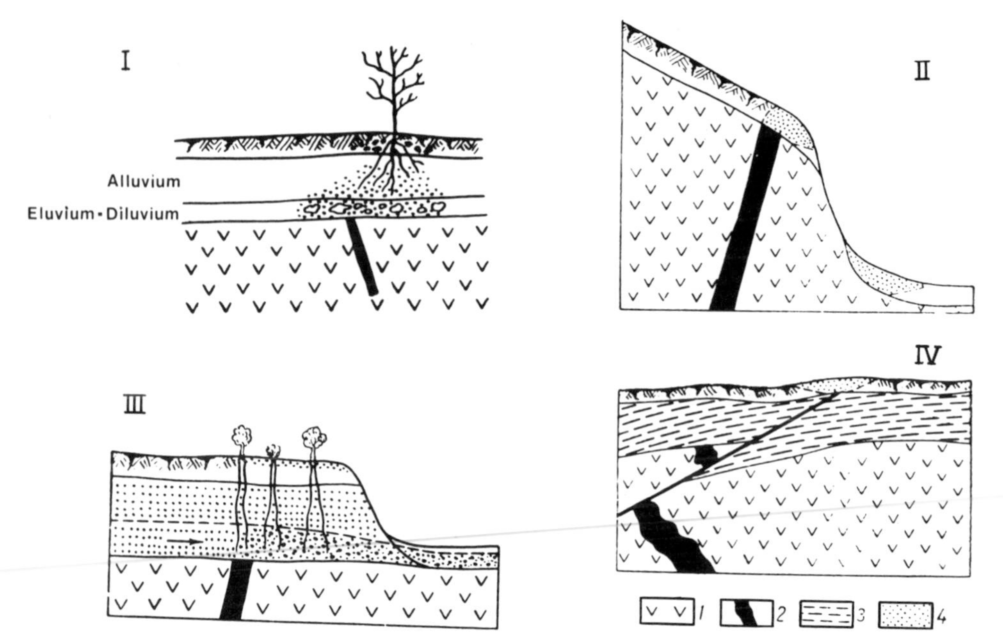

Fig. 68. Detached secondary dispersion halos.
I - supraore halo produced by biogenic accumulation in soils over a buried residual halo; II - detached halo in a bog due to the local relief (after Andrews-Jones, 1968); III and IV - detached halos associated with biogenic accumulation and migration of components in ground waters (after Andrews-Jones, 1968).
1 - enclosing rocks; *2* - ore body; *3* - rocks overlying the mineralization; *4* - dispersion halo.

Type III. **Superimposed exposed supraore halos of the accumulation type (Fig. 68, III).**

This type of halo differs from those described above in that they are separated from the ore body, which occurs at depth, by a horizon of unconsolidated allochthonous material which has approximately background contents of the indicator elements. These halos generally form as a result of an accumulation of the indicator elements in a soil layer due to biogenic accumulation by plants. The roots of such plants penetrate through a halo-free zone to the ore body or to its buried residual (or diffusion) dispersion halo. In some cases, supraore halos in arid regions are produced because of the presence of an active evaporation barrier at the surface of the unconsolidated deposits. This leads to the concentration, in these areas, of salts which are brought by subsurface waters as a result of capillary migration.

Interpretation of superimposed supraore halos may be difficult because of their resemblance to "false" halos (see below), which are in no way related to ore concentrations.

Type IV. **Superimposed exposed detached halos (Figs. 68 and 69).**

These are primary-residual, mechanical-hydromorphic, or primary-superimposed diffusion halos. They are shifted toward the presently existing runoff, and are separated from the parent source by a zone of barren rock in which the halo is not developed. Hydromorphic (chemical) halos of this type are sometimes associated with seepages of ground waters which flush a hidden deposit, or its primary or buried residual halo (Fig. 68, III). A specific type of exposed detached halo, which is quite difficult to interpret, is illustrated in (Fig. 68, IV).

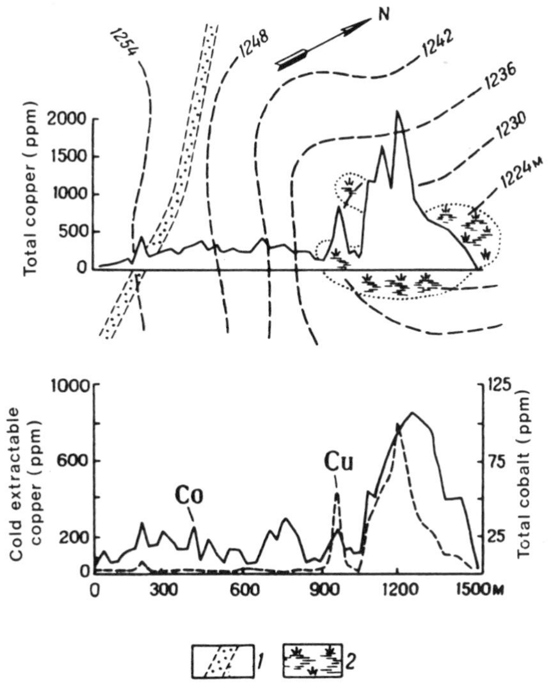

Fig. 69. Copper halo displaced into a marshy lowland along a slope in Zambia (after Tooms and Webb, 1961).
1 - ore horizon; *2* - marsh.

Type V. **Leached and extremely depleted halos.**

This type of halo forms under conditions which favor the removal of the mobile indicator elements from the surface horizons. As a result, the difference in the distribution of these elements in the halo and in the geochemical background becomes almost undetectable. Such conditions sometimes exist in humid zones in those environments in which strong leaching occurs in the upper soil horizons. The concentration of the indicator elements usually sharply increases at a certain depth below the surface. This depth, which determines the position of the representative sampling horizon, depends on the local landscape-climatic conditions, and may range from 1 to 2 meters. Dispersion halos of this type are similar to buried halos with respect to sampling and interpretation.

Type VI. **Buried residual halos (Fig. 70).**

These halos are analogs of the Type I residual halos which are overlain by allochthonous deposits, but the thickness of these deposits is too great to permit diffusion halos to be accessible from the surface. Buried residual halos generally develop in covered regions characterized by a two-stage evolution. Especially important, from this point of view, are regions which have experienced a prolonged period of continental erosion, prior to the deposition of allochthonous sediments which overlay the residual dispersion halos, which had been previously exposed by an erosion surface.

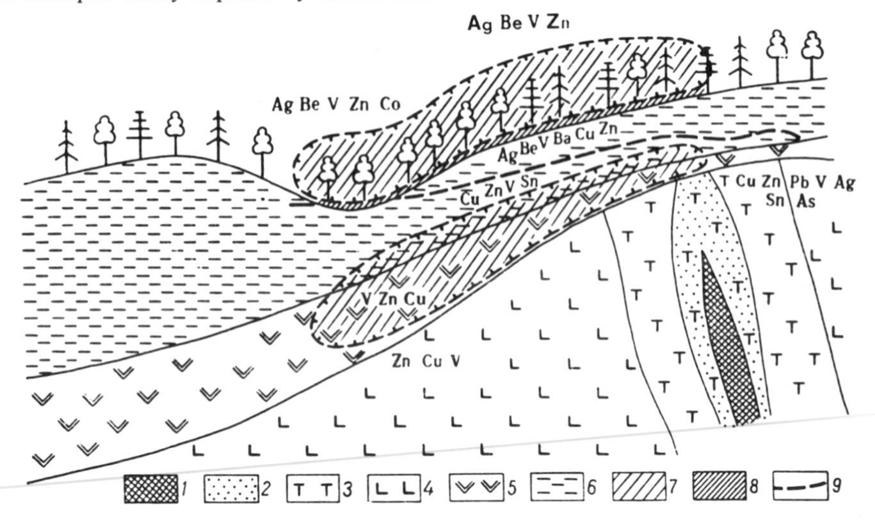

Fig. 70. Buried residual lithogeochemical dispersion halo with a detached biogeochemical halo in vegetation (after Glazovskaya).
1 - copper-pyrite ore; *2* - endogenic geochemical halo of the ore body; *3* - sericite-containing shales; *4* - diabase porphyrites; *5* - crust of weathering; *6* - Quaternary loams; *7* - buried residual dispersion halo; *8* - detached biogeochemical dispersion halo; *9* - perched water table.

Type VII. **Buried superimposed halos.**

These halos are often associated with buried residual halos, and they form "blind" diffusion caps over them. They are usually analogs of superimposed diffusion halos. However, halos leached and depleted at the surface are also similar to this group of dispersion halo. This type of secondary dispersion halo is particularly difficult to interpret and evaluate.

Geochemical Landscape Provinces in the USSR

It follows from the above classification of secondary dispersion halos, that their formation depends entirely on the natural environments in which the supergene destruction of mineral deposits occur. These environments, in turn, depend on the climate, geomorphology, and geological structure of the particular region. This implies that, in the selection and verification of the geochemical prospecting and sampling techniques to be utilized for secondary dispersion halos, the external regional and local factors responsible for the characteristics of the supergene migration and concentration of elements within a certain landscape must always be taken into account.

The theory of geochemical landscapes advanced by Academician B. B. Polynov, and developed by A. I. Perel'man for the purposes of geochemical exploration for supergene dispersion halos, may serve as the basis for an analysis of the environmental influences on the migration of the chemical elements in the supergene zone. When Polynov used the term *geochemical landscape,* he implied an area of the Earth's surface which is characterized by a particular type of supergene migration of the chemical elements. A combination of the regional bioclimatic and local geologic-geomorphological factors, is responsible for the formation of specific geochemical landscapes which are confined to certain landscape zones or provinces. For example, in a chernozem-steppe landscape zone, an area composed of granites would belong to one type of geochemical landscapes, whereas an area composed of limestones would belong to another type, etc. Moreover, within each of these areas various combinations of characteristic geochemical landscapes, due to variations in the local relief, may form. All this must be taken into account during geochemical exploration and sampling of secondary dispersion halos.

The alkaline-acidic (pH) and oxidizing-reducing (Eh) properties of soils and natural waters, as well as the nature of the colloidal migration of the elements, are the most important geochemical conditions for the migration and concentration of trace elements in the supergene zone. With this in mind, Perel'man and Sharkov (1957) made an attempt at zoning the USSR in terms of the geochemical environments of supergene migration of the chemical elements. They identified four major landscape-geochemical provinces within the boundaries of the USSR. These provinces include smaller subdivisions, specifically subprovinces and regions.

The first province, with a predominance of soils and waters having alkaline or neutral reactions, includes a subprovince composed of the arid steppes and deserts of Transcaucasia, Central Asia, and South Siberia, as well as the chernozem steppes and forest steppes of the Ukraine, Urals, Caucasus, and South Siberia.

The province, as a whole, is generally characterized by a weak migration of most trace elements, a lower mobility of colloids, and a low content of organic matter in surface and subsurface waters. The arid steppe and desert subprovince is characterized by an especially weak migration. The differentiation in the contents of the trace elements between the A and B soil horizons is practically absent (the A horizon in some places, does not develop at all). In this case, geochemical sampling at the surface may furnish representative results, provided the layer of unconsolidated deposits is not thick. The chernozem steppe and forest steppe subprovince is characterized by a larger amount of water and, consequently, by a greater mobility of the trace elements. The greater the rainfall in the region, the stronger the tendency for the accumulation of trace elements in the illuvial B horizon.

The second province embraces regions with a predominance of soils which have an acid or neutral reaction. It includes a subprovince composed of the taiga in the European North, the Urals, Siberia, and in the Far East, as well as subprovinces composed of the wooded mountains in the Carpathians and in the Transcaucasia. The entire province is characterized by strong leaching of the upper soil horizons and by the accumulation of trace elements in the illuvial B horizon. Vigorous migration of trace elements in natural waters within this province is due, in particular, to the increased mobility of organic and mineral colloids and the associated phenomenon of sorption. The mobility of the trace elements especially increases in some areas of the southern wooded mountain subzone which is characterized by a warm climate. The whole zone, especially its southern regions, is also characterized by the formation of diffusion dispersion halos, which either overlie the residual halos or are detached from them.

The third province differs from the second by the widespread development of permafrost. In addition, it has a considerable variety of local microlandscapes. Some types of soils in the taiga-permafrost landscapes differ very little from the taiga soils of the second province. They are characterized by an accumulation of trace elements in the illuvial B horizon. Also, soils in which concentrations of trace elements are confined to the upper A horizon which is strongly enriched with organic matter, are widespread. In the northern part of the zone (the tundra), the soils are rich in organic matter, and are characterized by a limited development of oxidation processes (i.e., reducing environments), and by a weak mobility of the trace elements. Sampling of such soils at the surface furnishes representative information on the presence of secondary dispersion halos. In regions where the soil cover is poorly developed, residual halos with a noticeable predominance of a mechanical component form.

The fourth province includes the high mountains of the Caucasus, Central Asia, Altai, and Sayan. This province is, in fact, a complex of landscapes which replace one another vertically. This takes place in a direction from the foothills of the mountain ranges (deserts in Central Asia, taiga landscapes in Sayan, etc.), to the landscapes characteristic of alpine meadows and glacial moraines in the high mountains.

Other schemes for describing the landscape zoning of the USSR exist. However, all are based on similar bioclimatic and geologic-geomorphological features of the landscapes, which are responsible for the character of the supergene migration of the chemical elements and, consequently, for the specific features in the formation of secondary dispersion halos from mineral deposits.

Methods of Geochemical Prospecting Using Secondary Dispersion Halos in Soils and in Unconsolidated Deposits.

Evaluation of Halos

Geochemical exploration using secondary dispersion halos are conducted by means of the systematic sampling of soils or unconsolidated deposits at optimal depths. This ensures that reliable information on the distribution of the indicator elements is obtained, and that the labor and time factors are reduced to a minimum. The depth of the representative sampling horizon, in cases of exposed dispersion halos, does not exceed several dozens of centimeters. The representative horizon is selected prior to the beginning of prospecting by means of lithological studies of cross sections through soils and unconsolidated deposits in the region of exploration. (Geochemical investigations of the cross sections are also quite desirable, although they are usually conducted later during the data processing stage in the office.)

In the same initial stage of prospecting, samples are collected in order to evaluate variations in the local geochemical background, and to determine the values of the minimum anomalous (threshold) contents of the indicator elements with the given probability.

The sampling is conducted with the help of special sampling devices, which make it possible to obtain the material from certain specified depths, or by means of pitting (the latter technique is much more labor-consuming). The fine sandy-clayey fraction from the representative horizon is sampled. Each sample weighs about 50 grams.

The sampling grid is planned with the aims and scope of the prospecting objectives in mind, and it takes into account the geological structure of the area being explored.

In the preliminary stage of the geochemical survey, the sampling grid within a promising area is selected on the bases of the expected size of the halos, and in particular, so that two or three samples will encounter a halo with a probability equal to 0.95 or higher. The *"Instructions on Geochemical Exploration Techniques"* (1965) recommend that the distance between the profiles of the major grid, oriented perpendicular to the expected trend of the halos, be no greater than 0.9 of the predicted length of the halos, whereas the sampling interval along the profile may be up to one-half their width.

In the detailed stage of exploration (at a scale 1:10,000 or larger), sampling is aimed at the ultimate delineation of the halos, and at studying their inner structure from the point of view of the distribution of the indicator elements. This is why the sampling grid at this stage of prospecting depends largely on the probable size of the ore bodies, i.e., on the type of mineralization. At prospective deposits which are composed of disseminated ores (for example, at porphyry copper deposits), at deposits with a sheet-like (layered) shape, or at large stockworks, the dispersion halos may be studied and delineated with the help of a relatively widely spaced network (with a grid of 40 by 100, or 20 by 100 meters). For small mineral deposits with complex outlines and with an irregular distribution of the useful components, the sampling grid must be smaller (10 by 50, or even 10 by 25 meters).

A comprehensive study of the geological and landscape-geochemical environments in the exploration region is essential. The resulting data may be helpful in the interpretation of the dispersion halos.

Geochemical maps of the secondary dispersion halos are drawn on a geological map (or, if there is no geological map, then on a topographic map) by means of plotting the results of the analyses of the geochemical samples. Subsequently, the areas with anomalous contents of the indicator elements are delineated. The delineation may be based on the minimum anomalous (threshold) contents determined with different significance levels. In order to obtain an objective idea as to the nature of the geochemical anomalies, it is advisable to evaluate the minimum anomalous contents and to delineate the anomalies at the 5, 1, and 0.3% significance levels.

Given below are the minimum anomalous contents of tin and lead (in ppm) calculated for various significance levels from the B soil horizons in the mountainous permafrost-taiga landscapes formed on arenaceous-metamorphic rocks in the Eastern Transbaikalia.

	5%	*1%*	*0.3%*
Sn	14	17	19
Pb	41	56	68

As shown in Figs. 71 and 72, various levels of concentration of the indicator element typical of a given zone are usually identified within the anomaly with the help of lines (contours) of equal concentration. Cross sections illustrating the indicator element distributions across the strike of the geochemical anomalies may also be useful (Figs. 73 and 74).

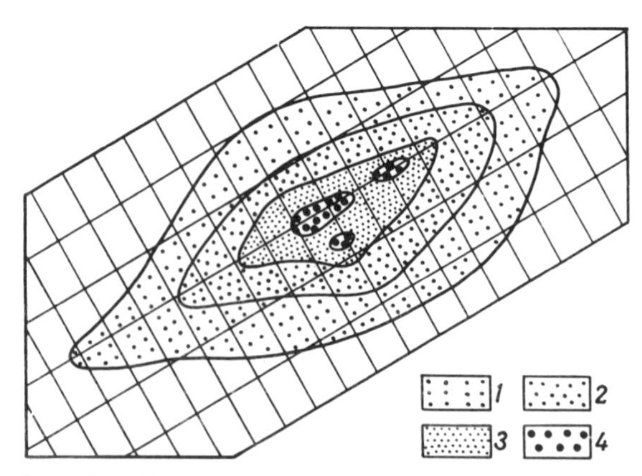

Fig. 71. Secondary dispersion halo shown in plan.
Contents, ppm: *1* - 50 to 250; *2* - 250 to 500- *3* - 500 to 1000; *4* - more than 1000.

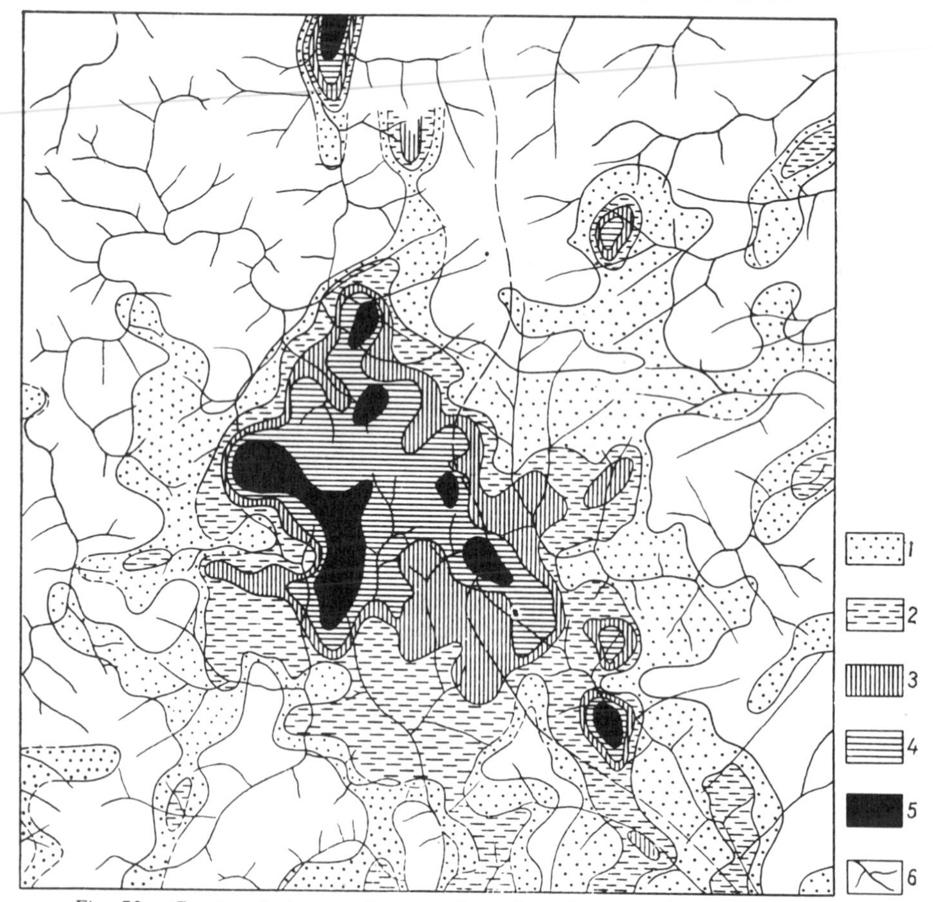

Fig. 72. Geochemical map of a secondary dispersion halo formed by chromium associated with a chromite ore deposit in Madagascar (after Borutskii).
Contents of Cr, ppm: *1* - 200; *2* - 500; *3* - 1000; *4* - 1500; *5* - 4000; *6* - streams.

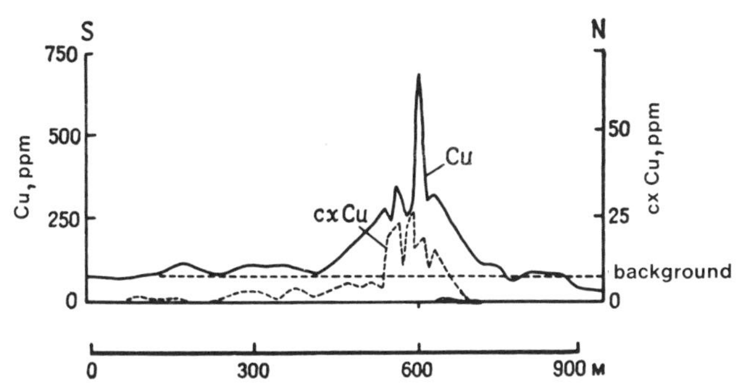

Fig. 73. Cross section through a secondary geochemical halo over a copper showing in Zambia. Shown are the distributions of the total copper (Cu) in the soil, and of the cold-extractable copper (cxCu) (after Tooms and Webb, 1962).

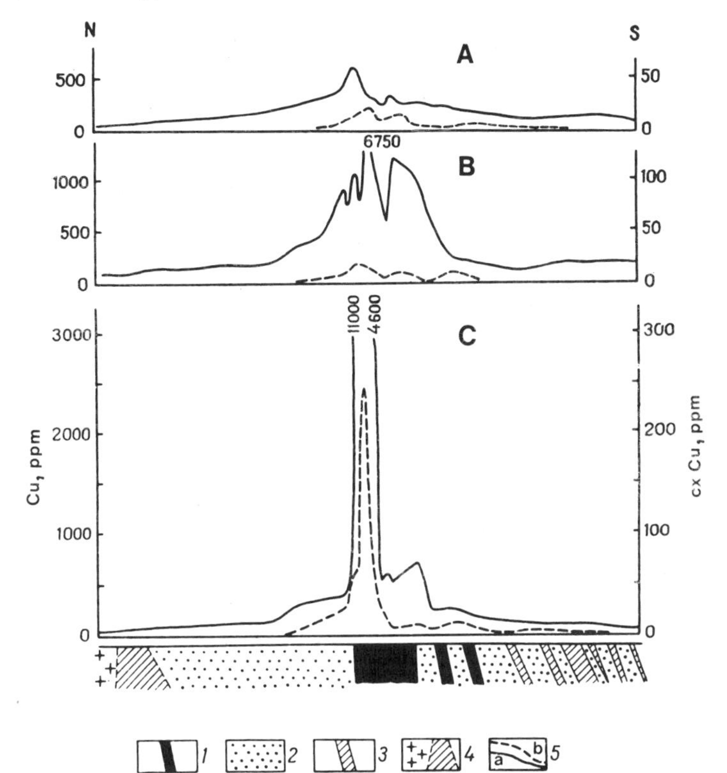

Fig. 74. Section across a secondary geochemical halo formed over a copper deposit in Zambia which illustrates differences in the distribution of copper in the A, B, and C soil horizons (after Tooms and Webb, 1962).
1 - copper-bearing ore zone; *2* - sandstones; *3* - shales; *4* - basement complex; *5* - plots of the copper distributions (*a*): total; (*b*): cold-extractable.

Years of experience gained during geochemical prospecting in the USSR and in other countries have shown that exploration for ore deposits, using secondary dispersion halos, is especially effective in regions where geological and landscape-geochemical conditions are favorable for the development of Types I to III exposed halos (described earlier in this chapter).

Solovov (1959) proposed a method for estimating the expected mineral reserves in bedrock based on the parameters of the secondary dispersion halos. He based his method on the productivity of the halo, expressed in tons of a specific metal for a 1-meter thick layer. However, expected reserves may be evaluated, in this case, only if data are available on the depth of the erosion surface of the mineral deposit or of the ore body being studied. To illustrate this point, Fig. 75 shows three variants of the erosion surface of an ore body

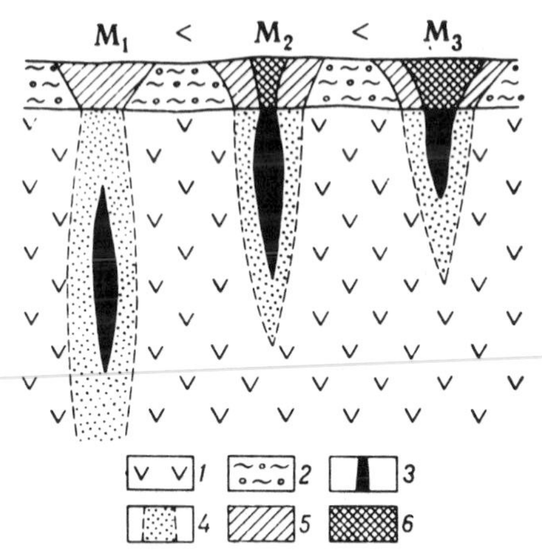

Fig. 75. Variations in the quantity of metal in secondary dispersion halos depending on the level of the erosion surface of the ore-bearing zones.
1 - bedrock; *2* - eluvial-diluvial formations; *3* - ore body; *4* - primary halos; *5* and *6* - secondary halos (*5* - with a low content of the metal; *6* - with a high content); M - metal reserves in the secondary halo.

[(1) blind, in which only the primary halo is exposed; (2) weakly eroded; and (3) an ore body more than half of which has been destroyed by erosion]. Productivities of the secondary halos for these ore bodies will be $M_1 < M_2 < M_3$.

The anomalies associated with the second and third ore bodies (M_2 and M_3 in Fig. 75), and which have larger productivities, may erroneously be assumed to be more promising than the anomaly representing the first body (M_1), unless their erosion surfaces are taken into account. This is why the method of evaluating expected reserves, based on the productivities of the supergene anomalies, may be used only if the dispersion halos of the ore bodies being compared have approximately equal levels of their erosion surfaces. In this connection, it is essential to develop criteria for evaluating the level of the erosion surfaces of ore bodies and their surrounding primary geochemical halos based directly on the distribution characteristics of the indicator elements in supergene dispersion halos.

That such an evaluation is possible was proved by Sochevanov and his co-workers (see Kablukov, Sochevanov and others, 1964) who found a close correlation (within a particular landscape-geochemical environment) between primary and secondary halos formed in some hydrothermal uranium deposits.

To illustrate the possibility of evaluating the erosion surface level using sampling data from overlying eluvial-diluvial deposits, the results from studying the correlation between primary and secondary dispersion halos within the Kanimansur-Almadon Fault, which is a major ore-bearing fracture in Central Karamazar, will be presented (based on unpublished data from Grigorian and Morozov).

This particular region is situated on the southern slopes of the Kuraminsky Range, and it is a strongly dissected massif with absolute elevations ranging from 1400 to 2000 meters. Individual summits rise above the local erosion surfaces as much as 200 to 400 meters. The mountain slopes are usually steep (15 to 30°) and flatten in the lower parts. The region is characterized by a considerable exposure of bedrock. However, within the Kanimansur-Almadon Fault zone, an eluvial-diluvial cover 0.5 to 2 m thick is present nearly everywhere. The soil cover in this area is represented by a dark-brown carbonate, and by a thin soil layer which are typically brown and gravelly on the steep slopes. The vegetation is quite sparse and is mainly characterized by grasses and low shrub associations.

Fig. 76 shows primary and secondary halos formed by the principal indicator elements which were delineated at the surface of the Koptarkhon deposit. A close spatial correlation between the primary and secondary halos of the main indicator elements can be seen not only within the external outlines, but within the areas of high concentrations as well. Despite the considerable steepness of the slopes (15 to 30°), there is no appreciable shift in the secondary dispersion halos down the slopes. Secondary halos of most of the indicator elements are somewhat larger than the primary halos, and they also differ in that they have a more uniform distribution of the indicator elements in their halo fields.

In order to ascertain the possibility and the reliability of using the distribution features of the indicator elements in the supergene dispersion fields for determining the level of the erosion surface of primary halos, the parameters of the primary and secondary dispersion halos were compared.

As previously pointed out, the level of the erosion surface of primary halos is usually determined by using ratios of the parameters for pairs of elements which are indicators of vertical zonality in the halos. Fig. 77 shows plots of the variations in the ratios of linear productivities for pairs of elements in primary and secondary halos, with increasing depth of the erosion surface. Some of the pairs exhibit a very close correlation between the primary and the secondary geochemical halos. However, this correspondence between the values of the pairs of indicator ratios is not always observed (in the case being considered, appreciable differences exist for the pairs silver-copper and arsenic-copper). The composite halos are characterized by a closer and more reliable correlation.

It follows from Fig. 77, that the plots of the variations in the ratios of the linear productivities of the primary and secondary multiplicative halos (elements typical of the above-ore (supraore) levels are Ba, As, Ag, and Pb; elements typical of the levels below the ore body are Cu, Bi, Co, and W) are similar in detail. They decrease very distinctly from the supraore cross sections of the halos toward the sections below the ore bodies. This shows that there are differences between the levels of the erosion surfaces in the halos.

Consequently, under the conditions found in this region of Central Asia, in areas which are covered by thin eluvial-diluvial deposits, the levels of the erosion surfaces of the primary geochemical halos may be revealed and evaluated using the results from studying the supergene dispersion fields of the indicator elements typical of the corresponding deposits. This increases the effectiveness of exploration for ore deposits. In particular, the depth range of exploration can be increased, because the evaluation of the level of the erosion

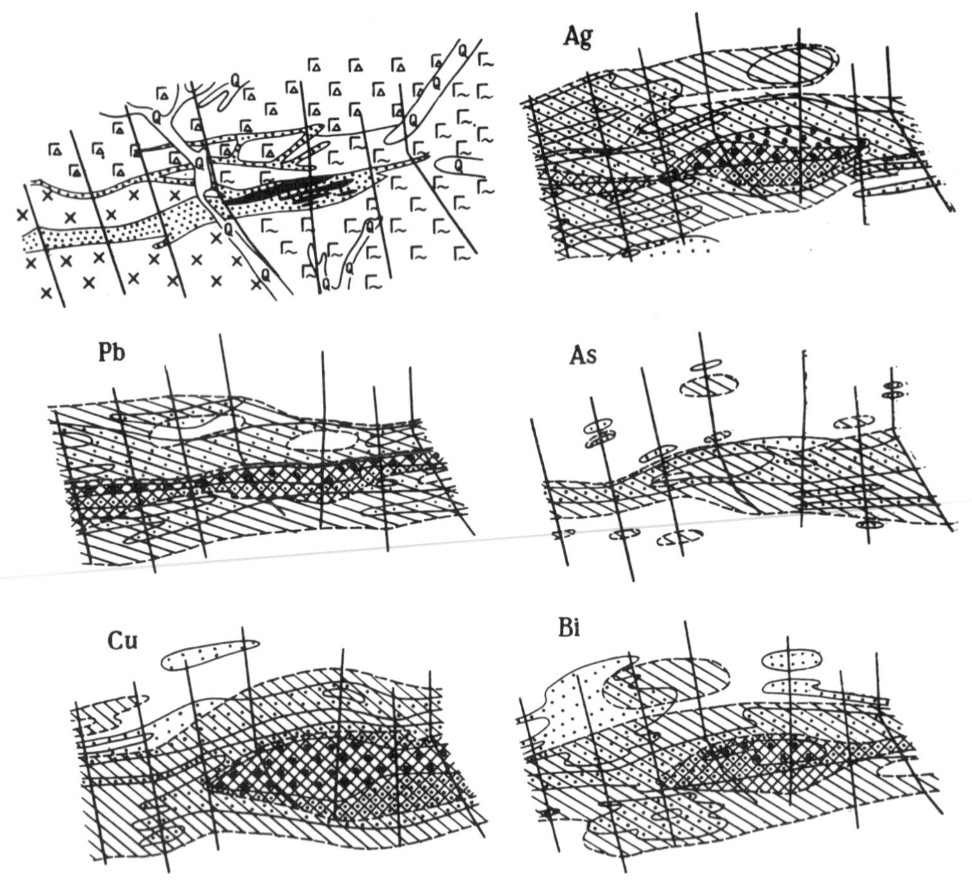

Fig. 76. Comparison between primary and secondary dispersion halos at the Koptarkhon deposit.
1 - tuffs of quartz-amphibole albitophyres; *2* - tuffaceous lavas of felsite-porphyries; *3* - felsite-porphyries; *4* - bleaching of rocks; *5* - ore body; *6* - Quaternary deposits; *7* and *8* - element contents in primary halos, % (*7* - 2 to $40 \cdot 10^{-5}$ Ag; 5 to $25 \cdot 10^{-3}$ Pb; 2 to $50 \cdot 10^{-3}$ As; 5 to $25 \cdot 10^{-3}$ Cu; 1 to $50 \cdot 10^{-4}$ Bi; 4 to $20 \cdot 10^{-4}$ Co; 1 to $50 \cdot 10^{-3}$ W; 4 to $50 \cdot 10^{-3}$ Ba; 5 to $50 \cdot 10^{-3}$ Zn; 3 to $50 \cdot 10^{-3}$ Mo; *8* - more than $4 \cdot 10^{-4}$ Ag; more than $25 \cdot 10^{-3}$ Pb; more than $25 \cdot 10^{-3}$ Cu; more than $5 \cdot 10^{-3}$ Bi); *9* and *10* - element contents in secondary halos, % (*9* - 5 to $40 \cdot 10^{-5}$ Ag; 15 to $50 \cdot 10^{-3}$ Pb; 5 to $50 \cdot 10^{-3}$ As; 5 to $25 \cdot 10^{-3}$ Bi; 15 to $50 \cdot 10^{-3}$ Co; 2.5 to $50 \cdot 10^{-3}$ W; 4 to $50 \cdot 10^{-3}$ Ba; 25 to $50 \cdot 10^{-3}$ Zn; 5 to $50 \cdot 10^{-4}$ Mo; *10* - more than $40 \cdot 10^{-5}$ Ag; more than $50 \cdot 10^{-3}$ Pb; more than $25 \cdot 10^{-3}$ Cu; more than $50 \cdot 10^{-4}$ Bi).

surface of the halos is the basis for the assessment of the extent of the prospective halo at depth.

It is clear that the above criteria may be used effectively only in the interpretation of residual exposed secondary dispersion halos. In regions covered with stable allochthonous or diluvial deposits which are more than 5 to 10 meters thick, the results from soil sampling at the surface may, at most, only help to solve the problem of localizing the anomalies in the bedrock. Further interpretation will require drilling to the bedrock, or through the

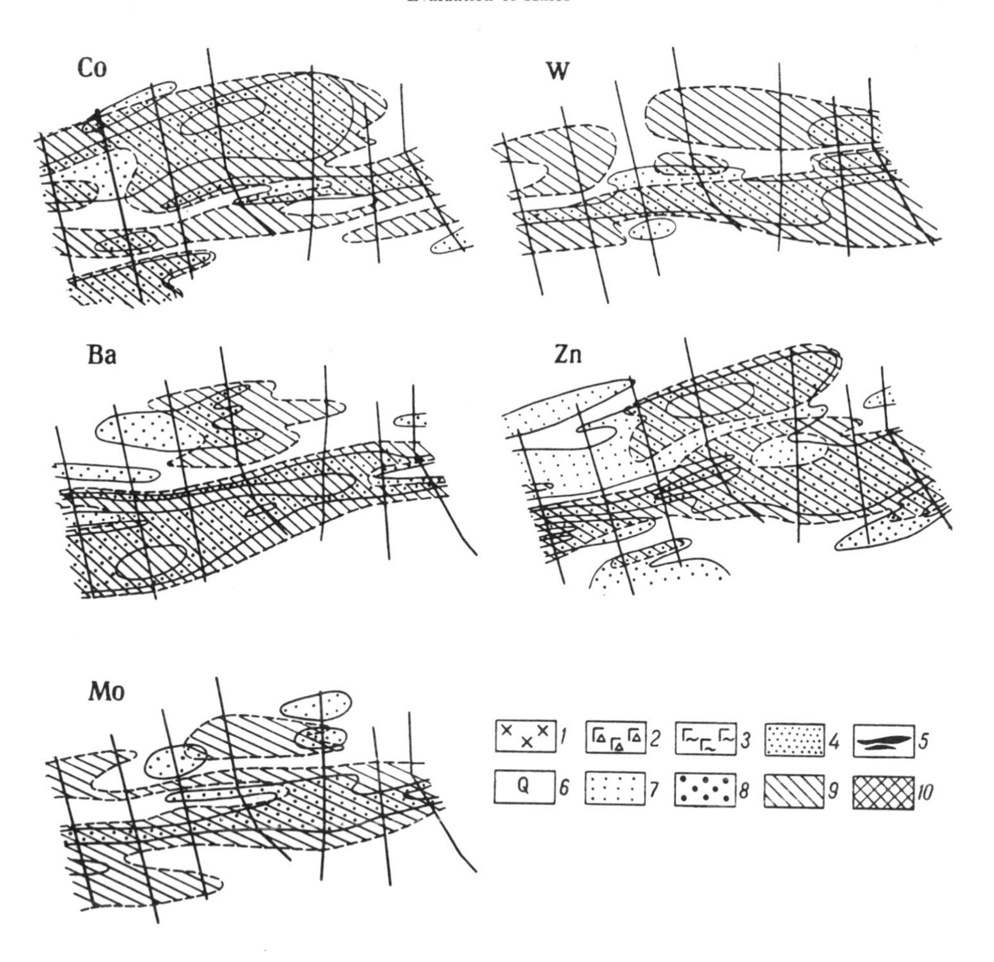

horizon of overlying unconsolidated deposits, which may well correlate with the halos in the bedrock.

Evaluation of secondary dispersion halos involves a complex of geochemical, geological, and landscape-climatic factors. The productivity, the contrast, and the size of halos, taken separately, cannot serve as unique criteria for the assessment of the ore potential indicated by the halo. Recognition of the so-called "false" halos is a very important problem which arises in the initial stage of exploration, and may result in the rejection of the exogenic anomalies which may have been detected.

The term *false halo* [generally called "false anomaly" in English; Editor] is used in exploration to designate supergene geochemical anomalies whose formation is not associated with processes of the supergene destruction of ore or non-metallic concentrations of the chemical elements. Their formation is generally due to the phenomena of concentration of trace elements, which are present in rocks or in natural waters, at different types of geochemical barriers. Such barriers are often produced in marshy landscape environments

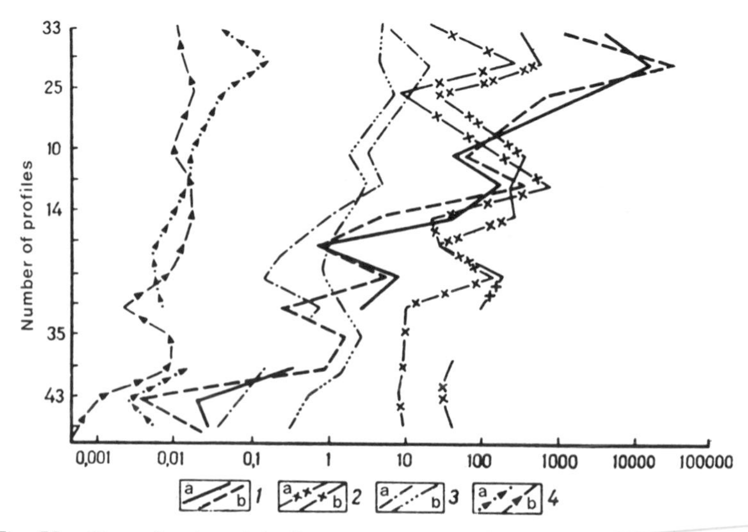

Fig. 77. Plots of ratios of the linear productivities (paired and multiplicative) of primary and secondary halos in areas with different erosion surface levels in the Kanimansur-Almadon Fault zone.
1 - multiplicative ratios; *2* - lead to copper ratio; *3* - barium to cobalt ratio; *4* - silver to copper ratio (*a* - in primary halos; *b* - in secondary halos).

where reducing swamp waters, and river or lake waters containing free oxygen, interact. It is, therefore, necessary to treat geochemical anomalies detected under marshy landscape conditions, with caution. In anomalies of this type, the ratios of the indicator elements typical of ore mineralization, are usually appreciably different from the ratios found in the dispersion halos of the particular deposits. This makes it possible to recognize "false" halos during laboratory analyses.* They often form in arid landscapes at the evaporation barrier, and this should be kept in mind during geochemical exploration in deserts, semideserts, and in the arid steppes. Formation of geochemical anomalies unrelated to ore mineralization in soils is, in some cases, due to biogenic accumulation processes. Anomalies which originate in this manner are usually weak, indistinct, and are characterized by specific combinations of the trace elements. For example, biogenic accumulation of trace elements in soils in the Transbaikalian taiga landscapes are characterized by increased contents of lead, zinc, and silver. However, accumulation of copper is not typical of these landscapes. Sampling from the most representative soil (B) horizon usually avoids any errors which may be due to biogenic accumulation of trace elements. Unconsolidated deposits enriched with certain trace elements may also indicate the presence in the region of some type of bedrock containing anomalous concentrations of these elements. These rocks may be some types of basic igneous rocks relatively enriched with certain ore minerals, or carbonaceous shales rich in organic matter, in which increased contents of copper, molybdenum, chromium, vanadium, and some other elements are often detected. The anomalous concentrations of trace elements in unconsolidated deposits and in soils, in this case, occupy areas much larger than could possibly be caused by anomalies from the disintegration of mineral

*The "false" halos of this type are not characterized by the simultaneous presence of elements with distinctly different mobilities, which is typically the case in secondary residual dispersion halos of mineral deposits.

deposits. This usually enables the correct conclusions to be made as to the sources of the anomalies. However, interpretation of such anomalies is not always an easy task. Analyses of paragenetic associations of elements within the anomaly, and of the most typical indicator ratios taking into account the geochemical features of the rocks in the region, may be very helpful in the interpretation of this type of "false" halo.

In some cases, errors in the interpretation of geochemical anomalies may be due to contamination of the atmosphere, natural waters, soils, and unconsolidated deposits, produced by human activity. The contamination of soils with metals through the introduction of the waste products from metallurgical and chemical plants, as well as from the use of fertilizers, usually only affects the A soil horizon. Such contamination usually decreases rapidly with depth, and this may be taken into account during sampling. However, contamination due to metal mining (especially to ancient workings) may reach considerable depths and are not easily recognizable during exploration. This may result in erroneous conclusions. It is very important to take into account possible contamination associated with human activity in all regions with advanced agricultural and industrial activity.

As previously mentioned, the most effective method for comparing and identifying supergene anomalies involves the comparison of the type of relationship which exists among the indicator elements within the anomalies. This implies that supergene geochemical anomalies which formed under similar conditions and have sources with similar compositions, must also exhibit similar quantitative relationships between the most typical indicator elements.

The character of the relationships between indicator elements within geochemical populations which have been sampled may be determined by various techniques. One such technique is the usual statistical evaluation of the presence, and of the degree of correlation, between the elements. These usually differ for anomalies of different origins, and also for anomalies and for the geochemical background. Table 41 lists data for the correlation be-

Table 41. Correlation between the contents of indicator elements which occur in different B soil horizons (after Berengilova).

	Taiga soils				Steppe soils		
	Geochemical background		Anomaly		Geochemical background	Anomaly	
Pairs of elements correlated	r	Character of correlation	r	Character of correlation	Character of correlation	r	Character of correlation
Li-Rb	0,363	+	0,799	+	WC	0,654	+
Li-Sn	—	WC	0,841	+	WC	0,783	+
Be-Sn	0,728	+	0,803	+	WC	0,553	+
Pb-Be	—	WC	0,584	+			
Rb-Be	—	—	—	—	WC	0,826	+
Rb-Sn	—	—	—	—	WC	0,815	+

Notes: r is a value of the correlation coefficient
+ designates a significant positive correlation
WC means weak correlation (at the 5% significance level)
— designates values which are not significant
blank indicates no data

tween indicator elements which are characteristic of rare-metal deposits, in different types of soils (Eastern Transbaikalia).

As can be seen from Table 41, correlations between the indicator elements, which are usually well expressed in exogenic dispersion halos, are generally absent in the background populations.

Supergene anomalies may also be classified with the help of indicator ratios of the elements which possess different geochemical characteristics. Solovov and Garanin (1968) proposed that the ratio of the products of groups of elements with differing mobilities between the bodies be compared. In order to identify these groups, logarithms of the contents of the indicator elements in the anomaly which is taken as the standard, are plotted on a graph at equal intervals.

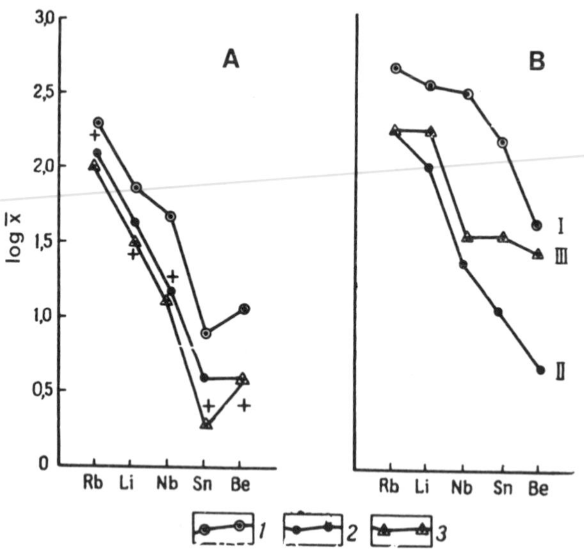

Fig. 78. Plots of geochemical spectra characterizing the distributions of the indicator elements in the rare-metal deposits with similar ore compositions (based on examples from Eastern Transbaikalia).

A - geochemical background; B - dispersion halos in soils: *1* to *3* - soils (*1* - mountainous taiga; *2* - forest steppe; *3* - steppe); I and II - rare-metal aplogranites; III - rare-metal pegmatites.

Graphs of the *geochemical spectra* (illustrated in Fig. 78) of the anomalies being compared are plotted in the same way, using the same order of the indicator elements as in the geochemical spectrum of the standard (known) anomaly. The standard graph (drawn on tracing paper) is superimposed on the graph being considered (the "target curve"), and the points representing the first indicator element of the standard series are superposed. If the graphs coincide completely, then the anomalies being compared may be considered geochemically identical, even though the absolute values of the indicator element contents show appreciable differences (the objects may have different concentrations of the elements). If geochemical differences are revealed between the anomalies being compared, then the indicator elements are divided into three groups. The first group (c) includes those indicator elements whose concentration is higher in the target anomaly than in the standard one. On superimposition of the graphs, the plots of the elements from this group will be located above the standard graph. The second group (n) consists of "neutral" elements whose plots coincide with the standard graph. The third group (d) encompasses elements whose content in the target anomaly is low relative to that in the standard anomaly. The plots of these elements will shift below the standard graph. Each of these groups may be described by a multiplicative index (M_c, M_n, and M_d) which characterizes the given type of the geochemical spectrum. The multiplicative index is a product of the logarithms of the average contents of the elements from each group, or of their concentration coefficients, relative to the geochemical background. The *geochemical index*, C_m, is an even better criterion. It is a quotient of the multiplicative index M_c of the first-group of indicator elements, divided by the M_d index, which characterizes the elements which are low in the target anomaly.

In order to normalize the geochemical index C_m, the number of elements in the first and the third groups must be equalized by raising one of the members of the smallest group to a corresponding degree, or by introducing into it the "neutral" elements of the second group.

The geochemical spectrum of the background population may also be used as a standard, and all anomalies detected in the region may be classified with respect to the geochemical background.

In the case of the secondary dispersion halos at the rare-metal deposits in Eastern Transbaikalia, the coefficient C_m was used (Fig. 78). This index expresses the ratio of the easily mobile (in the soluble form) indicator elements (Li and Rb) to the relatively immobile ones (Nb, Sn and Be), the majority of which accumulate in the halo mechanically. Fig. 78 shows curves which characterize the geochemical spectra of the indicator elements at the rare-metal deposits in Transbaikalian soils.

Solovov and Garanin (1968) gave an example of a geochemical coefficient (similar to C_m) which has been used for establishing differences between similar secondary halos from polymetallic deposits in the Transili Ala Tau:

$$V = \frac{As \cdot Ag \cdot Bi \cdot Cd \cdot B}{Zn \cdot Pb^4}$$

For a better classification of similar exogenic dispersion halos, these authors proposed using a linear discriminant function which makes it possible to take into account, simultaneously, the complete statistical characteristics of the objects (estimates of averages, dispersions, and covariances in the indicator element contents). The discriminant linear function is calculated with the help of computers. Its use is, therefore, recommended only when the objects cannot be differentiated by more simple methods.

In regions with a complex sedimentary series, where buried halos of different types predominate, geochemical exploration surveys have made use of so-called *deep lithogeochemical surveys,* which are based on drilling shallow boreholes along profiles in order to detect residual secondary dispersion halos from ore deposits, and to sample them at the level of the representative horizon.

The deep lithogeochemical surveys are much more expensive than geochemical exploration for exposed halos. Consequently, well-founded geological substantiation for such operations is required before they can be approved. This is why deep lithogeochemical surveys are usually conducted in regions where previous geological and geochemical studies have indicated the presence of specific types of mineral deposits. This method is especially effective in areas where buried ore-controlling structures with economic mineral deposits are known to exist.

As mentioned above, the most favorable conditions for the formation of secondary residual buried dispersion halos exist in regions which have experienced prolonged periods of peneplanation under continental conditions. The deeply reworked weathering crust on the eroded folded surface, which is characteristic of such environments, favors the formation of well developed residual dispersion halos which are subsequently overlain by veneers of deposits transported over long distances and of different thicknesses.

The distribution of the indicator elements in the profile of the weathering crust depends on the characteristics of its composition. When a kaolin horizon develops in the upper part of the weathering crust, the majority of the trace elements are partially leached away. This is accompanied by the accumulation of the trace elements at the level of the mottled zone, which is essentially a hydromicaceous weathering crust. The latter is usually the most representative horizon for sampling. When indicator elements are not, for some reason, leached out appreciably from the surface crustal horizon, superimposed diffusion halos may be produced in the sediments which overlie the residual halos. The vertical extent of these superimposed halos vary widely depending on many factors, such as the composition and intensity of the primary mineralization, as well as on the hydrological regime, and on the characteristics of the composition of the overlying sediments. Also very important is the ratio of the mechanical and hydromorphic (chemical) components in the residual dispersion halo. Eremeev (1963) noted considerable variations in the vertical extent of the dispersion of superimposed diffusion halos. They vary from 0.5 to 15 or 18 meters, but usually are no greater than 1 or 2 meters.

The size of the dispersion halo in deep horizons of the weathering crust near the surface of weakly altered bedrock decreases appreciably. It is, therefore, recommended these zones not be used for sampling in those cases characterized by the presence of uneroded upper horizons of the weathering crust.

The procedures for deep lithogeochemical surveys depend primarily on the thickness of the unconsolidated transported deposits. The technical means by which the representative sampling horizon is reached are chosen on the basis of this thickness (Table 42).

Table 42. Types of buried supergene dispersion halos from mineral deposits based on the depth of the cover which consists of unconsolidated material transported over long distances (after Krasnikov and Sharkov).

Category of difficulty in exploration	Thickness of unconsolidated deposits transported over long distances, mm	Type of dispersion halo	Methods and technical means of exploration
III	2-6	Relatively weak at the surface, and shallow	Deep lithochemical surveys conducted with the help of hydraulic cable tools
IV	6-25	Buried	Deep lithochemical surveys with the help of worm drills or other light drilling equipment
V	25-150	Deeply buried	Deep lithochemical surveys with the help of worm and core drills

Successful results from deep lithogeochemical exploration require information on: (1) the presence of ore in the region; (2) the lithological, stratigraphic, and genetic features of the sedimentary formations in the unconsolidated cover, as well as their thicknesses; and (3) the nature of the weathering crust of the bedrock. The preliminary stage of the exploration must be focused on a comprehensive study of all available geological materials for the region in which deep lithogeochemical surveys are planned. In this stage, it is most advisable to tentatively delineate those areas with different thicknesses of transported deposits. This is done with the help of data obtained by studying all available geological and geomorphological maps, as well as by generalizing the results obtained from previous drilling and geophysical studies. The latter are analyzed especially carefully, and all geochemical anomalies revealed previously in the region are indicated on a working map.

The materials processed during the preliminary stage of exploration are used as a basis for determining the optimal drilling depth and the depth at which the unconsolidated deposits will be sampled (the representative sampling horizon is usually selected following the first series of drilling operations). The technical decisions with respect to drilling are also made during this stage.

The pattern of exploratory boreholes is planned on the basis of the expected dimensions of the buried dispersion halos; it is approximately similar to the sampling grid selected in the search for exposed halos. Eremeev (1963) discussed the task of deciding upon the exploratory drilling grid to be used for deep lithogeochemical surveys, including the necessity of having the most widely spaced observation network which will provide a solution to the main exploration problem (i. e., intersecting the object with at least one point). Eremeev maintains that if one assumes certain reserves for the mineral deposits which are the target for the exploration, the contents of useful components in them, and the shape of the ore bodies by analogy with the known examples, then it may be possible to determine approximately the length and width of the exposure of the ore body at the basement surface. Possible sizes of the secondary dispersion halos may also be determined approximately, and the likely areas of the halos of indicator elements characterizing the deposits may be outlined.

Dispersion Trains From Mineral Deposits

Dispersion trains from mineral deposits which occur in the bottom sediments of a river system are produced by processes involving the supergene destruction of the deposits, as well as of their primary and secondary halos. The dispersion trains result from the removal of the chemical elements which are characteristic of a given type of deposit, by means of surface and ground waters. These elements later precipitate in the alluvium of the river system. The elements are removed both in the form of true or colloidal solutions, and mechanically in the form of solid disintegration products (Fig. 79). Precipitation of the solid particles takes place because of the gradual or discontinuous variations in the ability of the river to carry them. The distance over which the solid disintegration products are transported is inversely proportional to the particle size. Thus, the finest-sized particles are transported the longest distances from the source. Near the source they may precipitate in

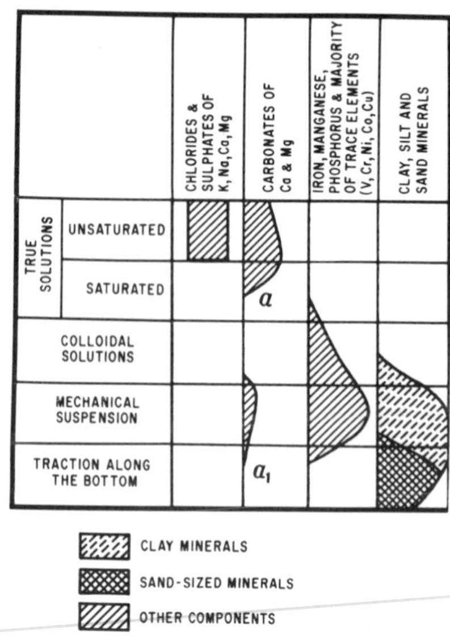

Fig. 79. Forms of transport by rivers of the principal components of sediments (after Strakhov).
1 - clay minerals; *2* - sand-sized minerals; *3* - other components; *a* and a_1 are characteristic only of mountain rivers and of arid climatic conditions.

those regions of the river bed where the velocity decreases sharply. The velocity of the river may decrease because of sharp bends in the river, obstructions or large boulders in the river bed, local depressions in the bottom, etc.

The majority of ore minerals which contain indicator elements are relatively fragile and chemically unstable. As a result, their migration in surface waters occurs mainly in the form of finely dispersed suspensions, as well as in the sorbed form. The few minerals which accumulate in placers (cassiterite, tantalum-niobates, cinnabar, etc.) are exceptions. However, the ability of minerals to accumulate in placers does not mean that they cannot also be transported by water in the finely dispersed state, which is subject to wider dispersion, in the bottom sediments of rivers. In this connection, the finely dispersed form of transport yields more information than large amounts of the same minerals which may accumulate in placers.

The ability of finely dispersed particles to adsorb on their surfaces those ions which are present in solution, further increases the amount of information which can be obtained from the fine fraction of bottom sediments. This fraction may be considered to contain two types of geochemical information, i.e., on the composition of the solid weathering products, and on the composition of the ions which are present in waters in the dissolved state.

The upper parts of dispersion trains are adjacent to secondary dispersion halos because they form as a result of the destruction of these halos. Consequently, dispersion trains are genetically associated with secondary dispersion halos, and through the latter, also with concentrations of ore in bedrock. This association has long been known and used in the search for the original sources of mechanically dispersed halos of those minerals which are resistant to weathering, such as cassiterite, cinnabar, gold, platinum, etc. (i.e., the heavy concentrate method). These well-known techniques were later included among the various geochemical prospecting methods.

However, as previously mentioned, many elements which are mobile in aqueous media (Cu, Zn, Pb, U, Mo, and others) are present in dispersion trains in the finely dispersed, soluble, or sorbed forms, and they cannot be detected by mineralogical methods. This is why studies of dispersion trains from mineral deposits employ geochemical methods based on the sampling and analysis of bottom sediments to detect indicator elements.

Polikarpochkin (1968), and others, found that dispersion trains may have very complex structures. The upper parts of the trains, directly adjacent to the deposit, are usually characterized by higher (from tens to hundreds of geochemical backgrounds) and nonuniform contents of the elements. The element concentrations decrease sharply downstream (where they may exceed the geochemical background by only 3 to 5 times); however, their distribution tends to become more regular as a result of intensive pulverization and mixing of the material by the river.

The indicator element composition found in dispersion trains depends on the composition of the primary halos and ores in the original source, as well as on the characteristics of the supergene migration of the elements in various landscape-geochemical environments.

Different investigators have proposed formulas, based on the concept of the dispersion train productivity, P_X, to quantitatively express the relationship between secondary halos and dispersion trains.

$$P_x = S_x (C_x - C_\Phi),$$

where C_X is the indicator element content in the dispersion train; C_Φ is the average background content; and S_X is the area of the basin being eroded (apparent drainage basin) for a given location in a river bed, in m^2.

Solovov proposed a simple relationship for this purpose: $P_x = k \cdot P$, where P is the productivity of the secondary dispersion halo which supplies the indicator element into the river system, and k is the coefficient which depends on a number of external conditions (which may be larger or smaller than unity).

Polikarpochkin (1968) recognized the difficulties arising in the determination of P_x, and suggested a more complex formula starting from a theoretical analysis of the processes which lead to the formation of mechanical dispersion halos in rivers:

$$P_x = \frac{1}{k_1} \cdot \frac{P}{L_1 - L_2} \cdot \frac{e^{-\delta(x-L_2)} - e^{-\delta(x-L_1)}}{\varphi_{(x)}} \cdot f_{(x)},$$

where P_x and P are productivities of the dispersion train and of the ore showing, respectively, in $m^2\%$; x is the distance from the top of the valley downstream; L_1 and L_2 are distances from the upper and lower edges of the ore showings, determined in a similar way; δ is the parameter which depends on the dynamic factors of dispersion train for formation; K_1 is the proportionality factor (found empirically); $\Phi_{(x)}$ and $f_{(x)}$ are functions whose forms depend on the shape of the river basin. For oval-shaped basins $f_{(x)} = 1/2\, x^2$ and $\Phi_{(x)} = x - 1/\delta \cdot (1 - e^{-\delta x})$. For basins with constant widths $f_{(x)} = x$ and $\Phi_{(x)} = 1 - e^{-\delta x}$. When δ approaches zero, the formula proposed by Polikarpochkin (1968) reduces to the simpler formula $P_x = k \cdot P$.

Other formulas exist which also describe relationships between secondary halos and mechanical dispersion trains (Bogolyubov, 1968). In view of the difficulties arising in the determination of the empirical coefficient k, all of them are essentially theoretical and cannot be applied in actual practice.

It might be useful to compare dispersion trains formed in similar landscape-geochemical environments in terms of their productivities, and contrast them with respect to the geochemical background.

The length of dispersion trains depends mainly on the landscape-geochemical environments in the region of the erosional basin, as well as on the dimensions of the source being eroded.* The shortest dispersion trains (a limit of only about one kilometer) are characteristic of mineral deposits which are situated in arid regions. Under the conditions characterized by a predominance of physical weathering, the proportion between the mechanical and hydromorphic components is markedly in favor of the former, which considerably restricts the length of dispersion trains in bottom sediments of a stream system. The temporary nature of rivers typical of arid regions, is responsible for the poor sorting of alluvial material and its contamination with windborne sands, which appreciably reduces the contrast of any anomalies. However, despite the above difficulties, geochemical prospecting for mineral deposits based on the use of dispersion trains in bottom sediments are effective even for normally arid regions. This has been proven by successful exploration which utilized bottom sediments from streams in the Eastern Desert of Egypt, and in the desert regions of Sudan, Morocco, etc. (based on data from United Nations exploration projects). Anomalous contents of copper in the Sudan desert could be traced 1.0 to 1.5 km downstream from vein deposits in dry water courses. In the arid regions of Mexico, dispersion trains from sulfide deposits sometimes reach 2.5 km. The lengths of dispersion trains in arid environments are similar for the trace elements which exhibit different mobilities. This is due to the essentially mechanical nature of the dispersion phenomena.

In contrast to arid regions, provinces which include forest areas with acidic or neutral waters and soils, are characterized by strong migration of organic and mineral colloids in natural waters. This is responsible for the increased mobility of the majority of the trace elements. Conditions especially favorable for aqueous migration of the trace elements are typical of humid tropical regions. In these regions, supergene removal of the indicator elements from the surface horizons of minerals deposits, and their subsequent transport in colloidal or sorbed forms by stream waters, is especially important. The longest dispersion trains recorded in humid tropical regions are associated with porphyry copper ore deposits which form relatively large (thousands of square meters) secondary dispersion halos in soils (Beus and Lepeltier, 1971). In Panama, for example, contents of copper up to 200 ppm were found in stream sediments at a distance of 12 to 15 km from a major soil anomaly overlying a porphyry copper ore deposit. Under similar climatic conditions in the Philippines (humid jungle), anomalous contents of copper in dispersion trains were traced for a distance of 16 km from a soil anomaly; this was also associated with a porphyry copper ore deposit.

---

*Parallels between the size of the initial erosional source, and the extent of the mineral deposits, should be used with definite reservations. Specifically, even large mineral deposits, in the initial stages of erosion, may be characterized by small areas of erosion.

Dispersion trains within the taiga regions of the USSR, forest regions of Canada, and other places within the temperate zones being discussed, may extend from 1.5 to 4 km. The extent of the dispersion train depends on the specific landscape-geochemical conditions, and on the composition of the mineral associations within the deposits which are the sources of material being eroded. It is quite natural that, under such conditions, the longest dispersion trains will be composed of indicator elements which migrate easily in acidic and neutral waters, in the form of true or colloidal solutions (Cu, Mo, U, etc.). In the case of these elements, the soluble components of the dispersion trains are usually of considerable importance, and this importance tends to increase as the distance from the erosional source of the elements becomes larger.

The data obtained from prospecting surveys which have been conducted in various landscape-climatic environments in the last decade, show that dispersion trains from mineral deposits of various compositions and dimensions, ranging from large to medium, are at least 0.8 km long even under conditions unfavorable for their formation. As a consequence of this feature, it is possible to choose distances between sampling points on the basis of the specific landscape-geochemical conditions.

From studies of the distribution of the trace elements in various fractions of the bottom sediments (Tables 43 and 44), it is recommended that the finest silty-clayey fraction from

Table 43. Total zinc content, and the total cold-extractable nonferrous metal content (Zn, Pb, and Cu), in different size fractions of stream sediments from Canada (after Hawkes and Webb, 1962).

Fraction, mesh (British standard)	Content %	Total content of zinc, ppm*	Cold-extractable content of the total nonferrous metals (ppm)
−8 + 32	65	500	6
−32 + 80	25	500	15
−80 + 115	4	800	25
−115 + 200	3	800	45
−200	3	1000	75

* The extraction was performed with nitric acid.

Table 44. Contents of copper in different size fractions of stream sediments from Zambia (after Webb).

Fraction, mesh (British standard)	Content in 2-mm fraction, %		Total content of copper, ppm		Content of cold-extractable copper	
	Background	Anomaly	Background	Anomaly	Background	Anomaly
−20 + 35	5.4	24.1	80	180	8	80
−35 + 80	64.4	43.5	40	160	2	35
−80 + 135	21.1	21.4	40	210	3	70
−135 + 200	5.4	4.6	80	250	12	110
−200	1.7	0.6	110	360	22	170
−80			50	220	4	80*

* The −80 mesh fraction furnishes reliable results for both total copper (contrast equal to 4.4), and for the cold-extractable copper (contrast equal to 20.0).

the sediments be sampled because they are usually the most informative. However, certain departures from this conventional procedure are also possible. It is, for example, advisable to sample fractions up to 0.1 mm in the cases of gold, platinum, in some cases for tin, as well as tantalum and niobium minerals. These samples should be collected from places in the stream bed which favor deposition of heavy minerals in the bottom sediments.

Samples are collected directly from the bottom of the stream, or from its dry floodplain. They are collected at the surface or from depths ranging from between 10 and 20 cm, depending on the specific local conditions. Difficulties in sampling the most informative silty-clayey fraction arise in mountainous regions which have very rapid streams, as well as in desert regions where alluvial deposits in the dry water courses are strongly contaminated with windborne sands.

Silty and sandy-clayey material in mountainous rivers should be sought between large boulders and at sharp bends in the river. In the dry water courses of deserts, the fine fraction of the bottom sediments (−0.25 mm, or −60 mesh) may be separated from the sand fraction by sieving during the process of sampling. This fraction, likewise, may be appreciably impoverished as a result of windborne contamination. If contamination is suspected, Bugrov (1974) recommends using the 1.0 to 0.25 mm fraction. This recommendation is based on experiences gained from experimental studies in the Eastern Desert of Egypt.

The distances between sampling points in streams depend both on the landscape-climatic conditions and on the features of the drainage basin. Sampling is performed in an upstream direction, possibly from the mouth, and is continued into the mouths of all the tributaries. The lengths of the dispersion trains make it possible to select distances between sampling points on the basis of specific local conditions. These sampling intervals may range from 0.5 km (in regions with limited mobility of the trace elements) to 1 km or even to 2 km (in regions with increased mobility of the trace elements). However, if tributaries whose mouths are closer spaced than the selected sampling interval discharge into the

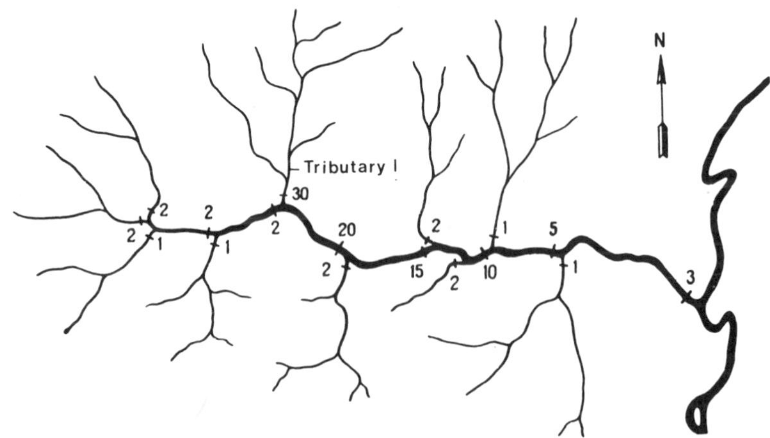

Fig. 80. Schematic map showing results of a reconnaissance geochemical survey using bottom sediments in the Philippines. An anomalous content of copper was found at the mouth of tributary I in the fine fraction of bottom sediments. Bars and numbers designate sampling sites and the contents of cold extractable copper (in ppm) (after Govett et al., 1966).

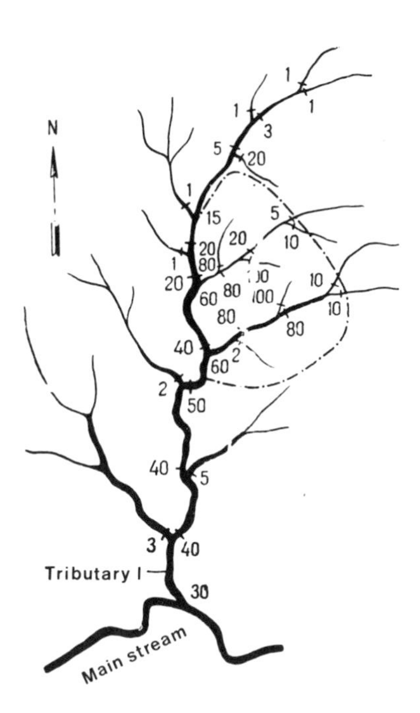

Fig. 81. Detailed sampling of bottom sediments along tributary I, which led to the detection of a secondary dispersion halo associated with copper mineralization (after Govett et al., 1966).
(The secondary halo in the soils is encircled with a dot-and-dash line. For other conventional symbols see Fig. 80).

valley of a river which is being used as the main route of the survey, then these tributaries should always be sampled at their mouths (Figs. 80 and 81).

In the course of reconnaissance surveys, one often has to confine himself to collecting samples from the mouths of shallow tributaries.* If anomalous contents are detected, then the upper reaches of such a tributary are investigated during the more detailed stages of exploration with the purpose of identifying the source of the anomaly. This may turn out to be an exposed ore occurrence or its geochemical halo (primary or secondary). Worldwide geochemical exploration experiences involving the use of bottom sediments, which have been summarized by Webb and others, suggests that a sampling density of 1 sample per km^2 should make it possible to detect all anomalies from dispersion trains originating from large and medium sized mineral deposits.

When sampling bottom sediments from streams, one should take into account that there are various possible sources of contamination of the river water by man. Contamination of waters and bottom sediments by mining operations and metallurgical plants is particularly important. Their presence in the upper reaches of drainage basins sometimes rules out the possibility of obtaining reliable results by means of the technique being discussed. In order

*Dispersion trains are usually strongly diluted in deposits of major rivers below the place where a tributary joins the river.

to avoid erroneous results from false anomalies, one should take into account, and note in their field records, the nature of the material being sampled, especially its size, approximate content of organic matter, etc.

Geochemical prospecting based on stream sediments from drainage basins results in the construction of dispersion trains maps which show data on the indicator element distributions in bottom deposits. Figs. 82 and 83 illustrate results which can be obtained from sampling bottom sediments from stream systems.

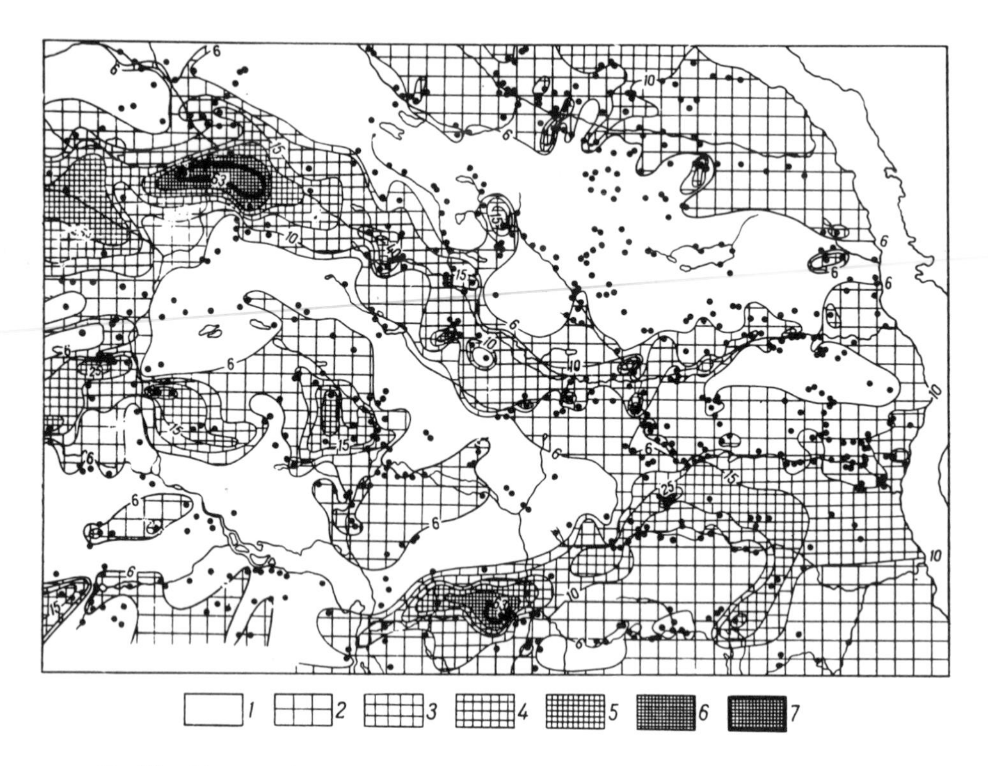

Fig. 82. Map showing lines of equal concentrations in the distribution of copper in bottom sediments (Southern Norway). Sampling locations are marked by dots (after Bolviken).
Content: ppm; *1* - less than 6; *2* - 6 to 10; *3* - 10 to 15; *4* - 15 to 25; *5* - 25 to 40; *6* - 40 to 63; *7* - more than 63.

During the reconnaissance stage of exploration, when the metallogenetic features of a region are still inadequately known, it is recommended that geochemical stream sediment samples be analyzed by the semi-quantitative emission spectrographic method in order to detect the largest number of elements. As soon as the geochemical anomalies are detected, and the associations of the trace elements ascertained, the number of elements which need to be analyzed may be reduced. Then the most effective analytical method can be employed to obtain sufficient accuracy for quantitative interpretation, rapidly and inexpensively.

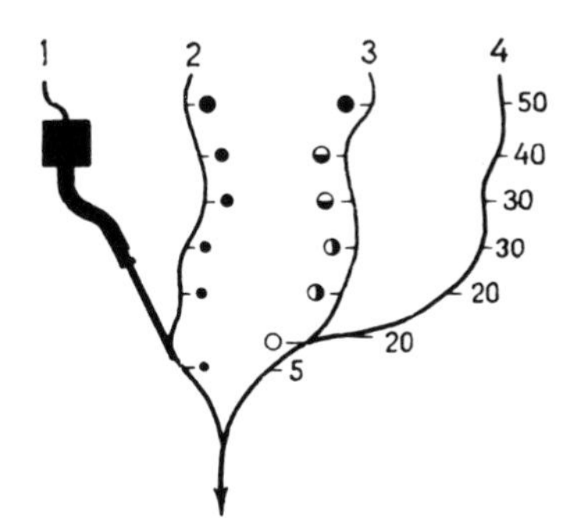

Fig. 83. Different ways of showing results obtained from sampling bottom sediments from river systems on geochemical maps (after Andrews-Jones, 1968).
1 - the thickness of the line along the water-course expresses the content of an indicator element; *2* - the content is expressed by the size of the circle; *3* - the content is expressed by the type of shading within the circles; *4* - the contents are shown by numbers in ppm.

Geochemical exploration techniques using bottom sediments from streams, which have been proven to be very effective under various landscape-climatic conditions, have now become very popular, especially during the initial stages of prospecting. For example, during the period from June, 1970 to June, 1971, for geochemical exploration in the developing countries of Asia and Latin America, 176,000 stream sediment samples (64.4% of the total) were collected and analyzed.

7

Hydrogeochemical Halos From Mineral Deposits

General Information

Water is known to have the unique capability to dissolve certain amounts of the majority of chemical elements which have been components in natural compounds, and to transport them over considerable distances. As a result, water which comes into contact with ore concentrations (mineral deposits and ore occurrences, as well as their geochemical halos) become enriched in chemical elements which are typomorphous for specific ores. This results in the formation, around ore concentrations, of certain zones in which the subsurface waters contain increased amounts of various ore components. These zones are called *aqueous dispersion halos* of the chemical elements.

The hydrogeochemical method of exploration for mineral deposits is based on detecting aqueous dispersion halos which accompany ore accumulations, and on studying other indirect criteria (hydrogeological and hydrochemical) which indicate specific geochemical processes taking place during the weathering of ore bodies, mineral deposits, and their primary halos.

The detection of aqueous dispersion halos of the indicator elements, which always accompany ore occurrences, makes it possible to outline areas which are promising for ore deposits.

The theoretical and procedural principles of this method were developed by a large number of Soviet and non-Soviet investigators. These include Brodskii, Belyakova, Germanov, Goleva, Krainov, Ovchinnikov, Smirnov, Saukov, Udodov, Hawkes, Webb, White, and others, who determined methods to be used for studying features concerned with the formation and use of aqueous halos for various types of mineral deposits.

Based on our present level of understanding, it is possible to use hydrogeochemical methods in a wide range of natural environments in the search for mineral deposits, both exposed and hidden. This method may be used in exploration for various ore deposits primarily because the analytical techniques are now available which make it possible to detect the contents of a wide range of chemical elements in water which are indicators of ore deposits.

It has now been proved that hydrogeochemical methods are effective in the search for many different types of mineral deposits (boron, beryllium, lithium, fluorine, cesium, tungsten, uranium, etc.). However, the best results are likely to be achieved if this method is used in the exploration for sulfide deposits, around which extensive aqueous dispersion halos of the chemical elements form as a result of the weathering of readily oxidizable ores.

The size and contrast of aqueous halos depend on various natural factors. The major ones are:

1) morphology and composition of the ores, as well as of the primary halos surrounding them;

2) physical nature, including permeability, of the enclosing and overyling rocks;

3) hydrogeological and paleogeochemical environments;

4) intensity and duration of the erosion processes;

5) landscape-geochemical conditions and the migration characteristics of the indicator elements in a given environment.

Hydrogeochemical halos are divided into *permanent* and *temporary* in terms of their stability in natural waters. The former are typical of deep aquifers where the environmental conditions are constant. The latter form in shallow ground waters or in surface waters, and their anomalies may temporarily disappear during periods of strong seasonal precipitation.

Aqueous dispersion halos are always extended in the flow direction of the surface or subsurface waters.

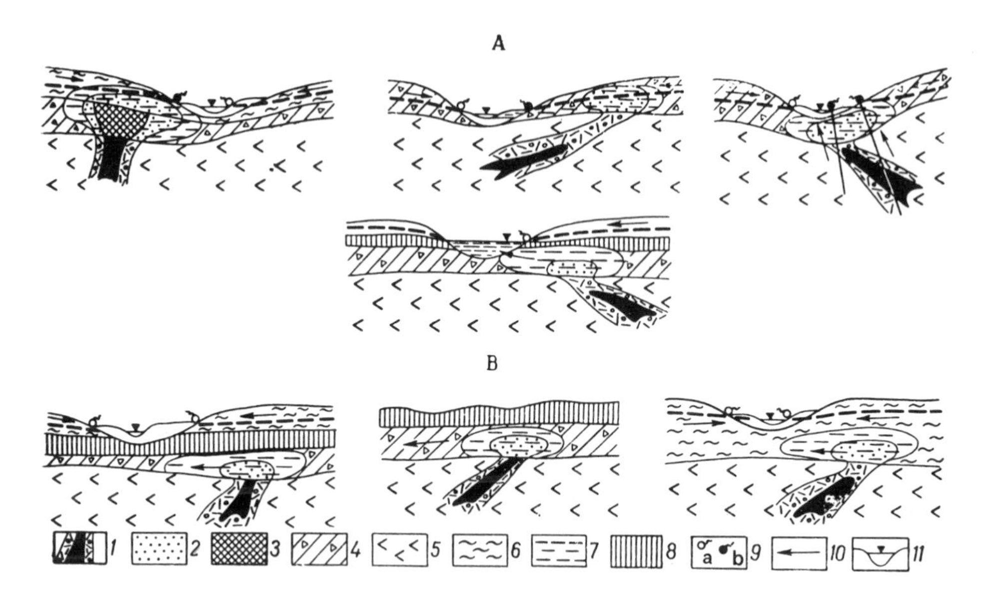

Fig. 84. Types of aqueous dispersion halos around buried mineralization (A: exposed; B: hidden).
1 - ore body and its primary dispersion halo; *2* - secondary dispersion halo; *3* - oxidized ores (gossan); *4* - crust of weathering (water permeable, fractured rock); *5* - ore-enclosing bedrock; *6* - sedimentary water-bearing rocks; *7* - aqueous dispersion halo; *8* - rock impervious to water; *9* - sources of ground water with background (*a*), and anomalous (*b*), contents of hydrogeochemical indicator elements; *10* - direction of water movement; *11* - local erosion level.

Hydrogeochemical anomalies are usually classified as *exposed* or *hidden* depending on the mode of their surface expression.

The former include anomalies which have expression at the surface of the Earth as various types of water discharges in which halos may be found (springs, swamps, streams, etc.). Fig. 84 shows various types of exposed aqueous halos (from Goleva, 1965).

The latter are associated with ore bodies situated below the present local levels of erosion in poorly dissected, and flat foothill and plain regions (Fig. 84). In contrast to exposed anomalies, these are not exposed at the present-day erosion surface, and they are not detected in surface waters.

The dimensions of aqueous dispersion halos around strongly oxidizing ore deposits range from 0.5 to 3.0 km. Under favorable conditions, water from streams with increased concentrations of certain metals may be traced over distances of up to 5 - 8 km.

Table 45. Hydrogeochemical indicator elements for different types of deposits.

Type of deposit	Indicator elements of ore bodies	
	Strongly oxidizing	Weakly oxidizing
Copper-pyrite	Cu, Zn, Pb, As, Ni, Co, F, Cd, Se, Ge, Au, Ag, Sb	Zn, Pb, Mo, As, Ge, Se, F
Polymetallic	Pb, Zn, Cu, As, Mo, Ni, Ag, Cd, Sb, Se, Ge	Pb, Zn, As, Mo, Ni
Molybdenum	Mo, W, Pb, Cu, Zn, Be, F, Co, Ni, Mn	Mo, Pb, Zn, F, As, Li
Tungsten-beryllium	W, Mo, Zn, Cu, As, F, Li, Be, Rb	W, Mo, F, Li
Mercury-antimony	Hg, Sb, As, Zn, F, B, Se, Cu	Ag, As, Zn, B, F
Gold ore	Au, Ag, Sb, As, Mo, Se, Pb, Cu, Zn, Ni, Co	Ag, Sb, As, Mo, Zn
Tin ore	Sn, Nb, Pb, Cu, Zn, Li, F, Be	Sn, Li, F, Be, Zn
Titaniferous magnetite	Ti, Fe, Ni, Co, Cr	Ni, Fe
Spodumene	Li, Rb, Cs, Mn, Pb, Nb, Sr, F, Ga	Li, Rb, Cs, F
Copper-nickel	Ni, Cu, Zn, Co, Ag, Ba, Sn, Pb, U	Ni, Zn, Ag, Sn, Ba
Beryllium-fluorite	Be, F, Li, Rb, W	Be, F, Li
Baritic-polymetallic	Ba, Sr, Cu, Zn, Pb, As, Mo	Be, Sr, As, Mo

Table 45 lists the most important hydrogeochemical associations (from Goleva, 1965) of indicator elements which are found in aqueous halos from different types of mineral deposits.

Studies of aqueous dispersion halos from various deposits have shown that they have zonal structures, i.e., as one goes away from the ore bodies, the concentration of the elements varies, and the zones have different gradients (concentrations). This is caused by the processes of dilution, hydrolysis, sorption, etc. This leads to the "precipitation" of some elements from the above associations in different parts of the aqueous dispersion halos.

Practically all elements from the associations listed above are found in the zone immediately surrounding the particular type of ore body. The intermediate zones of aqueous dispersion halos are most frequently characterized by increased concentrations of copper, lead, selenium, and germanium. The peripheral zones of the aqueous dispersion halos are characterized by anomalous concentrations of a more limited number of indicator elements, such as molybdenum, zinc, arsenic, and sometimes mercury.

Successful application of hydrogeochemical methods in exploration depends, to a large extent, on the variables controlling the supply of water in the regions being studied. Naturally, in regions with abundant precipitation the results of hydrogeochemical sampling will be more representative owing to larger numbers of ore occurrences which can be detected (in comparison with arid regions).

A characteristic feature of hydrogeochemical anomalies (as compared to lithogeochemical anomalies) is the rather uniform distribution of the indicator elements within anomalous fields. This is why each individual hydrogeochemical sample may give relatively reliable information for an area of considerable size. This feature of hydrogeochemical anomalies ensures that this method can be used effectively during small-scale (regional) exploration, when large areas are being investigated in order to detect promising ore fields and regions.* For this particular reason the hydrogeochemical method is always considered as a component among the complex of methods which can be used during the stage in which regional geological surveys are conducted.

The above-mentioned uniform distribution of the indicator elements within their aqueous dispersion halos, which is an advantage in regional exploration, becomes a disadvantage when one considers the value of the method during the stage in which more detailed surveys are conducted (1 : 10,000 or larger). The hydrogeochemical method is less effective during this stage of exploration due to the above reason, as well as because it is difficult to reveal correlations between the ore accumulations and the behavior of the indicator elements in the aqueous dispersion halos. This correlation is complicated, appreciably, by the influence of various factors (hydrogeolocial, structural-geological, and others) on the formation of hydrogeochemical anomalies.

Efficiency in the application of the hydrogeochemical method at the detailed stage of exploration depends mainly on the number of water occurrences which exist within the area being prospected. In many cases, the lack of water at the surface prevents the delineation and conclusive interpretation of hydrogeochemical anomalies.

Consequently, lithogeochemical methods (using primary geochemical halos and secondary dispersion halos of the indicator elements) are the major geochemical methods to be used during the detailed stage of exploration.

*This method is less effective economically (i.e., more expensive) than the equally informative stream sediment method. In this connection, the use of hydrogeochemical sampling is recommended during regional exploration particularly to study the discharges from subsurface waters.

8

Biogeochemical Halos from Mineral Deposits and Their Use in Exploration

General Information

The biogeochemical methods used in mineral exploration are based on an intimate, intrinsic interrelationship between the biosphere and its environment; this includes the hydrosphere, atmosphere, and surface of the lithosphere.

The immediate link between the biosphere and the environment is effected through vegetation. Vegetation uses certain components of the atmosphere, as well as of minerals which are extracted in the form of aqueous solutions from the surface of the lithosphere, as nutrients. As a result, plants accumulate numerous trace elements in their tissues whose composition and quantitative relationships, depend to a large extent on the characteristics of the nutrient-supplying substrate. These include both the soil and underlying bedrock.

Biogeochemistry as a science was founded by Vernadsky early in the 20th century. Later, further advances, which were essential for geochemical exploration, were developed by Vinogradov, Malyuga, Glazovskaya, Grabovskaya, and others. Both the scientific principles, and the methods for utilizing the biogeochemical (more specifically phytogeochemical; "phyton" is a Greek word for plant) features of the vegetable kingdom, were developed specifically for mineral exploration.

Vinogradov, who studied the ability of plants to accumulate certain trace elements, identified two types of concentration (which we call phytogenic): *group* and *selective*. In the case of group concentration, all plants growing in a certain region contain increased amounts of particular chemical elements, which depends on the presence of these elements (high contents) in the soil and in the underlying rocks. During selective concentration, the accumulation of the individual chemical elements or their associations, is effected by certain species, more frequently by genera of plants. In the vegetable kingdom, plant concentrators of certain elements exist, so that one can refer to the existence of the specific lithium, boron, calcium, aluminum, etc., flora. Vinogradov developed the theory of *biogeochemical provinces*. These provinces are characterized by definite abundance levels of one or several chemical elements (deficient, normal, excessive), and definite biological reactions.

There are no exhaustive, statistical data on the distribution of the chemical elements in plants. However, the general situation may be illustrated by the data in Table 46 which were calculated by Vinogradov (1950).

Table 46. Average contents of the chemical elements in the "granitic" layer of the Earth's crust, in soils, and in plants, in ppm (after Vinogradov).

Element	"Granitic" layer of the Earth's crust	Soil	Plants (in ash)	Concentration ratio (plant/soil)	Element	"Granitic" layer of the Earth's crust	Soil	Plants (in ash)	Concentration ratio (plant/soil)
Li	30	30	6*	0,2	Cu	22	20	20	1,0
Be	2,5	3*	3*	1,0	Zn	51	50	900	18
B	10	10	400	40	Ga	19	20	20	1,0
F	720	200	10	0,05	As	1,6	5	0,3	0,06
Na	22 000	6300	20 000	0,3	Se	0,14	0,01	?	?
Mg	12 000	6300	70 000	11	Br	2,2	5	150	30
Al	80 000	71300	14 000	0,2	Rb	180	130*	120*	0,9
Si	309 000	330000	150 000	0,5	Sr	230	300	300	1,0
P	800	800	70 000	88	Mo	1,3	3	9*	3,0
S	300	850	50 000	59	Ag	0,05	0,1*	1	10
Cl	170	100	100*	1,0	Cd	0,15	0,5	0,01	0,02
K	26 400	13600	30 000	2,2	Sn	2,7	4*	5	1,2
Ca	25 000	13700	30 000	2,2	I	0,5	5	50	10
Ti	3 300	4600	1 000	0,2	Cs	3,8	4*	4*	1,0
V	76	80*	61	0,8	Ba	680	500	600*	1,2
Cr	34	50*	90*	1,8	Au	0,001	0,001*	0,007*	?
Mn	700	850	7 500	8,8	Hg	0,03	0,01	0,001	0,1
Fe	36 000	38000	6 700	0,17	Pb	16	10	10	1,0
Co	7,3	10	15	1,5	U	2,6	1	0,5	0,5
Ni	26	30*	20	0.7					

*More accurate data

In order for the contents of the trace elements in plants to be used as indicators of the distribution patterns of these elements in the bedrock underlying the soil, a clearly expressed correlation must exist between the distribution of the chemical elements in the media being studied. That this correlation does exist is shown in Table 47, which is based on data obtained from materials used in biogeochemical exploration in Eastern Transbaikalia.

Table 47. Correlation between the contents of lead and tin in plant ash and in bedrock from Eastern Transbaikalia (after Grabovskaya, 1965).

Pairs being correlated	Correlation coefficient	
	Pb	Sn
Larch-bedrock	+ 0,41	+ 0,35
Birch-bedrock	+ 0,40	+ 0,81
Red bilberry-bedrock	+ 0,32*	+ 0,45
Wild rosemary-bedrock	+ 0,54	—
Rhododendron-bedrock	+ 0,46	+ 0,55

* Correlation is insignificant (at 5 % significance level).

Indicator plants should be selected only from those in which the distribution of the ore trace elements, and their associated elements, correlate well with the distribution in the bedrock.

In some cases (especially for grassy vegetation), significant correlations may exist only with the B soil horizon (Table 48). As is shown in the Table, geochemical correlation be-

Table 48. Correlation between the contents of lithium in the ash of legume grasses, in the soil, and in the bedrock from the forest-steppe zone of Eastern Trans-baikalia (after Grabovskaya, 1965).

Pairs being correlated	Correlation coefficient
Legume grasses — A soil horizon	+ 0,27*
Legume grasses — B soil horizon	+ 0,48
Legume grasses — bedrock	+ 0,40*

* Correlation insignificant (at 5 % significance level).

tween the plant and the bedrock, as well as between the plant and the A soil horizon, may be absent. (The absence of a correlation in the distribution of indicator elements in plants and in the A soil horizon, is typical of several geochemical landscapes, and emphasizes that this soil horizon is not usually geochemically representative.)

Since correlations exist between the distribution of trace elements in plants and in the

Table 49. Accumulation of ore trace elements in the ash of plants, ppm.

Element	Over barren areas (local biogeochemical background)	Over mineral deposits
Cu	15-30	From 50-60 to 100-200 (depending on the concentration of copper in ores); in individual cases more than 1000
Pb	0.2 to 30; seldom to 50	40 to 500, in places up to 1000 or higher
Zn	20 to 100 (in grasses); 100 to 1000 (in wood species)	More than 300 in grasses, 500 to 3000 in wood species
Ag	0,01 to 3	0,04 to 10 (depending on the background), maximum contents 30 to 60
Au	Less than 0,01	More than 0,01; rarely up to 10
Mo	0,1 to 20	20 to 350 (depending on the background); in individual cases up to 1000
Sn	1 to 6	More than 10, usually 40 to 150 sometimes approaching 550
Ni	5 to 30	More than 100, sometimes up to 2000
Co	Up to 30	More than 50
Bi	Less than 5	10 or more (up to 100)
Li	From less than 1 to 15	More than 10 to 15, usually 40 to 200, in individual plant species more than 1000
Be	0,5 to 4, rarely up to 7 or 8	5 to 20, in individual cases reaching 30 to 40
Nb	Less than 10	10 to 50
Zr	Up to 10	More than 10, sometimes approaching 100

bedrock, anomalous contents of the indicator elements in plants must correspond to geochemical anomalies of these elements in the bedrock and in soils. Extensive information is now available confirming the increased contents of ore elements and their associates in plants which grow over the corresponding mineral deposits. Table 49 gives a general idea of this relationship.

One should keep in mind, when using biogeochemical halos from mineral deposits for exploration purposes, that different plant species usually differ appreciably in the degree of accumulation of particular indicator elements. This may be illustrated by *coefficients of concentration* of the trace elements in different plant species relative to their content in the soil (Table 50).

Table 50. Coefficients of concentration of some indicator elements in the ash of different plant species relative to their content in soils (from Eastern Transbaikalia).

Plant species	Li		Be		Pb	
	In background areas	Over rare-metal deposits	In background areas	Over rare-metal deposits	In background areas	Over rare-metal deposits
Birch	0,02	0,09	0,10	0,06	1,3	0,16
Larch	0,02	0,16	0,20	0,18	1,1	0,19
Rhododendron	0,02	0,11	0,20	0,10	1,0	0,13
Red bilberry	0,02	0,38	0,50	0,30	1,7	0,23
Wild rosemary	0,02	0,22	0,20	0,04	1,3	0,17
Dwarf stellaria	0,40	0,55	0,59	0,30	0,3	—
Gmelin wormwood	0,37	0,55	—	—	—	—
Spruge olive	0,30	0,43	—	—	—	—
Adonis	0,22	0,31	0,59	0,25	—	—
Legume grasses	0,20	0,33	0,09	0,12	—	—
Cereals	—	—	0,40	0,20	—	—
Meadow-rue	5,30	20,0	0,40	0,18	—	—

As seen from Table 50, soils within geochemical anomalies are generally characterized by larger accumulations of the trace elements. Plants which concentrate particular trace elements are an exception. These include meadow-rue *(Thalictrum)* which accumulates lithium in its tissues. In background areas, the usual contents of lithium in the ash of plants ranges from 3 to 17 ppm, whereas the concentration of this element in the meadow-rue averages 160 ppm, and ranges between 80 and 280 ppm (the probability for the limiting values being 0.05). Thus, sampling of meadow-rue among other grasses may easily mislead the inexperienced investigator because it exhibits "sharply anomalous" contents of lithium in ashed specimens. The difference in the accumulation of lithium between meadow-rue and other plants, is even greater within geochemical anomalies containing increased contents of lithium. For example, in areas with rare-metal pegmatites, the content of lithium in the ash of trees and grasses varies between 1 and 61 ppm. Meadow-rue growing in the same area contains, on the average, 1500 ppm of lithium (range 200 to 2800 ppm). Other examples of concentrator plants include various types of astragalus which contain up to tenths of a percent of selenium per live weight of the plant; foxglove *(Digitalis purpurea)*, whose ash contains up to 9% of manganese; wood horsetail *(Equisetum sylvaticum)* which accumulates tantalum (up to hundredths of a percent in the ash); and Siberian patrinia *(Patrinia sibirica)* which concentrates beryllium and other elements.

Table 51 gives a general idea of the concentrations of some indicator elements in the ash of various plants growing within the same areas.

Table 51. Relationships between the contents of beryllium and lithium in the ash of different plant species, in ppm (after Grabovskaya, 1965).

	Li		Be	
Plant species	Background	Geochemical anomaly	Background	Geochemical anomaly
	Mountainous taiga landscapes			
Larch	1,0	1,0	1,0	1,0
Birch	0,64	0,63	0,33	0,25
Rhododendron	0,72	0,72	0,55	0,70
Wild rosemary	1,36	1,40	0,67	1,78
Red bilberry	1,97	2,02	1,67	1,67
	Forest-steppe landscapes			
Wood small-reed	1,0	1,0	—	—
Siberian patrinia	—	—	1,0	1,0
Legume grasses	1,5	1,6	0,11	0,33
Grasses	1,8	1,8	0,50	0,55
Dwarf stellaria	3,0	2,5	0,75	0,83
Meadow-rue	40,0	167,0	0,50	0,50
	Steppe landscapes			
Feather grass	1,0	1,0	0,67	0,63
Siberian tansy	1,7	1,5	1,0	1,0
Cold wormwood	1,7	1,6	1,0	0,95
Small meadow-rue	45,0	75,0	—	—

It follows that biogeochemical sampling of plants for exploration purposes requires not only an individualized approach to various plant species, but also a knowledge of the major features of concentration of the indicator trace elements by specific genera and species of plants. This knowledge is also necessary because different parts of plants accumulate different quantities of the trace elements. It is, therefore, of relevance as to which parts of the plants (leaves, branches, or roots) are sampled. Table 52 presents data on the preferential concentration of some trace elements by different plant organs. Table 53 contains quantitative information on the distribution of Be and Zr in the leaves, branches, and roots of birch, aspen, and larch.

Table 52. Distribution of some trace elements in different plant organs.

	Plant organ	
Trace element	With maximum concentration of the trace elements	With minimum concentration of the trace elements
Pb	Roots	Bark
Mo	Leaves	Bark
Zn	Leaves	Wood
Cu	Roots	Bark
Li	Leaves of trees and epigeal parts of grasses	Bark
Ta and Nb	Coniferous needles and leaves of trees, fruit and seeds of grasses	Bark

Table 53. Average contents of beryllium and zirconium in the ash of different organs of wood species within geochemical anomalies, ppm (after Grabovskaya, 1965).

Species and organs of trees	Be	Zr
Common birch *(Betula verrucosa)*		
Leaves	6	14
Branches	2.6	7
Bark	1.5	4
Dahurian larch *(Larix dahurica)*		
Coniferous needles	8	60
Branches	7	56
Bark	5	12
Aspen *(Populus tremula)*		
Leaves	6	—
Branches	5	—
Bark	2	—

The age of plants also has a considerable effect on the accumulation of trace elements. The processes of biosynthesis are especially vigorous in young plants, and these processes slow down with age. Therefore, in old plants the capacity for absorption of trace elements from the environment is often more weakly expressed than in young plants. According to Grabovskaya (1965), the difference in the accumulation of trace elements by plants of different age varies by a factor of 1.5 to 2, sometimes it reaches 8 to 10 (Table 54).

Table 54. Contents of some indicator elements in wood species of different age over rare-metal deposits in the forest-steppe zone of Siberia (after Grabovskaya, 1965).

		Average content in ash, ppm		
Species of plant	Age, years	Be	Li	Zr
Common birch (leaves)	4-5	12	40	10
	40-50	6	70	40
Aspen (leaves)	4-5	8	200	70
	50-60	6	12	10

It should also be remembered that requirements for trace elements, and consequently the level of their accumulation, vary with the phases of plant development. Some plants vigorously concentrate certain trace elements in the spring, others in autumn, whereas others have periods of vigorous accumulation of trace elements twice a year. The tendency in many plants to concentrate particular trace elements increases noticeably during periods of flowering or fruiting. For example, in the steppe zone of Siberia beryllium, yttrium, and ytterbium are accumulated by the flowering plants of *Pulsatella patens* in May, and by *Potentilla tanacetifolia* at the end of July to the beginning of August; lanthanum and cerium are accumulated by flowering grasses of the *Leguminosa* family in July. Tantalum and niobium are concentrated by gramineous grasses, and lithium, by plants of the genus *Thalictrum* (meadow-rue), during the fruiting period (Grabovskaya, 1965). In some cases, differences in the exposure to solar radiation have an effect on the extent of trace element accumulation. The difference in the content of trace elements in the same plant species, growing under the same conditions but illuminated differently by the Sun, may vary by 3 to 4 times. As an example, Grabovskaya cited a case where the ash of coniferous needles (of larch), which was collected on the side well illuminated by the Sun, contained 5 ppm of beryllium, whereas on the darkened side, the content of beryllium was 1.5 ppm.

Variations in the extent of assimilation of various trace elements by plants, within background and anomalous regions, make it possible to use not only the absolute variations in the contents of ore elements and their associates in the ash of plants, but also certain typical relationships as indicator criteria. Warren and Delavault (1949) pointed out the fact that the Cu/Zn ratio in vegetation from barren regions is constant, whereas geochemical anomalies due to the presence of copper or zinc mineralization cause regular variations in this ratio. According to these authors, a Cu/Zn ratio greater than 0.23 suggests the presence of copper mineralization. If this ratio is smaller than 0.07, there may be zinc mineralization. The Th/U ratio may be used successfully in the search for uranium deposits; its decrease in the ash of plants usually indicates the presence of uranium mineralization in the bedrock.

Malyuga (1963) found that the ratio of Co/Ni/Cu in the ash of plants, which is 1/2.5/10 under normal conditions, becomes 1/15/3 in regions with nickel silicate deposits. Another example is a massif which contains beryllium-bearing aplogranites, above which the Be/Mo/Sb/Pb ratio in vegetation was 1/1/0.2/0.3, whereas this ratio is typically 1/1.4/7/4.3 in background regions.

Parameters of the phytogeochemical (biogeochemical) background and minimum anomalous (threshold) contents of the indicator elements are estimated during biogeochemical exploration with the help of the same methods which were discussed above during the description of primary and secondary lithogeochemical halos. By way of example we will consider parameters of the distribution of lead in the coniferous needles of larch within the geochemical background of the mountainous permafrost-taiga landscapes of Transbaikalia (Etyka region). These parameters are: normal distribution; average (35 ± 4) ppm; standard 16; coefficient of variation 44.8; minimum anomalous contents are 61 (5% significance level), and 73 ppm (1% significance level).

The probability (P) of encountering the calculated minimum anomalous and higher contents within phytogeochemical halos from known deposits in this region are (for 61 and 73 ppm) 0.60 and 0.48, respectively. In order that at least three specimens with critical contents of lead would be included in the set of geochemical samples from the area of a halo, it is necessary to collect eight (for $P = 0.60$) or eleven (for $P = 0.48$) samples (see Table in the Appendix). By using an analogous method, and by taking into account the possible size of the geochemical anomaly, one can easily calculate the optimal sampling grid for the coniferous larch needles during exploration.

Along with biogeochemical criteria proper, the so-called *geobotanical* criteria must be taken into account when plants are being used as indicators in mineral exploration.

The geobotanical characteristics of plants, which may be of relevance in exploration, are divided into several groups. The first of these includes the affiliation of specific plant species with specific geochemical environments. Plants exist which are *universal indicators* of concentrations of particular trace elements in soils, and others which are *local indicators* of certain trace elements.

The first type of indicators (i.e., universal) grow only in areas with increased contents of particular trace elements and are direct indicators of the presence of ore occurrences and deposits. These include *Viola calaminaria* and *Thlaspi calaminaria* (indicators for zinc), as well as some species of the genera *Astragalus*, *Stanteya*, and *Xylorrhiza* (indicators for selenium).

The second type of indicators (i.e., local) are quite common, however, they are more frequently encountered in areas enriched with particular trace elements. Consequently, plants of this group can, under certain conditions, indicate zones of ore mineralization concealed beneath soil or alluvium.

Plants which are local indicators of copper and which belong to the families Caryphyllaceae and Labiatae, as well as to mosses (Bryophita), have been the most extensively studied. In Sweden, three sulfide deposits containing copper were found with the help of "cupriferous" mosses. Several copper deposits were found in Zambia with the help of the plant *Osimum nomblei*. Some plants are local indicators of tin, nickel, cobalt, uranium, and rare metals.

Another group of geobotanical features of plants associated with increased contents of certain trace elements includes: variations in the shape and coloring of flowers and leaves; the character of plant pubescence; and the density of stems, branches, and foliage. The high contents of some elements may even cause various diseases in plants which are expressed as deformities. In some cases, geochemical anomalies of certain trace elements may be partly, or totally, devoid of vegetation. Table 55 lists the nature of the changes which have been observed in plants in connection with increased contents of trace elements in soils.

Table 55. Changes in plants due to increased contents of some trace elements in soils.

Element	Character of changes
U, Th, Ra	When present in small amounts, cause acceleration of growth in plants. High concentrations, lead to the appearance of deformities in vegetative shoots, dwarfing, dark-colored or blanched leaves.
Fluorine (topaz greisens)	Premature yellowing and falling of leaves.
B	Slow growth and ripening of seeds, dwarfing, procumbent forms. Dark-green leaves, deustate at edges. High concentrations in the soil causes total or partial disappearance of vegetation.
Mg	Reddening of stems and leaf-stalks, coiling and dying of leaf edges.
Cr	Yellowing of leaves, in some cases thinning of vegetation until its total disappearance.
Cu	Blanching of leaves, necrosis of their tips, reddening of stems, appearance of procumbent, degenerating forms. In some cases total disappearance of vegetation.
Ni	Degeneration and disappearance of some forms, appearance of white spots on leaves, deformities, reduction of corollar petals.
Co	Appearance of white spots on leaves.
Pb	Thinning of vegetation, appearance of suppressed forms, development of abnormal forms in flowers.
Zn	Chlorosis of leaves and dying of their tips, appearance of blanched underdeveloped dwarfed forms.
Nb	Appearance of white deposits on the blades and leaves of some types of plants.
Be	Deformed shoots in young individuals of pine.
Rare earths	Sharp increase in the size of leaves in some wood species.

Note: This table is based on data from Cannon (1960), Grabovskaya (1965) and other investigators.

Application of the Biogeochemical Method in Mineral Exploration

The most important advantage of the biogeochemical (phytogeochemical) method of mineral exploration is the increased depth range which can be attained as a result of the development of the plant root system. However, the biogeochemical method is more labor-consuming and more expensive than other methods of geochemical exploration. This is because of the specific requirement for the sampling and processing of phytogeochemical samples. (It is also desirable to have a geobotanist or botanist on the exploration team during biogeochemical exploration.) Therefore, the biogeochemical method may be used most effectively in cases where the less expensive lithogeochemical methods cannot be used for some reason or other.

Such conditions usually arise within specific landscapes. These conditions are characterized by: (1) the development of secondary lithogeochemical dispersion halos which are buried or leached at the surface; (2) the presence of a cover of allochthonous (transported) deposits 3 - 5 to 10 - 20 m thick; or (3) a zone of vigorous leaching of the trace elements from the upper soil horizons. It is, for example, appropriate to use the biogeochemical method in desert and semidesert landscapes covered by windblown sands up to 30 m in thickness. Plants growing in distinctly arid regions usually have deep root systems reaching depths down to 25 m. They include *Acacia raddiana, Acacia ehrenbergiana, Haloxylon aphyllum, Alhagi pseudoalhagi, Astragalus aureus, Medicago sativa,* etc.

Sampling of tree foliage is sometimes effective under conditions characterized by flat covered landscapes with allochthonous deposits reaching 3 to 15 m in thickness. The depth range of the biogeochemical method usually exceeds, to a certain extent, the depth to which the roots penetrate, owing to the capillary ascent of ground waters, connection between the root system and interstitial waters, etc.

The biogeochemical method is effective in marshes and peatbogs, where soil sampling usually involves difficulties. As noted by Grabovskaya (1965) biogeochemical halos under these conditions outline ore bodies rather distinctly, whereas soil halos are usually very vague. Difficulties in lithogeochemical sampling usually also arise in areas with large-boulder eluvium, diluvium, or moraines, where unconsolidated soil materials cave in between large fragments, thus preventing the use of soil sampling. The sampling of plants solves the problem of geochemical exploration within such areas. Finally, the biogeochemical method may be used in tropical and subtropical regions where soils and upper horizons of the weathering crust are subject to vigorous leaching. Such leaching is accompanied by the removal of the trace elements which are indicators of the presence of ore.

The sampling grid used during biogeochemical exploration depends on the scope of the exploration, as well as on the specific objectives of the surveys. Biogeochemical exploration is usually conducted in regions where previous small-scale regional surveys have revealed potential mineral reserves. In this stage of exploration (scale 1:50,000 or larger), at least tentative information is available on the possible types of mineral deposits which may be present in the region being studied, and on the possible size of the related biogeochemical halos. These data form the basis for selecting an optimal sampling grid. If the probability of encountering minimum anomalous (threshold) or higher contents within a biogeochemical halo is no smaller than 0.5, then in order to obtain three anomalous contents within a geochemical set representing the halo, one should collect a maximum of 10 samples. If only one "hit" is required, then 4 samples will be sufficient. At higher (than 0.5) probabilities of encountering selected critical values of anomalous contents in the halo, the number of samples will decrease accordingly (see Appendix 1).

The sampling of vegetation, as well as other types of geochemical sampling, is conducted along profiles and traverses perpendicular to the prospective strike of the ore bodies or the ore-controlling structures.

Sites for biogeochemical samples are selected so they represent 25 m^2 of terrain, in the case of forest vegetation, and 1 to 5 m^2 for steppes, semideserts, and deserts. This area is sufficiently representative to characterize the vegetation covering a region being studied, and it makes it possible to collect enough material for a sample. Sampling and processing (drying and ashing) of biogeochemical samples are performed in accordance with the current edition of *Instructions On Geochemical Exploration Techniques* (1965, pp. 127 to 137). It is recommended that a certain amount of experimental work be carried out in order to select the optimal conditions for conducting biogeochemical sampling.

In view of the above-mentioned biogeochemical features of trace element accumulation by plants, a particular species or several species of plants may be selected. One should base his choice on the abundance of the given species (the selected plant must occur quite uniformly within the area being sampled), on the depth to which the root system reaches (the deeper the better), as well as on the ability of the plant to accumulate indicator trace elements (this problem may best be solved during experimental studies). Plants known as concentrators of indicator elements typical of a particular type of ore mineralization, are usually not sampled. This is because of the complex nature of the variations within them of the contents of the element being concentrated. Because of this feature, it is frequently not possible to correlate the trace elements in such plants with their distribution in the soil. If it is necessary to sample different species of plants in different areas of the region being studied, because of variations in plant associations, it is necessary to evaluate the parameters of the biogeochemical background for each particular species. The features of the trace element accumulation by a particular species are also studied.

Different organs of plants concentrate trace elements differently. For this reason, a specific part of a plant is sampled (leaves, stems, branches, etc.). Plants of approximately equal age must be sampled. In view of the seasonal variability in the trace element contents of plants, it is recommended that sampling be restricted to as short periods of time as is possible (no more than two or three weeks). If wood species are present in a plant assemblage, preference should be given to them, rather than to grasses or brushwood. In most cases, leaves, coniferous needles from branches of trees or brushwood of the same age, as well as two- or three-year old branches with leaves or coniferous needles, are sampled. The age of trees is determined, approximately, on the basis of the thickness of the trunk and on its height. Representative average data will most likely be obtained if samples are taken uniformly from all sides of the tree or brushwood. Stems and leaves are usually also collected when grasses are sampled.

The size of a biogeochemical sample depends on the amount of ash required for an emission spectrographic analysis (about 30 mg). The weight of the original sample collected must be approximately 20 g. For more complicated analyses, the weight of the sample may have to be increased to 100 - 400 g.

Results from biogeochemical surveys are presented as maps or as graphs which show the biogeochemical (phytogeochemical) anomalies, along with corresponding cross sections showing the sampling data (Figs. 85 and 86). In addition to monoelement maps, it is

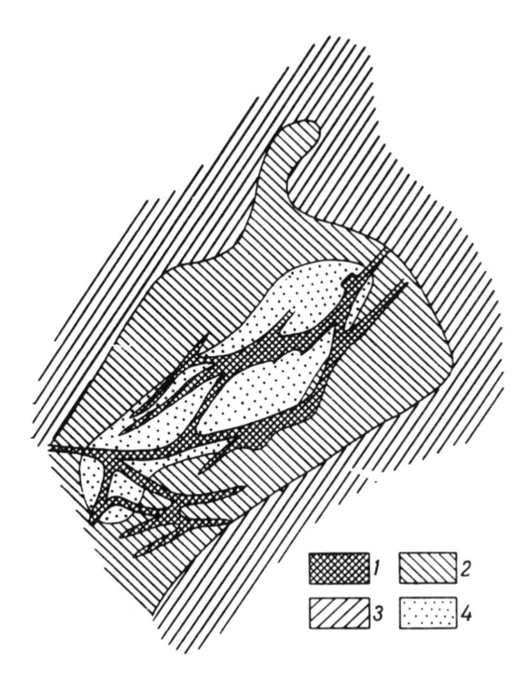

Fig. 85. Biogeochemical dispersion halos of beryllium in a region with rare-metal aplogranites (after Grabovskaya, 1965).
1 - ore body; *2* to *4* - content of Be in ash, % (*2* - 0.0001 to 0.0003; *3* - 0.0004 to 0.0009; *4* - 0.0009 to 0.0019).

possible to prepare maps of typical indicator ratios between elements with similar properties, whose ratio in the soils and underlying rocks changes abruptly as one passes from the geochemical background to the halo of a mineral deposit. The ratio Zn/Cu is an example of just such a ratio which may be used in the search for copper deposits; Zn/Sn, Zr/Sn, and V/Sn may be used in exploration for tin deposits; Mg/Li, for rare-metal deposits; etc.

Evaluation of any biogeochemical anomaly which may be detected reduces to ascertaining the relationship between its origin and the geochemical halo of the deposit, or with the phenomena of the local accumulation of particular trace elements in a specific type of landscape. False biogeochemical halos may be due to abrupt changes in the composition of the underlying rocks (for example, an ultrabasic stock among acid rocks), sharp changes in the pH of the soils, as well as to some other factors.

Biogeochemical anomalies are verified, depending on the particular situation, with the help of any one of several geophysical methods, and subsequently by drilling to expose an ore body or its lithogeochemical (primary or secondary) halo.

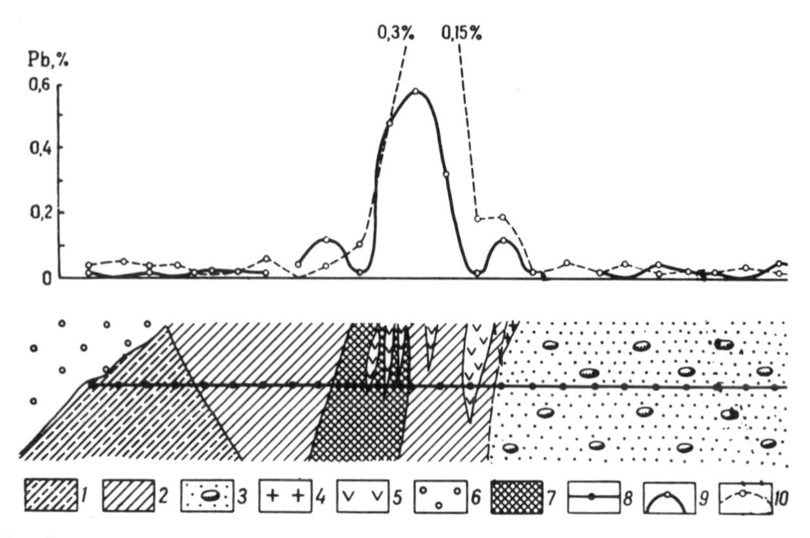

Fig. 86. Biogeochemical profile across a polymetallic metal deposit (After Grabovskaya, 1965).
1 - sandy-clayey shales and hornfels; *2* - bleached sandy-clayey shales; *3* - sandstones and conglomerates; *4* - granite porphyries; *5* - lamprophyres; *6* - recent alluvial deposits; *7* - mineralized tectonic zones; *8* - sites for the sampling of birch; *9* - element contents in leaves; *10* - element contents in branches.

The Use of Geobotanical Halos in Mineral Exploration

Geobotanical halos have great potential use in mineral exploration in view of the growing role of airborne geological and remote sensing methods. The ability to obtain various types of colored photographs of the vegetation of the Earth (with the help of filters) from aircrafts and man-made satellites, makes it possible to record the minutest changes in the color of vegetation, including shades indistinguishable to the naked eye. As a result, geobotanical color anomalies may be detected, including those due to increased concentrations of ore elements, and their associated elements, in soils and in plants.

The U.S. Geological Survey carried out experiments on the comparison between the reflection spectra of some species of trees *(Picea rubens* and *Abies balsamea)* which grew within a geochemical background, as well as within rather strong soil anomalies caused by copper and molybdenum. A difference in the infrared region of the spectrum was detected in the reflection spectra of *Picea rubens,* whereas in the case of *Abies balsamea,* the reflection spectra proved to be essentially different in all regions.

There are reports on the successful utilization of airborne geobotanical methods in the exploration for copper in humid, almost impassable jungles. In these areas, promising porphyry copper occurrences are reported to have been found with the help of specific associations of trees, which apparently does not depend on any other landscape factors (including elevation of the region).

During surface exploration in covered or semicovered regions, geobotanical criteria are used on various scales, starting from the reconnaissance stage, up to the detailed stage of exploration. Criteria to be considered, which were summarized by Polikarpochkin (1968) in *Instructions on Geochemical Exploration Techniques* (1965, p. 132), include the following:

(a) changes in the plant association, with deviations of the typical species composition, which cannot be explained by variations in the environment;

(b) appearance of specific indicator plants;

(c) changes in the external appearance of plants (the shape of leaves, flowers, etc., which are generally not characteristic of the given plant; unusual coloring; unusually luxuriant development; etc.);

(d) deviations in the rhythm of plant development (premature or, on the contrary, late flowering, or late falling of leaves, etc.);

(e) indications of the suppression, or the unaccountable absence, of vegetation.

In regions in which the effectiveness of geobotanical criteria have been established, systematic geobotanical surveys are conducted during mineral exploration. This amounts to preparing special geobotanic maps showing areas where more detailed exploration is warranted by means of a number of other geochemical and geophysical methods.

9

Atmochemical Halos from Mineral Deposits

General Information

It has long been known that gaseous halos exist in the atmosphere above various types of mineral deposits. However, their compositions and structural features are not yet adequately understood. This is due, primarily, to the absence of effective methods which might help in measuring the extremely low natural concentrations of the majority of the gaseous components found in atmogeochemical halos which originate from mineral deposits. [The terms "atmochemical" and "atmogeochemical" are used interchangeably in the Russian literature. Editor]

Gas surveys which detect volatile hydrocarbons and helium have now become an integral part of geological exploration for oil and natural gas. Gaseous halos of radon, which are present in the atmosphere and in the soil gases which overly volcanic, metamorphic, or sedimentary rocks containing increased concentrations of uranium, have been widely used in the search for uranium deposits since the early 1940's. At that time, the use of other gaseous components of atmochemical halos for the purposes of exploration had just been started, and in most cases, they are still in the experimental stages. About 43,000 atmochemical samples were collected in the entire world in 1971 (this number does not include gas samples collected during emanation [radon] surveys in the exploration for uranium deposits).

Atmochemical halos from mineral deposits include all components present in the gaseous form which may originate from the ore in mineral deposits, or in the products of supergene destruction of these deposits. The components may be present in liquids, as well as in solids. Under the atmospheric conditions existing at the surface of the Earth's crust, they exhibit high vapor pressures and this promotes their transition to the gaseous state at the temperatures prevailing in the lower layer of the atmosphere (Table 56).

Table 56. Composition of atmochemical halos from mineral deposits.

Source of gaseous components in halos	Composition of gases
Gaseous components of rocks	Br, Cl, F, CO_2
Gaseous products of radioactive decay	Rn
Gaseous products of supergene reactions between ore and gangue minerals	SO_2, CO_2, H_2S, F
Solid and liquid components of rocks and ores with high volatility	Hg, I

Below, data are given on the background contents of possible indicator gases in the atmosphere, in micrograms per m³ (from McCarthy, 1972):

Hg	SO_2	H_2S	F	Cl	Br	I
0, 02	2	3 to 30	<0, 05	1 to 5	2	0, 1

(The data for mercury seem to be high. Background contents of mercury in air do not generally exceed 0.01 micrograms per m³.)

At present, the gaseous halos of mercury associated with mercury deposits proper, as well as with deposits of sulfides containing mercury in the form of extremely minute impurities, have been the most intensively studied. Mercury vapors were first detected by Sergeev in 1961 over ore zones of the cinnabar-containing Khaidarkan deposit.

Comprehensive studies carried out by Fursov (1970) on some mercury deposits of the USSR, by Karasik and Bol'shakov (1965) in the region of the Nikitovsky cinnabar deposit, as well as by Khairetdinov (1971) on a mercury deposit in Tuva, suggest that the average background concentration of mercury in the soil gas ranges between 100 and 200 nanogram per m³ (1 nanogram equals $1 \cdot 10^{-9}$ grams), and it increases to thousands, tens of thousands, or even hundreds of thousands of nanograms per cubic meter of air over cinnabar deposits.

Fursov (1970) calculated that the amount of mercury vapor in equilibrium in soil gas, if the soil contains 80 to 200 ppm of mercury, depending on the temperature, ranges from 100 (at 2° to 10°C) to 500 nanograms per m³ (at 20°C). Mercury vapors may infiltrate up to dozens of meters through porous unconsolidated formations. Gaseous dispersion halos of mercury were detected in Canada in a moraine more than 20 m thick overlying a deposit of polymetallic sulfide ores. Sampling of soil gas on a 30 x 60 meter grid in the region of the Cortez mine (Nevada) made it possible to delineate a deposit of disseminated gold ore overlain by a gravel layer which was up to 30 meters thick (Gott et al. 1969). This was done using data on the content of the mercury vapors.

Gas surveys for mercury in regions covered with overburden may help in studying the fracture systems in the area being mapped. Fursov et al. (1968), who studied the effects of the Tashkent earthquake, noted that the air pumped from boreholes drilled over hidden fractures contained 15 times as much mercury as the soil gas sampled away from the fracture zones. Khairetdinov (1971) also detected a considerable increase in the content of mercury vapors in the soil gas over active seismic zones. He emphasized that this characteristic of atmochemical mercury anomalies must be taken into account during the interpretation of results obtained from geochemical gas surveys.

The concentration of mercury in atmospheric air decreases sharply with height. For example, based on data from the U.S. Geological Survey, the concentration of mercury vapor in the air near the Ord mine (Arizona), reached 20,000 nanograms per m³ directly at ground level. At an altitude of 120 meters the content of mercury decreased to 120 nanograms per m³. This anomaly showed a distinct characteristic dependence on the wind direction. On the side coinciding with the direction of the wind, the anomaly could be followed for 30 meters away from the mine, whereas on the other side, the background contents of mercury in the air were detected at a distance of only 10 meters from the source of the mercury vapors. At lower contents of mercury in the soil gas, the decrease in the content of this element on passing from the soil to the atmosphere may be even stronger.

For example, 2000 nanograms per m^3 of mercury were found in the soil gas in the region of the Cobalt mine (Ontario, Canada). At an altitude of 1 meter above the ground, the content was just 10 to 12 nanograms per m^3 (McCarthy, 1972). The use of atmogeochemical anomalies based on mercury above the soil as exploration criteria, involves difficulties in interpretation due to the effect of the weather and the related phenomena. Experiments carried out during the 1971 field studies at the Pershing mine (Nevada) described by McCarthy (1972) are typical. A study of the air samples collected at altitudes of up to 120 meters above the soil in the region of this mine showed that the mercury content is quite variable at different altitudes, depending on the atmospheric conditions. When the mercury content at the Earth's surface was 400 nanograms per m^3, it reached 350 nanograms per m^3 at 120 meters height, but only occasionally. Following rain, the concentration of mercury in the air decreased appreciably. In contrast to the anomalies revealed in the soil gas, the atmogeochemical anomalies over sulfide deposits which do not contain high concentrations of mercury, are frequently very weak (up to 10 to 20 nanograms per m^3). This suggests that this particular method is not very effective during detailed exploration.

At the same time, in the case of reconnaissance geochemical surveys aimed at detecting promising regions to be covered by more detailed studies in the future, the method based on the direct determination of mercury from aircraft, helicopter or truck, deserves very serious consideration. Positive results obtained with the help of helicopters in the search for mercury anomalies in the air have been described by Fursov (1970). The U.S. Geological Survey has been developing airborne atmochemical exploration methods for mercury since 1965. Successful experiments on detecting mercury anomalies in the air from a moving truck are described by McCarthy (1972). In Australia, mercury anomalies in the air were detected over copper, nickel, and zinc deposits. At background levels, the content of mercury in the air is equal to 1 nanogram per m^3, whereas the content of mercury in the air over occurrences of sulfide mineralization increased to 11 nanograms per m^3. The size of one atmochemical anomaly was several times larger than the dimensions of the deposit in plan. Gaseous halos of mercury over deposits containing native silver and tetrahedrite were detected in Ontario as a result of atmochemical surveys carried out from a moving truck. Increased amounts of mercury in the air were also detected in the winter, when the temperatures had decreased to minus 10°C.

Gaseous SO_2 halos are typical of mineral deposits containing sulfides. Sulfur dioxide, which forms during the oxidation of sulfur-containing minerals, accumulates in the soil gas over mineral deposits in quantities ranging from 25 to 50 parts per billion, whereas background concentrations range from 2 to 10 ppb. Atmochemical halos of SO_2 have been detected over very different types of mineral deposits which contain sulfides in the ores, and under various climatic conditions. SO_2 migrates through thick unconsolidated deposits.

There is much less information on the presence of H_2S in gaseous halos originating from mineral deposits which contain sulfides. Anomalous quantities of hydrogen sulfide were specifically found by Frederikson et al. (1971) in the air over gold ore deposits in the Cripple Creek region of Colorado. They exceeded the background content of this gas in the air, at some distance from the deposits, by a factor of almost two.

The oxidation of sulfides and the formation of SO_2 during weathering, are first accompanied by the production of sulfurous, and then by sulfuric acid, in the surface solutions which leach altered minerals. If fluorite or carbonates are present in the ore, sulfuric acid reacts with them, and the formation of hydrofluoric acid or carbon dioxide result, respectively. The presence of both fluorite and sulfides in an ore may cause anomalous concentrations of fluorine in the air above the ore zones. Accordingly, increased concentrations of carbonic acid must indicate accumulations of sulfide-carbonate ores.

Gaseous-liquid (fluid) inclusions in the gangue and in the ore minerals, which contain soluble salts of chlorides and fluorides, may serve as additional sources of fluorine and chlorine in the air above ore deposits which are being weathered.

Atmogeochemical anomalies of fluorine and chlorine were detected over some greisen deposits, as well as over gold ore deposits in Colorado. Anomalous contents of the halogens were three times as large as the average value of the geochemical background for these elements in the region.

Very limited data on the distribution of CO_2 in atmogeochemical halos of ore deposits are available. The most interesting information on this subject was reported by Netreba et al. (1971) who detected high (more than 5%) concentrations of CO_2 in the soil gas over fault zones which control a mercury mineralization.

Very little information is available on the gaseous anomalies formed by bromine and iodine. McCarthy (1972), in his review of gaseous halos from ore deposits, presented data obtained from Barringer Research Ltd. (Rexdale, Ontario, Canada) on the presence of anomalous concentrations of bromine and iodine over known porphyry copper ore deposits in Arizona, as well as over petroliferous regions in California.

Sampling of Atmogeochemical Halos and Analysis of the Gases

Studies of atmochemical halos include the sampling of soil gas, or of the air at a certain height above the ground. In each particular case, the method of sampling is closely associated with the technique subsequently used to analyze the gaseous components in the halo. The low contents of indicator gases in halos from mineral deposits complicate the analytical procedure and is the main factor which inhibits the development of the exploration method being discussed.

The most useful analytical procedures are the direct methods of gas determination by means of highly sensitive instruments which enable measurement to be made of the background contents of the individual gases in the air. These instruments include: (1) atomic absorption instruments for the determination of trace quantities of mercury, which are used in the Soviet Union, and which are also manufactured by some companies in the United States and Canada (Barringer, Scintrex, etc.); (2) mass spectrometers specifically designed for the direct measurements of minute quantities of chlorine, fluorine, mercury, and hydrogen sulfide in the air (Frederikson et al., 1971); and, (3) the so-called correlation spectrometer which measures extremely small amounts of sulfur-containing gases, bromine, and iodine in the air. The use of instruments of this type which are mounted on airplanes, helicopters, or trucks, makes it possible to continuously sample the air.

When less sensitive analytical methods are used, it is necessary to sample large amounts of air and to concentrate the required gas by appropriate methods. For this purpose, the air is passed through a solid filter within a chamber, or through a liquid in which the required gaseous component is accumulated.

Gold or silver, which amalgamate mercury vapors, may be used as concentrators of mercury; sulfuric gases, in turn, are quantitatively concentrated when air is passed through the water; etc. Gas chromatography, selective ionic electrodes, and other methods are used for the subsequent analyses of the concentrated gases. The mercury concentrated in a solid (e.g., on gold or silver), or in an aqueous solution, may be analyzed by means of various methods, including colorimetric and spectrographic analyses.

The sampling of air above the surface of the ground is usually done by directly introducing air into a highly sensitive instrument mounted on a moving vehicle (airplane, helicopter, or truck), or into a special chamber equipped with a gas collector. Increased concentrations of mercury and iodine vapors in the air can be determined optically from aircraft of various types, including artificial Earth satellites. Preliminary experiments carried out in the stratosphere with the help of sounding balloons by Barringer Research Ltd. have yielded positive results.

Soil gas sampling is done by either the active or passive pumping of the gases from the soil. Active pumping of the soil gas is carried out with the help of special pumps from small boreholes or directly from the surface of the soil. The passive method is widely employed by the U.S. Geological Survey. For this purpose, a plastic hemisphere is used, in the upper part of which there is an opening with a silver mesh. The hemisphere is placed on the surface after the upper soil layer (1 to 2 cm in thickness) is removed. After two hours, the silver filter is removed and sent for analysis (McCarthy, 1972).

Sampling of air above the soil, at different heights above the surface, from a moving truck, airplane, or helicopter, accompanied by a continuous recording of the variations in the quantities of the gases being analyzed in the air, is very effective during reconnaissance surveys. This is especially so in poorly studied regions which are well-removed from industries which may cause contamination of the atmosphere.

Geochemical sampling of soil gas may be used during detailed exploration, especially if there are no (or few) exposures. This type of sampling makes it possible to detect atmochemical halos from mineral deposits hidden below alluvium.

It should be stressed that progress in the field of atmochemical exploration depends, to a large extent, on further improvements in analytical instrumentation.

10

The Application of Geochemical Methods Depending on the Scope and Objectives of Exploration

Each stage of geological exploration is characterized by a combination of specific problems which may be solved with the help of geochemical methods, as well as by the combined application of these methods with other techniques, particularly geophysical methods. The maximum effectiveness of geochemical exploration methods may be achieved only if they are used in a logical sequence.

Small-Scale (Regional) Geological Surveys and Exploration (scale 1:200,000 to 1:100,000).

The main objectives in this stage of geochemical studies are very clearly defined in the "Instructions On Geochemical Exploration Techniques" (1965). They are:

(1) Concurrent with studies of the features characteristic of the regional geochemical specialization of igneous, metamorphic, and sedimentary rocks, attempts are made to detect potential ore-bearing geological complexes, and to obtain geochemical data for the solution of several geological problems associated with magmatism, metamorphism, lithology, conditions of sedimentation, the level of the erosion surface in the region, etc. Such studies are made along with geological surveys.

(2) Reconnaissance lithogeochemical surveys are conducted, simultaneously, with geological surveys and using the same scale, based on secondary dispersion trains in the present-day drainage system. It is sometimes recommended that these studies be carried out in combination with hydrogeochemical surveys.

The main objective of regional geochemical prediction and reconnaissance exploration within the given scale, is the detection and delineation of regions to be covered by more detailed exploration for particular minerals. In some cases, reconnaissance surveys using bottom sediments in drainage basins in poorly studied regions may precede geological surveys.

In uncovered regions with well developed drainage systems in which the bedrock is exposed, at least in part, preference should be given to the stream sediment method. Uncovered regions include areas where the ore-bearing bedrock is exposed at the surface, or is overlain by thin eluvial-diluvial deposits, in which exposed secondary dispersion halos form as a result of the destruction of ore bodies or of their secondary halos. These conditions are mainly typical of mountainous regions with dissected relief, as well as of flat areas where the bedrock is overlain by relatively thin allochthonous or autochthonous deposits. In uncovered regions with little exposure of the bedrock, geochemical surveys using stream sediments in drainage basins are the only method which furnishes systematic geochemical information on the region being studied.

This method is much more informative than the more labor-consuming heavy mineral concentrate method, because a complex of the indicator elements in the dispersion trains is much more representative than the mineral association in the heavy mineral concentrates. It can be shown from examples in the mountainous taiga landscapes of Transbaikalia, that the stream sediment method makes it possible to detect mineral deposits and ore showings which are not detected by heavy mineral sampling, even in the case of rare-metal deposits (Bezverkhnii, 1968). Combined use of both these methods is recommended in the search for gold and platinum.

Sampling of exposed bedrock in uncovered regions has an auxiliary role as a source of additional geochemical information on the patterns of distribution of the indicator elements in the rock types, and is of particular interest from the exploration or geological points of view. In any case, all potentially ore-bearing formations found along the traverses should be subjected to geochemical sampling; these include zones of mineralization, fault zones, altered rocks, veins, etc. These formations may prove to be interesting ore showings or primary geochemical halos of hidden promising mineralization. On the whole, however, small-scale (regional) geochemical mapping using bedrock furnishes valuable information on the geochemical specialization of geological complexes. Nevertheless, this method, when used in small-scale surveys, cannot be a means for detecting either primary or secondary geochemical halos of mineral deposits in the area covered by the mapping. At the same time, geochemical exploration using stream sediments enables one to judge both the geochemical specialization of various types of rocks, and the presence in the exploration region of geochemical anomalies which may turn out to be primary or secondary halos from mineral deposits.

Geochemical mapping using bedrock during reconnaissance exploration in uncovered regions acquires a more important, and sometimes even a major role when the sampling of stream sediments is difficult for some reasons. Such conditions usually arise in arid regions, where the proper combination of stream sediment sampling and geochemical mapping using bedrock is especially effective. Sampling of stream sediments may involve difficulties in mountainous regions with high stream velocities. This must be remembered during the organization of small-scale (regional) geochemical exploration in highly dissected mountainous regions.

In flat regions with poor exposures and with poorly developed stream systems, additional information can be obtained with the help of hydrogeochemical sampling from wells and springs, in combination with the sampling of bottom sediments.

In covered regions, where the bedrock is overlain by thick series of allochthonous sediments and where rivers do not expose the bedrock, the stream sediment method is, in most cases, ineffective.

The only other more or less adequate source of geochemical information during small-scale surveys in covered regions are cores obtained from mapping boreholes drilled on a large grid during geological surveys in the area. Consequently, geochemical mapping using bedrock in covered regions is the main method which makes it possible to evaluate prospects. It is based on the features of the geochemical specialization of the rocks obtained from boreholes. It thus enables one to direct exploration in the most logical way.

The possibility of reconnaissance evaluation of the ore potential in covered regions, using the results of geochemical studies of core samples from boreholes, significantly increases the effectiveness of exploration in covered regions and sharply reduces the amount of drilling in subsequent stages of exploration. Exploration for copper-nickel ores associated with massifs containing ultrabasic rocks within covered ancient platformal areas, may serve as a good example. Massifs containing ultrabasic-basic rocks buried beneath thick series of allochthonous deposits have been successfully outlined in the course of geophysical surveys. However, the presence of ore in each particular massif may be evaluated only with the help of systematic deep drilling to the covered massif, unless geochemical methods are applied. The possibility of evaluating the ore potential in the massif, based on data from geochemical studies of core samples from isolated boreholes, makes it possible to reduce the amount of exploratory drilling and to concentrate drilling operations on potentially ore-bearing massifs and complexes.

The samples collected in the process of small-scale (regional) geochemical exploration from stream sediments in drainage basins, as well as from the bedrock, are analyzed for a wide range of chemical elements after the necessary processing. (See, "Instructions On Geochemical Exploration Techniques" (1965), for the requirements on the processing of geochemical samples.)

It is necessary to analyze for a large number of elements because of the possibility that within the area being studied, ore deposits with different compositions, and multi-element halos which accompany many mineral deposits, may have developed. For this reason, as well as because of the large quantity of the geochemical samples being collected, the least expensive analytical procedures are required; the semi-quantitative emission spectrograph method is by far the best.

The list of elements for which geochemical samples must be analyzed depends primarily on the metallogenetic features of the region being investigated and must, by all means, include the major indicator elements which are typomorphous for the ore formations in the region. The samples must also be analyzed for indicator elements of other possible types of ore formations which have not previously been detected in the region.

Kvyatkovskii et al. (1972), who studied the complex of indicator elements recommended for geochemical exploration, pointed out that at least one of the major indicator elements listed below is always present in sufficiently large quantities in ores of practically any endogenic mineral deposit. These indicator elements are: beryllium, fluorine, sulfur, vanadium, chromium, nickel, copper, arsenic, molybdenum, tin, mercury, lead, uranium.

Table 57 gives a more complete list of the indicator elements. This list includes mainly the elements which can be determined by the semi-quantitative emission spectrographic method. Only a small number of elements require the more labor-consuming special analytical methods. These elements include the rare alkalis and fluorine which are indicators of rare-metal deposits, as well as gold, uranium, and mercury which are direct indicators of gold, mercury, and uranium deposits. Analysis of all geochemical samples for the elements which can be detected by the emission spectrographic method is indispensable. On the other hand, analyses for rare alkalis, fluorine, uranium, mercury, gold, sulfur, and

Table 57. Indicator elements of mineral deposits with different compositions.

Type of deposit	Major indicator elements identifiable by the semi-quantitative emission spectrographic method	Additional indicator elements identifiable by special methods
Copper-nickel in basic and ultrabasic rocks	Cu, Ni, Co, Zn	S
Rare-metal aplogranites and pegmatites	Be, Sn, Nb, Ta, W	Li, Cs, Rb, F
Tin-tungsten of quartz-greisen origin	Be, Sn, W, Mo, Bi	Li, Rb, F
Rare-metal carbonatites	Nb, Zr	U, P
Tungsten-molybdenum in skarns	W, Mo, Sn, Cu, Be, Bi	—
Bismuth in skarns	Bi, Cu, Pb, Zn	—
Tin-sulfide	Sn, Sb, Pb, Cu, Ag, Zn	—
Polymetallic	Pb, Zn, Cu, Ag, Ba	—
Gold ore	Sb, As, Ag	Au
Porphyry copper	Cu, Zn, Mo, W, As, Sb	—
Copper	Cu, Pb, Zn, Ag, Mo	—
Hydrothermal uranium	Pb, Mo, Zn, Ag, Cu	U
Antimony-mercury and mercury	As, Sb, Ba	Hg
Celestite	Sr	—
Phosphorites	—	P
General list	Be, Sn, Mo, W, Nb, Zr, Bi, Ta, Ni, Co, Cu, Pb, Zn, Ag, Sb, As, Ba, Sr	Li, Cs, Rb, Au, Hg, U, F, S, P

phosphorus can be performed only on samples from areas where the emission spectrographic analysis has established anomalous concentrations of other indicator elements in these types of mineral deposits (Table 57).

Geochemical surveys at the scale of 1:200,000 to 1:100,000 are conducted in combination with airborne and surface geophysical surveys at corresponding scales. This is particularly important for covered regions where core samples obtained from boreholes are the only source of geochemical information. Under these conditions, geophysical surveys must precede drilling performed for the purposes of geochemical bedrock mapping.

Geochemical mapping and exploration at scales from 1:200,000 to 1:100,000 result in geologic-geochemical prediction maps, which determine the further direction of exploration in the area. These maps show, first of all, areas which deserve more detailed study to detect probable areal geochemical anomalies. Indications for more detailed studies may be found in the distribution patterns of anomalous contents of the indicator elements in stream sediments.

In regions where geochemical mapping has been carried out with the help of the bedrock samples, the maps also show massifs and rock complexes which, in terms of geochemical specialization, are promising for a specific type of ore mineralization and, therefore, deserve more detailed exploration.

Experience accumulated throughout the world over the last two decades indicates that geochemical studies at a scale of 1:200,000 to 1:100,000 are the best for small-scale (regional) geological exploration.

Geological Mapping and Exploration at a Scale of 1:50,000

Geochemical exploration on this scale differs from the above in that, in this case, geochemical mapping based on stream sediments (or bedrock) is more detailed. In addition, areal geochemical surveys aimed at revealing primary and secondary geochemical halos around ore deposits, are carried out in promising regions.

The types of geochemical techniques used in these cases depends entirely on the landscape-geochemical conditions in the region being studied, and in particular, on the ratio of the exposed to the hidden area. Under certain conditions, when the effective use of lithogeochemical methods at this stage of exploration have been ruled out, it is possible to conduct biogeochemical surveys, as well as hydrogeochemical surveys by means of sampling subsurface waters in promising areas.

Such conditions sometimes arise in arid regions where the bedrock is overlain by sands transported over long distances, as well as in moorlands and woodlands, where the thicknesses of the sediments and soils are such that the root systems of trees can reach the bedrock, or at least to the zone of weathering.

For example, acacia in North African deserts, as well as saxaul and some other types of desert plants in the arid regions of Central Asia, can be used effectively where sampling of stream sediments involves difficulties. Biogeochemical methods can be used successfully in covered and boggy regions within the Ukrainian Crystalline Shield, in Eastern Transbaikalia, and in other regions.

Hydrogeochemical sampling from wells and springs is, in some cases, the only source of geochemical information under the conditions characteristic of flat arid and semi-arid regions overlain by unconsolidated allochthonous deposits as much as 10 - 15 meters in thickness.

More detailed sampling of dispersion trains which contain anomalous amounts of indicator elements in stream sediments, helps in delineating areas which are the sources of the indicator elements. Areal surveys, which include sampling on a grid corresponding to the scope of the exploration, are conducted within the promising regions thus delineated. The objective of areal sampling, in this case, amounts to detecting primary and secondary geochemical anomalies (depending on the extent of exposure) so that they can be studied in greater detail in the future. In order to ensure the greatest value from the isolated samples which are collected at this scale of exploration (which is a rather low sampling density), it is necessary to use analytical procedures, as well as sample processing, which "enhance" weak anomalies. These procedures should primarily include the composite (additive and multiplicative) halos which were discussed in great detail earlier in this book. The use of these methods is especially important in preparing the geochemical framework for prediction maps of the ore-bearing basement in covered regions, because monoelement anomalies, which are indistinct and limited in size, are practically useless for these purposes.

The method of selectively analyzing different fractions, in particular the heavy fractions, may be used successfully to expand the effective range of borehole mapping in covered regions of a particular metallogenetic type (regions promising for sulfides, tin, or other types of mineral deposits).

Practice has shown that geochemical sampling of potentially ore-bearing geological formations, carried out in the course of geochemical surveys in uncovered regions at the scale of 1:50,000, makes it possible to detect the largest geochemical anomalies and to establish their nature with a certain degree of reliability (their association with ore formations of various types), the approximate level of the erosion surface relative to possible ore deposits, etc.

It is sometimes helpful to conduct exploration in greater detail (at a scale of 1:25,000 to 1:10,000) within individual promising areas identified during regional geochemical exploration. The geochemical methods in this stage of exploration are used in combination with regional geophysical surveys.

Geochemical mapping at a scale of 1:50,000 makes it possible to prepare corresponding geologic-geochemical prediction maps which show primary and secondary (including biogeochemical and hydrogeochemical) geochemical anomalies delineated in the course of exploration, as well as massifs of igneous rocks, and metamorphic and sedimentary geological complexes which may be promising in terms of geochemical specialization.

Detailed Surveys at a Scale of 1:25,000 - 1:10,000 or Larger

Detailed geochemical surveys are planned on the basis of the results obtained from smaller-scale geochemical surveys and are conducted in areas which have been recognized as promising in an earlier stage of exploration. Geochemical exploration at a scale of 1:25,000 - 1:10,000 or larger may be recommended for systematic studies of previously detected promising areas in order to delineate, and give tentative evaluation to, al geochemical anomalies present in the region. Lithogeochemical methods of exploration, using primary and secondary geochemical halos, are applied in combination with biogeochemical and hydrogeochemical methods, depending on the landscape-geochemical conditions in the region. The principal method used in the detailed stage of geochemical exploration is the systematic sampling of the surface on a grid whose parameters are chosen in such a way as to provide a complete understanding of the extent of the geochemical features of the anomalies developed within the area covered by the mapping.

In areas where the bedrock is sufficiently exposed, surveys using primary geochemical halos are effective.

The following principal problems may be solved, consecutively, by the interpretation of geochemical anomalies detected in bedrock.

Determination of the Formational Nature (Composition) of the Anomalies

This problem is, in most cases, solved by taking into account the metallogenetic features of the region in question (types of mineral deposits occurring in the region), as well as the results of mineralogical observations carried out during geochemical sampling or during subsequent study of the anomalies in the field (the presence of mineral associations which are typomorphous for particular types of mineral deposits, etc.). The method of analogy is also used; that is, comparison is made with anomalies revealed by halos of other mineral deposits in the region. If such mineral deposits are absent, the sequences of transverse zonality of the indicator elements may be used.

Evaluation of the Erosion Level of Geochemical Anomalies

An evaluation of the erosion level is an important factor in the interpretation of a geochemical anomaly. Specifically, we mean the level of the erosion surface of the anomaly relative to the ore body with which the given anomaly is associated. It is easy to see that, in the general case, this level will not coincide with the erosion surface level in the area being studied. The echelon arrangement of ore bodies in cross section may be responsible for cases where supraore geochemical halos are developed in an area with a maximum depth to the erosion surface (over blind mineralization), whereas in other areas halos formed by the elements below the ore bodies may be present (deep erosion surface of anomalies).

Vertical zonality of primary halos, discussed in detail earlier, is used for the evaluation of the level of the geochemical anomaly with respect to the erosion surface. The surveys usually detect complicated anomalies which have formed as a result of the coincidence in space of the halos which accompany, more or less spatially, separated ore bodies. The levels of the erosion surfaces of each particular body may differ sharply. Consequently, probable *elementary anomalies* (anomalies corresponding to individual ore bodies) must be identified within the general anomaly. After this, the level of the erosion surface of each particular anomaly must be evaluated. Elementary anomalies are usually readily identified based on the epicenters of the indicator-element anomalies, primarily the major components of the given type of mineralization. This differential approach to the interpretation of geochemical anomalies is necessary because halos formed below ore bodies, or at the level of root portions of practically totally eroded ore bodies, are often encountered in the immediate vicinity of blind or weakly eroded mineralization.

Discussion

The results of bedrock sampling in regions with limited exposures frequently misrepresent the intensity and other characteristic features of primary halos. Intensive destruction of the bedrock, and the formation of unconsolidated strata, take place primarily in those regions which are particularly tectonically active and which exhibit rock alteration, etc.; such regions contain potentially metalliferous geologic formations. In such cases, primary halos may be partly or completely overlain by unconsolidated deposits and overlooked if sampling is done from the bedrock only. Consequently, during exploration at the above scale in areas with only partial rock exposure, it is recommended that areal sampling of the eluvial-diluvial deposits be conducted first in order to reveal, delineate, and study the secondary dispersion halos. The so-called combined areal-geochemical sampling, when both unconsolidated and solid rocks are sampled within partly exposed areas, is incorrect. In these cases, areal sampling of the unconsolidated material may be accompanied by bedrock sampling along certain profiles only. Primary and secondary halos must be identified and compared taking into account the parameters of the background distribution of the indicator elements separately for the bedrock, and for the unconsolidated deposits, since the values of these parameters usually differ appreciably.

Systematic sampling from unconsolidated deposits using a definite grid is also a principal exploration method in poorly exposed or completely covered areas, where landscape-geochemical conditions suggest the presence of exposed residual or superimposed secondary dispersion halos. In this case, sampling is done in order to collect sufficient amounts of geochemical information to enable the reliable determination of the type and formational affiliation of any supergene geochemical anomalies. In the case of geochemical anomalies which are poorly developed at the surface or deeply buried, biogeochemical methods are used. These methods expand the depth range of the surveys. Hydrogeochemical sampling of subsurface waters is also employed. However, information obtained with the help of these techniques in the detailed exploration stage is, as a rule, far from sufficient. Therefore, the deep lithogeochemical survey is the principal method for the detailed geochemical investigation of covered promising areas revealed during earlier stages of exploration. When these investigations are under way, it is extremely important to use special procedures to gain as much information as possible from the boreholes. This is achieved by using the above-described method for summing the halos of the indicator elements.

The results of detailed geochemical surveys often form the basis for conducting preliminary drilling. It is, therefore, especially effective at this stage to use both geochemical and geophysical techniques in combination. Geophysical methods, which generally include various modifications of electrical prospecting, make it possible to determine more accurately the optimal parameters of the boreholes and mine workings, as well as to choose the most logical locations for the boreholes. The possibility of revealing structural features within covered promising areas with the help of geophysical methods must, by all means, be taken into account when geochemical and geophysical methods are used in combination for the purpose of detailed investigation of geochemical anomalies.

The detailed stage of geochemical exploration is completed after the geochemical anomalies which might be detected are checked. In the case of exposed primary halos, the recommendations for further prospecting are based on the data concerning their formational affiliation, and the interpretation of the geochemical zonality of the halo in order to evaluate the level of the erosion surface. The latter, which proceed from interpretation of the paragenetic features of the indicator elements in an endogenic geochemical halo, must also be evaluated for manifestations of ore revealed in the process of exploration. Examples which illustrate the effectiveness of this evaluation were discussed earlier in this book.

In the interpretation of supergene geochemical anomalies, the landscape-geochemical conditions in the region being studied must be analyzed very thoroughly. These conditions frequently play a leading role in the distribution and parageneses of the indicator elements in secondary dispersion halos. It may turn out, in complicated cases, that the features of the distribution of the indicator elements in their supergene dispersion field, do not make it possible to determine the origin of the geochemical anomalies with sufficient reliability. In such cases, the results of bedrock sampling along certain profiles (partly in exposed areas), or from artificial exposures (ditches, etc.), may be needed. It was previously mentioned that the criteria of vertical zonality of primary halos may be used in the interpretation of supergene geochemical anomalies only on the condition that a satisfactory correlation between primary and secondary dispersion halos of indicator elements was previously established. In this connection, studies of secondary anomalies in soils and in eluvial-diluvial deposits at this stage of exploration must be accompanied by the digging of shallow prospecting pits or boreholes in order to ascertain whether the given anomaly is a secondary dispersion halo of an ore deposit (or its endogenic halo), and to determine unambiguously its formational affiliation.

The data obtained from studies of secondary dispersion halos are further confirmed by sampling of the bedrock uncovered by mine workings or boreholes and by studies of endogenic geochemical halos. The latter is especially important if the anomaly detected in unconsolidated deposits is due to destruction of the endogenic halo of a hidden ore body.

Detailed geochemical surveys result in the preparation of geochemical maps of the halos at the corresponding scales. Cross sections of the halos along mine workings or boreholes where the anomaly is detected, are also prepared. This documentation, as well as the analyses of geochemical and geophysical data, are used as the basis for specific recommendations on mining operations, or for drilling to check the detected anomaly, and for exposing the ore body.

Drilling Stage of Mineral Exploration

Drilling is known to be extremely expensive. It is, therefore, essential to develop techniques which might increase its effectiveness. Recent studies have shown that one of the most promising ways to increase the effectiveness of drilling is the practical implementation of geochemical methods, primarily the technique of studying primary geochemical halos. The role of primary halos at this stage of geological prospecting increases owing to the fact that drilling operations open up ore bodies and create favorable conditions for: (1) a three-dimensional study of primary geochemical halos; (2) determining the features of their development in space; and (3) evaluating ore reserves in individual areas.

Geochemical methods can be used for the solution of various other problems in mineral exploration at this stage of exploration. Several are discussed below.

Evaluation of the Ore Potential in Metalliferous Zones Recommended for Detailed Exploration at Depth

Solution of this problem reduces to determining the level of the erosion surface in the ore-bearing zones being studied. In other words, it is necessary to find out which parts of these zones (upper, middle, or root) are exposed at the surface. Procedurally, the solution to this problem is essentially the same as the evaluation of the ore potential discussed in detail above. It is necessary to stress that considerable expenses are associated with detailed studies of these ore-bearing zones because the evaluation of the prospective ore reserves at depth must be highly reliable. It is, therefore, necessary to carry out detailed areal geochemical sampling of the bedrock in areas recommended for drilling.

Specific examples of the evaluation of ore potential at depth, using the results of bedrock sampling at the surface, were considered earlier in this book (see discussions on the Maidantal and Kamazak-Marazbulak areas in Figs. 58 to 60).

These, as well as other examples, show that geochemical sampling prior to detailed exploration may contribute greatly to the accurate evaluation of ore potential and may greatly reduce expenses, because they eliminate deeply eroded ore showings from expensive drilling.

Exploration for Blind Mineralization

As was discussed earlier, the echelon arrangement of ore bodies within metalliferous intervals means that the ore body (ore zone) which appears in an outcrop, and is of prime concern in the drilling stage, is not the only such ore body. It may be accompanied by blind bodies, whose successful detection and exploration, increases appreciably the effectiveness of exploration as a whole.

Geochemical sampling from mine workings and boreholes, drilled during exploration for previously known metalliferous zones, makes it possible to detect primary halos not only in the known zones being explored (to evaluate their ore potential at depth), but also halos due to blind ore bodies occurring next to, or below, the known deposits. To put it differently, geochemical sampling of mine workings and core samples from boreholes makes it possible to significantly increase the effective range of exploration.

The results of bedrock geochemical sampling at the surface of the Northern Kurusai II skarn-polymetallic mineral deposit are a typical example illustrating the above statement. An anomaly of the indicator elements, without any traces of ore mineralization, was revealed at the surface in this area alongside anomalies which indicate known outcrops from the metalliferous zones. This anomaly is located within the northwestern margin of the mineral deposit in question, and occurs in dolomites which are pure on the outcrop and do not contain any traces of ore mineralization or skarning. Increased contents of barium, antimony, silver, lead, and other elements were detected within the anomaly (Fig. 87). The extensive occurrence of indicator elements in the supraore cross sections of the

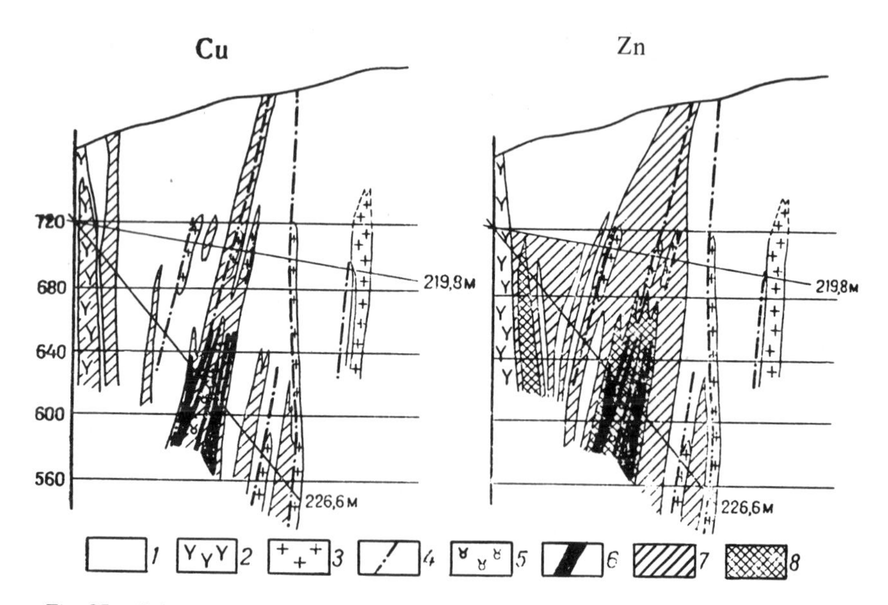

Fig. 87. Primary halos around ore bodies of the Northern Kurusai II mineral deposit. *1* - dolomites and limestones; *2* - diorites; *3* - quartz porphyries; *4* - faults; *5* - skarns; *6* - ore body; *7* and *8* - element contents in the halos, % (*7* - $5 \cdot 10^{-3}$ As and Sb; 0.01 to 0.1 Ba; 0.005 to 0.01 Pb; $1 \cdot 10^{-4}$ to $3 \cdot 10^{-4}$ Ag; 0.005 Cu; 0.005 to 0.01 Zn; *8* - more than 0.1 Ba; more than 0.01 Pb; more than $3 \cdot 10^{-4}$ Ag; more than 0.01 Zn).

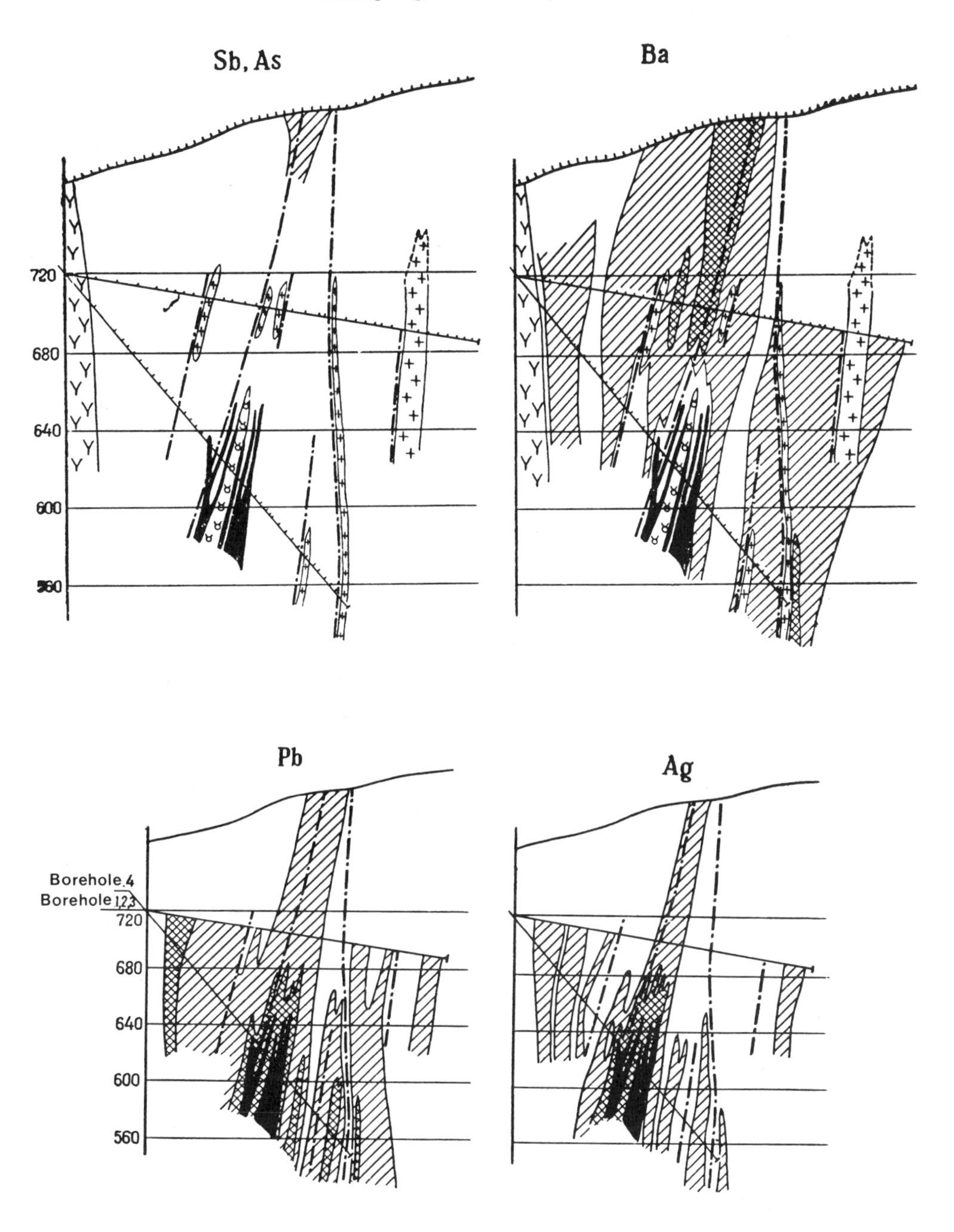

polymetallic mineralization halos (barium, arsenic, antimony, and silver), and the practical total absence of the anomalies formed by indicators below the ore bodies (copper and molybdenum), suggested that this anomaly was of a supraore nature. This made it possible to conclude that the area is promising for blind mineralization, and to recommend it for drilling. A predominance of lead in the suspected blind ore bodies was suggested by the distribution patterns of the indicator elements. This followed from the high lead-to-zinc ratio (the ratio of the average contents at the surface was 4). In order to test the anomaly, a borehole was drilled in a mine working nearest to the area to a depth of 100 meters. No ore bodies were revealed. After this, an additional two boreholes were drilled to the same depth. They did not give any positive results either. Geochemical core sampling from the boreholes revealed a pronounced expansion of the halos formed by lead, zinc, and silver, which are the major ore components, at depth. This suggested that the suspected ore mineralization occurred at even greater depths. A borehole drilled in accordance with this recommendation revealed a rich ore body of economic importance at a depth of 200 meters.

Another example taken from the experience of Boliden, a Swedish mining company, will be illustrated. Geochemical sampling from subsurface boreholes was performed at a skarn-polymetallic mineral deposit (Harpenberg) which is situated in metamorphosed rocks of Archean age. The characteristics of the development of primary geochemical halos around these ore bodies were described earlier. From Fig. 23, two anomalies represented by supraore halos of the given type of mineralization (strong halos formed by silver, lead, etc.) appear at a depth, despite the complete wedging out of the ore bodies. This conclusion is fully confirmed by a plot of the ratio of the linear productivities of the multiplicative halos versus depth. Fig. 88 shows that a monotonical and steady decrease in the indicator ratio continues to the level at which the known ore bodies wedge out, after which this ratio sharply increases. This phenomenon marks a new "spike" in the supraore-element halos. The conclusion was made that blind ore bodies lie at a depth. Deep drilling of one of the anomalies revealed a large blind ore body at a depth of 500 meters (see Fig. 88).

The above examples show that geochemical exploration for blind mineralization involves: (a) detection of primary halos from blind ore bodies; and (b) establishment of the supraore, or below ore, nature of these halos, i.e., the identification of the most promising geochemical anomalies represented by supraore geochemical halos. Solution of this

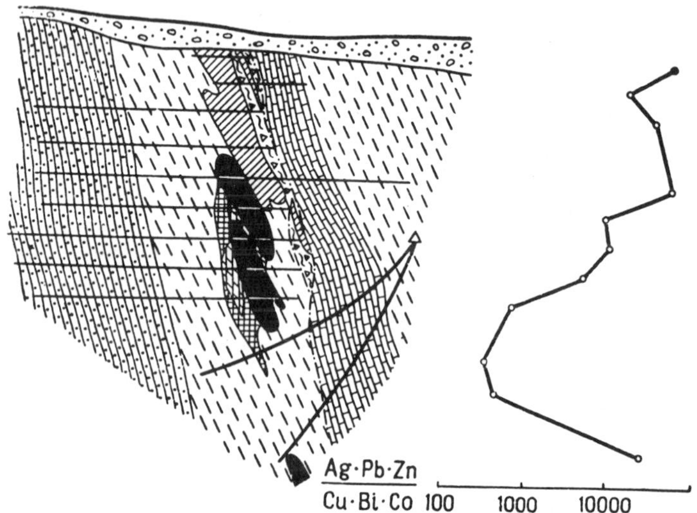

Fig. 88. Cross section illustrating the variation in the multiplicative index with depth at the Harpenberg mineral deposit (Sweden). See Fig. 23 for conventional symbols.

problem is essential in the search for blind mineralizations because in ore regions, especially in those intensively eroded, the anomalies represented by halos formed below the ore bodies are numerous, and their reliable rejection is indispensable to the successful exploration for blind mineralization.

Zonalities of primary halos, discussed in great detail in the preceding chapters of this book, are used to solve this problem.

Adjustment in the Direction of Exploratory Work

Studies have shown that primary halos may be used quite successfully to adjust the direction of exploration as a result of geochemical sampling from mine workings, and from core sampling from boreholes. This may be illustrated by the above example of the study of the anomaly at the margin of the Northern Kurusai II mineral deposit (Fig. 87). A geochemical sampling of the cores from the first exploratory boreholes established that there was a deeper occurrence of the blind mineralization at that locality. It is quite possible that the ore potential in the area, determined by the results from the first boreholes which did not reveal any ore bodies, might have been evaluated as nonexistent, unless the exploration work had been accompanied by geochemical sampling.

Let us consider the results of geochemical studies carried out in the Skarn Zone mineral deposit (the Kurusai ore field, Tadzhik SSR) as an example of an adjustment in exploration work.

This mineral deposit is situated in an extrusive fault zone west of the Kurusai II mineral deposit. The ore mineralization is located in skarns developed in marble within a tectonic block between a quartz porphyry dike and dacite porphyries. A fault zone in the area of the deposit has a southwestern dip at an angle of 70 to 75°. Fig. 89 shows an ore body uncovered by mine workings at the third horizon.

In order to determine the direction of further exploratory work, and in particular to identify ore potential at depth, mine workings at this horizon were sampled. As a result, strong halos of the indicator elements characteristic of this type of mineralization were found. From Figs. 89 and 90, it can be seen that the known ore body is distinctly indicated by anomalous concentrations of the elements. A strong anomaly formed by lead and other indicator elements was also detected, against a background which included general contamination of these elements from the mine workings, at the northwestern continuation of the known ore body. The network of the available mine workings did not make it possible to delineate the entire anomaly, so that average values of element contents in the anomalies (rather than the productivities) were used for the interpretation of this anomaly.

Table 58 lists the parameters (ratios of the average contents of element pairs) of the anomaly situated to the northwest of the known ore body. For comparison, parameters for the halos around the Northern Kurusai II blind ore body are given (these are the standard cross sections in Fig. 87).

These data suggested that the anomaly is due to the supraore levels, and that another ore body exists at depth which is close to the known ore body. It was also concluded that the suspected ore body occurred at shallow depths. This was suggested by the similar values of the indicator ratios of the anomaly in question, and by the halos developed in the upper

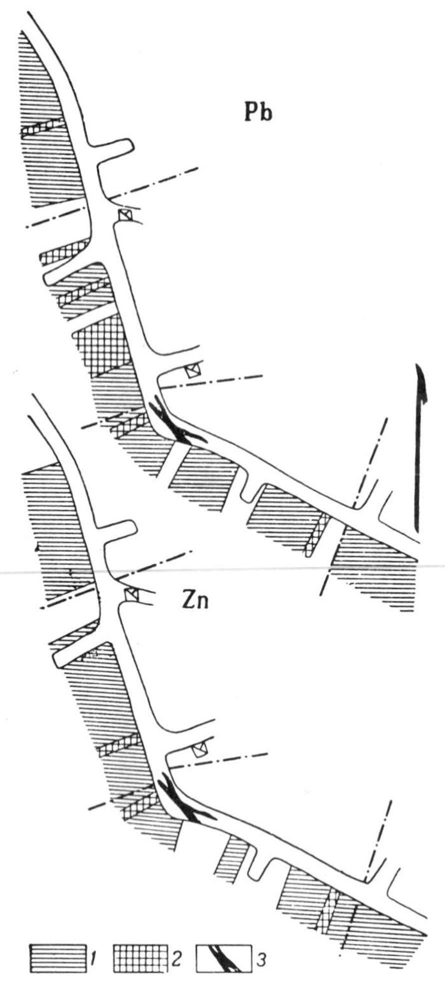

Fig. 89. Primary halos formed by lead and zinc at the third horizon of the Skarn Zone mineral deposit.
Element contents, %: *1* - 0.01 to 0.1; *2* - more than 0.1; *3* - ore body.

levels of the ore bodies within the standard cross section (borehole 65).

The ore potential was also evaluated positively on the basis of the results of geochemical sampling. This conclusion was based on the relatively high values of the indicator ratios.

It follows from Table 58 that appreciable differences in the indicator ratios between the ore body and the western anomalies exist only for the ratios in which tungsten, the least mobile of the indicator elements listed in Table 58, is in the denominator.

Later drilling confirmed that the recommendations were correct. A drift in the fourth horizon at the western margin of the ore body uncovered a large blind ore body with high contents of lead and zinc.

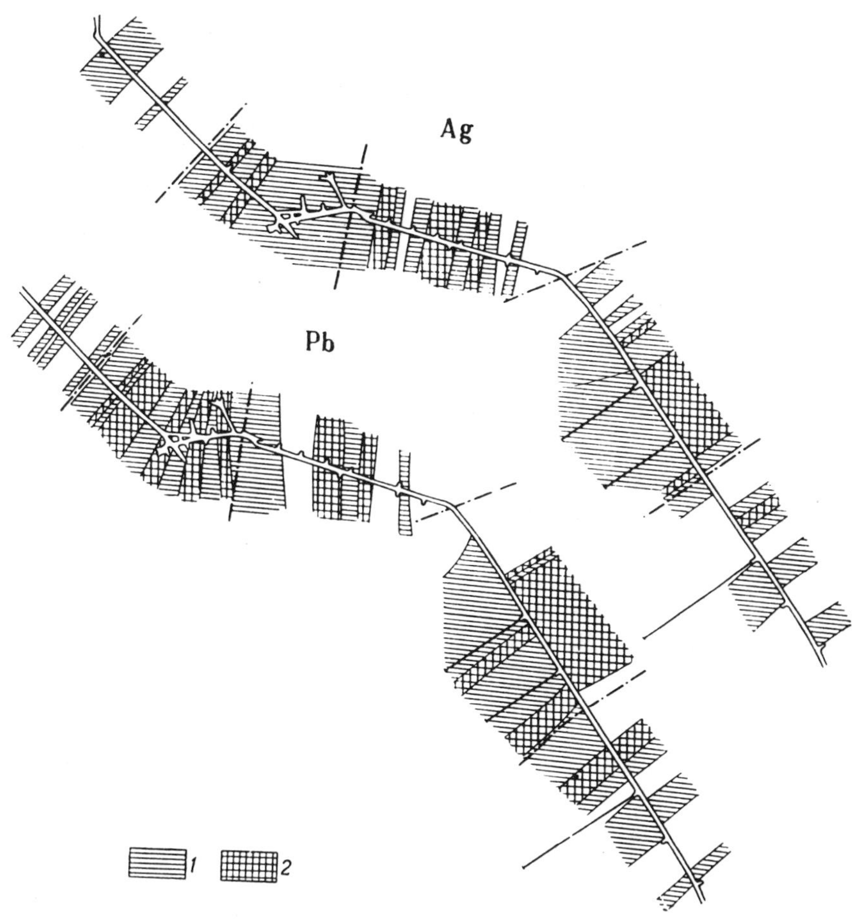

Fig. 90. Primary halos formed by silver and lead at the second horizon of the Skarn Zone mineral deposit.
Element contents, %: *1* - $5 \cdot 10^{-5}$ to $1 \cdot 10^{-3}$ Ag; 0.01 to 0.1 Pb; *2* - more than $1 \cdot 10^{-3}$ Ag; more than 0.1 Pb.

Table 58. Ratios of average contents of elements.

Pair of elements	Standard cross section	Horizon III	
		Western anomaly	Anomaly due to the ore body
As/Zn	0,3 - 0,01**	0,05	0,05
Ba/Zn	9,7 - 0,12	0,28	0,39
Pb/Zn	3,6 - 1,5	2,8	2,0
As/W*	—	4,4	0,8
Ba/W	—	27,0	6,5
Pb/W	—	275	32,6

*Tungsten could not be detected in the halos because of the low sensitivity of the analytical procedure used (for the standard cross section).

**Range of values from the cross sections above the ore body, to the upper levels of the ore body.

Anomalies were also detected at the second horizon, based on the results of geochemical sampling from the drift. One of them (anomaly I) represents supraore halos of the bodies considered above. Another anomaly is confined to marbled and skarned limestones and also corresponds to the supraore cross sections of the halos, in terms of the average contents of the elements (anomaly II).

As/Zn	0.1
Ag/Zn	0.008
Pb/Zn	1.0

These data suggested that the anomaly is promising for a blind mineralization, and it was recommended for drilling. To test this anomaly, two oblique (at an angle 55 to 60°) boreholes were drilled from the drift at the fourth horizon. They uncovered an economically important ore body.

In conclusion, we will consider a case where an exploratory working was cut along the strike of an ore body (metalliferous zone), in the absence of transverse workings which are usually used for geochemical sampling. In these cases, duplicates of the trench samples (splits), which are collected at working faces along lines directed across the strike of the metalliferous zones, can be used. These samples are generally analyzed for a limited number of elements which are the major components of the ores. Therefore, the sample duplicates must also be subjected to spectrographic analysis for a wide range of elements which are typomorphous for the given type of mineralization.

The results of the emission spectrographic analyses are extremely important, because this particular additional set includes the most valuable indicators of halo zonality, particularly those elements typical of the levels above and below the ore bodies. The contents of these elements are relatively low, so that they do not have any economic significance. Sampling from the working faces furnishes information on the distribution patterns of the elements directly in the ore zone. However, it does not make it possible to delineate primary halos. Consequently, average contents of the elements may be used for an evaluation of the levels of the ore-bearing zones.

In interpreting the results of sampling carried out in ore-bearing zones, one should keep in mind that owing to the enormously high element concentrations in ore bodies (as well as in the immediately adjacent areas), and especially because of the extremely high scatter in the element concentrations, the fluctuations in the indicator ratios are considerable, and this reduces the reliability in actual practice. In this connection, it is advisable to use the multiplicative index (Solovov, 1966).

Table 59 presents ratios of the average contents of the major indicator elements calculated for the ore zone and for the primary halos (Fig. 46).

As one can see from Table 59, the multiplicative index gives the most consistent and distinct zonality both for the ore zone and for the primary halos (which include the ore zones); the elements characteristic of the levels above and below the ore body were taken as indicated by the vertical zonality of the halos.

Revision of Previous Exploration Results

Not all mineral deposits are minable, mainly because of the limited proven reserves which do not make the mining profitable. Geochemical studies on such deposits make it

Table 59. Indicator ratios for the ore zone and the primary halos (from an Eastern Kanimansur mineral deposit).

Sampling horizon	Ag/Pb	Ag/Cu	Pb/Cu	Pb/Bi	$\frac{Pb,Ag}{Cu,Bi}$
		Ore zone			
Surface	0,42	10	24	240	2400
Horizon I	0,008	0,35	35	240	1015
Horizon II	0,02	0,036	1,8	9,0	0,32
Horizon III	3,5	0,19	0,06	0,33	0,063
Coefficient of contrast	0,12	53,2	400	72,7	38090
		Primary halos			
Surface	0,007	0,18	21,8	215	38,7
Horizon I	0,0009	0,022	24,0	1020	22,4
Horizon II	0,0005	0,01	19,3	288	2,88
Horizon III	0,0027	0,0014	0,53	0,56	0,001
Coefficient of contrast	2,6	1286	41,1	384	38700

Note: Average contents of the elements were used for the ore zone, and linear productivities for the halos.

possible, in some cases, to identify areas promising for blind (weakly eroded) ore bodies, whose detection and exploration may add to the proven reserves and make them economically profitable.

An example of such a revision are results obtained from investigations in the Promezhutochnyi (Kurusai ore field) area where previous studies revealed an isolated blind ore field.

Geochemical sampling at the surface and from boreholes, carried out many years after the exploration, made it possible to detect and delineate strong halos formed by barium, arsenic, lead, silver, and zinc above the known ore body. These halos have appreciable vertical extents (in most cases more than 500 meters).

An unrelated zone with anomalous contents of barium, arsenic, lead, and silver was detected at some distance from the ore body after processing the geochemical sampling results. This anomaly is, in many ways, similar to the halos over the known ore body. As evidenced by core sampling, this anomaly expands with depth, where its intensity increases. Table 60 shows the values of the linear productivities in the newly detected zone by comparison with those from the halos over the known ore body. Some other parameters of the halos which are developed over the known ore body (I) are also given, as well as those calculated for the new zone (II). The ratios of the linear productivities for the pairs arsenic-copper, barium-copper, silver-copper, and bismuth-copper are seen to be very similar. Considerable differences were found only for the pairs lead-copper and zinc-copper.

The halos of both zones show a decrease in the productivity of arsenic and an increase for the other elements with depth.

These data suggested that the anomaly which was detected is from the levels above the ore body, so that it was recommended for drilling. The close proximity of the known "standard" ore body enabled a rough determination to be made of the depth to the suspected ore body. It was assumed that the geologic-structural environments of the ore bodies and sur-

Table 60. Linear productivities of primary halos and their ratios (from the Promezhutochnyi area).

Element	Number of halo	Level of halo	Average content, %	Mineralization coefficient	Width of anomaly	Linear productivity, m %	Ratio to productivity of copper
As	I	Surface	0,001	0,125	100	0,0125	0,25
	II	Surface	0,001	0,1	130	0,013	0,25
	II	Borehole	0,0043	0,183	130	0,105	0,037
Ba	I	Surface	0,01	0,125	100	0,125	2,5
	II	Surface	0,01	0,10	130	0,13	2,5
	II	Borehole	1,05	0,187	130	25,525	8,97
Ag	I	Surface	0,00004	0,125	100	0,0005	0,01
	II	Surface	0,00005	0,10	130	0,00065	0,013
	II	Borehole	0,00043	0,75	130	0,0419	0,015
Pb	I	Surface	0,003	0,125	100	0,0375	0,75
	II	Surface	0,004	0,40	130	0,206	4,0
	II	Borehole	0,25	1	130	32,50	11,4
Zn	I	Surface	0,0042	1	100	0,42	8,4
	II	Surface	0,004	0,428	130	0,2226	4,3
	II	Borehole	0,14	0,875	130	15,925	5,6
Cu	I	Surface	0,004	0,125	100	0,05	—
	II	Surface	0,004	0,10	130	0,052	—
	II	Borehole	0,035	0,625	130	2,844	—
Bi	I	Surface	0,00015	0,125	100	.0,0019	0,038
	II	Surface	0,00015	0,1	130	0,00195	0,033
	II	Borehole	0,00043	0,375	130	0,02096	0,0074

Notes: I. Halos over the known ore body
II. Anomaly revealed

rounding halos are identical. However, the above-mentioned "anomalous" behavior of the halos formed by lead and zinc caused problems. It could be due either to a more shallow occurrence of the suspected ore body in comparison with the known one, or to a relatively high content of lead. The second possibility seems more likely, because in the case of the shallow occurrence of the blind mineralization, differences would exist in the productivity ratios of the halos for all pairs of elements, rather than only two. Later drilling confirmed this conclusion. A relatively large blind ore body of economic importance with a high content of lead (10.7% Pb; 8.8% Zn) was found at a depth of 450 m in close proximity (50 m) to the known ore body.

The above results obtained from geochemical studies in the Central Orlinaya Gorka ore field (Rudny Karamazar) are, in fact, an example of the revision of previous exploration data. As shown above, geochemical bedrock sampling at the surface, as well as from the exploratory mine workings and of the borehole cores, confirmed the existence of polymetallic mineralization which wedges out at depth.

11

Methods of Mathematical Statistics Used in Geochemical Exploration

Introduction

The qualitative and quantitative improvements in the technical means and methods of investigating the chemical composition of materials, which is characteristic of modern geochemistry, also leads to a considerable increase in the amount of geochemical information. Special mathematical procedures are necessary for processing and interpreting this information.

The methods of mathematical statistics, as used in geology and geochemistry, are based on the theory of probability. A knowledge of the main concepts of this theory is an essential prerequisite for the proper utilization of statistical methods, because this will enable one to apply the laws of statistics to the analysis of natural phenomena.

The use of statistical methods in geology and geochemistry must lead primarily to better reliability and accuracy of the conclusions, based on the comprehensive analysis of the factual material. Statistical methods cannot be applied unless one is well aware of the meaning of the geological or geochemical problems to be solved with the help of these methods. Whereas it is a powerful tool in the hands of a qualified geologist or geochemist, statistics may lead to serious errors when used routinely without due consideration to the entire complex of natural factors.

The distribution function of a random variable is a major concept in the theory of probability. The distribution function is a universal characteristic of a random variable and makes it possible to give a complete mathematical description to any population of random variables which are united by a common qualitative or quantitative feature. Populations which have identical properties are characterized by identical distribution functions. Therefore, the latter can be used as objective criteria to evaluate the homogeneity or heterogeneity of various populations.

In order to apply the methods of mathematical statistics to geochemical studies and mineral exploration it is also necessary to introduce the concepts of a geochemical population and a geochemical set.

A *geochemical population* is a set of values of the contents of a chemical element, which represents statistical regularities in the distribution of this element in specific natural formations. In practice, individual sets from the general geochemical population are studied.

These sets include limited (finite) quantities of members (results of the determination of the content of a chemical element) and are called *geochemical sets.*

Depending on the statistical law of distribution of a random variable, its distribution function is described by a definite set of statistical parameters. The determination of the law (type) of distribution and the calculation of the parameters of a distribution of a random variable in the geochemical population being studied, is the major problem in the statistical processing of the geochemical information. Since the investigator always deals only with geochemical sets of values, which represent the geochemical population being studied with a certain degree of reliability, only approximate estimates (characteristics) of the corresponding distribution parameters can be obtained. The smaller the geochemical set, the more these estimates differ from the real ones. However, the possibility of determining, in each particular case, the accuracy of the calculated distribution parameters enables one to use them for a general characteristic of the population as a whole. Consequently, if one knows the distribution function one can solve, with the help of the distribution parameters which have been found, some geological and geochemical problems based on the laws of the distribution of the chemical elements in natural formations. Listed below are some of the problems which are directly relevant to the problem of mineral exploration.

1. Determination of the statistical parameters of a distribution of the contents of chemical elements for a mathematical substantiation of the geochemical conclusions resulting from studies of specific natural objects.
2. Evaluation of fluctuations in the contents and the determination of maximum (critical) contents of indicator elements within a geochemical background.
3. Detection of anomalous contents in the geochemical sets being studied.
4. Mathematical substantiation of exploration geochemical criteria.
5. Substantiation of a geochemical similarity or difference between specific geochemical objects in the development of exploration criteria or in mineral exploration.
6. Determination of statistical relationships between varying geochemical criteria which characterize a specific natural object (the correlation and regression problems).

Only an elementary knowledge in the theory of probability and in mathematical statistics is necessary for the solution of these problems which are quite elementary from the point of view of mathematical statistics.

We assume the reader is familiar with the basic postulates of probability theory usually taught within the framework of higher mathematics. Therefore, we can omit definitions of elementary statistical concepts in examples of the practical application of the methods of mathematical statistics applied to geochemical problems.

Statistical Estimates of Parameters of Distribution of Chemical Elements in Natural Formations for Substantiation of Geochemical Conclusions

In studies of specific geochemical sets which characterize specific natural formations, the first stage of the statistical analysis reduces to the determination of the type of distribution of the chemical element contents suggested by the data.

The distribution of contents can be represented graphically. The contents are plotted along the X-axis at an appropriate scale; the number of samples associated with a given content is plotted along the Y-axis. The resulting points are connected with line segments. The figure produced is called a *distribution polygon*. The distribution of contents can be represented also by a *histogram*.

Now suppose we increase the number of observations. In such cases the polygons, as well as histograms, will begin to approximate a smooth curve. The limiting curve is called a *distribution curve* of the population being studied, whereas the analytical expression describing the curve in the corresponding coordinate system is called the *distribution function* or *distribution law* of the content X.

The distribution function may be used to determine the probability of occurrence of any content of interest, including the probability of contents not found in the sample group studied. In such a case, it is necessary only to calculate the corresponding ordinate of the distribution function. The calculations involved are not complicated as special tables are available usually in texts on mathematical statistics and probability theory.

Each statistical law or distribution is characterized by a corresponding function containing parameters which makes it possible to define a population. The first two types of curves shown in Fig. 91 are the most common types of distributions encountered in geochemical data. The symmetrical curve, type I (Gauss curve), describes the normal law of probability shown in Fig. 92. This is perhaps the distribution most often used to describe a given geochemical population.

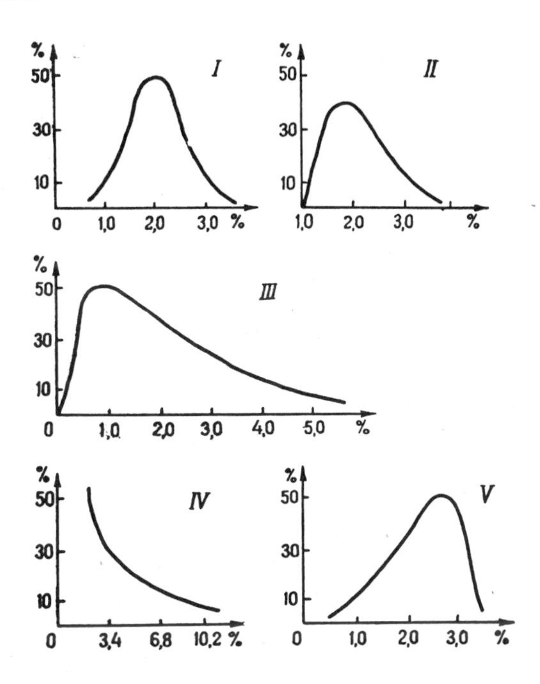

Fig. 91. Types of variation curves (frequencies (%) along the X-axis; contents (%) along the Y-axis).

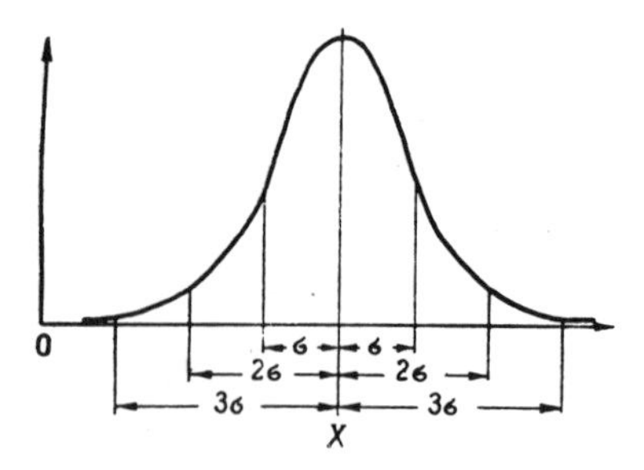

Fig. 92. Normal curve of probability distribution.

The two parameters of the normal distribution are the mean, μ, and the standard deviation, σ. The equation which describes the normal distribution has the form:

$$y = \frac{1}{\sigma\sqrt{2\pi}} \cdot e^{\frac{-(x-\mu)^2}{2\sigma^2}} \tag{1}$$

The random variable x can take on any value within $-\infty$ and $+\infty$. The probability of x occurring between two fixed values, x_1 and x_2, equals the area below the curve bounded by the ordinates of the values, x_1 and x_2.

For the normal distribution, it is possible to determine the number of members in a statistical population, contained within the interval

$$\mu - z \cdot \sigma_x; \qquad \mu + z \cdot \sigma_x, \tag{2}$$

where μ is the mean value of a normally distributed variable;

σ_x is the standard deviation;

z is a coefficient which depends on the desired probability level.

For example, the interval in (2) for $z = 1$ contains 68.3 percent of the members of the statistical population; for $z = 2$, the percentage is 95.4; and for $z = 3$, the percentage is 99.73.

The so-called three sigma rule is based on this property of the normal distribution, that is, the range of fluctuations of a given normally distributed variable x must not exceed, with probability $p = 0.9973$ or 99.73 percent, three times the standard deviation or $3\ \sigma_x$. In accordance with this rule, any value from a statistical population for which its deviation from the mean value is less than three standard deviations, is considered a probable value.

In practice, approximate estimates of the aforementioned parameters are used, that is, the sample mean, x, and the sample variance, s^2.

If one wishes to establish the confidence interval for the fluctuation of a random variable $\bar{x}$ in which essentially all values of x would be included, one can write

$$x - 3\sigma; \qquad x + 3\sigma. \tag{3}$$

Of course, the expression "all values of x" is not quite accurate (99.97 % of all x values, rather than all of them). In most cases, confidence intervals are determined with a level of probability of 0.05, that is, a 5 % significance level. The coefficient z in (2) for a 5 % significance level is 1.96.

The normal distribution may be realized in the case when the distribution of a random variable is determined by a sufficiently large number of mutually independent and approximately equally effective factors. However, these conditions are not always satisfied in nature. For instance, empirical curves which characterize the distribution in specific geochemical sets are in most cases asymmetrical and different, therefore, from the normal distribution curve. These distributions cannot be described based on formulas for the normal law or by the corresponding statistical tables.

Curves of type II in Fig. 91, which exhibit positive asymmetry, are the types which are especially common in geochemistry. A characteristic property of such distributions is the change to a symmetrical shape if the contents which make up the distribution are replaced by their logarithms. With such a transformation, the data can be treated the same as for normally distributed data. The original data, therefore, is considered to be lognormally distributed.

The parameters of the lognormal distribution (in contrast to distributions which obey the normal law) are based on the mean and variance of the logarithms of the contents (or other numerical values).

The normal and lognormal distribution laws do not exhaust the possible range of distributions of chemical elements in natural objects. For example, in populations which are characterized by strongly non-uniform distributions of contents, curves of type III in Fig. 91 may result. In such cases, the asymmetry is greater than that of the lognormal distribution.

There exist distributions also whose curves exhibit negative asymmetry. It is more difficult to describe the distributions characterized by curves of types III and IV in Fig. 91, for instance, than the normal or lognormal distributions. We will not concern ourselves with this problem here.

The operations on the statistical processing of specific geochemical sets reduce to establishing an agreement of empirical distributions with a particular distribution law, and to calculating the parameters of a distribution of a random variable x for a given geochemical population. The known estimates of the distribution parameters are used further as the basis for the solution of geological and geochemical problems, and for the evaluation of the representativeness of the conclusions suggested by analyses of specific geochemical information.

It is recommended that in the first stage of the mathematical processing of analytical data, the data be classified on the basis of the contents of chemical elements in the samples by content intervals, that is, to construct frequency distributions. The following considerations must be adhered to in the selection of the intervals:

(1) the possibility of reducing all values of the element contents in the samples referred to the given interval to an average value for the interval; this must not involve appreciable errors;

(2) to make the processing convenient, the interval must be taken reasonably large without, however, violation of the above condition.

In practice it has been shown that, in most cases, the best number of the intervals of the contents is about 10 to 15 (at least five, however). The intervals of the different contents may be equal or unequal. It is naturally simpler to have them equal.

Each interval comprises a lower (x_{min}) and an upper (x_{max}) limit; the size of an interval is

$K = x_{max} - x_{min}$. The midpoint is determined by

$$X_i = \frac{x_{max} - x_{min}}{2} \tag{4}$$

To facilitate the calculations it is recommended that intervals are chosen so that the midpoints are represented by whole numbers.

The limits of the intervals of contents are usually written down a column, and the data are enumerated within each interval. Contents whose values are equal to any limit should be referred to the lower interval. Values which exceed even slightly an upper limit are included in the next higher interval. Let us take for example two intervals, 62 - 64, and 64 - 66. All contents equal to or lower than 64, but higher than 62 are included into the first interval. The values greater than 64, but smaller than or equal to 66 are included into the second interval, and so forth.

In calculating the number of samples within the given interval it is convenient to make notes in the form of an envelope. The first four points are arranged to form a rectangle (four samples). These are connected by four straight lines and two diagonals (another six samples). This produces a figure resembling the reverse side of an envelope corresponding to 10 samples.

The total number of samples included within each interval represents the frequency. Dividing by the total number of samples represents the relative frequency, $W_i = n_i/N$, where n_i is the number of samples within the ith interval and N is the total number of samples in the given set. The sum of the relative frequencies is equal to unity and may be used as a control for the correct classification of the samples into intervals according to contents.

Table 61 is an example of classifying data based on the sampling of granites for lithium from different regions of the Earth.

Table 61. Classification of results of granite analyses for lithium.

Average value, ppm	Number samples	Average value, ppm	Number samples
5	1	50	8
10	7	55	4
15	13	60	8
20	20	65	1
25	15	70	6
30	26	75	2
35	8	80	1
40	9	95	2
45	12	100	3
		120	3

The above geochemical set is small and could be calculated without further grouping of the data. However, for the sake of simplicity, as shown below, it can be grouped into larger intervals without any significant sacrifice in the accuracy.

Let us take an interval equal to 16 ppm. We obtain eight intervals (Table 62).

Table 62. Calculation of parameters of normal distribution of lithium contents.

Interval, ppm	Middle point of interval x_i	Number of samples n	$x_i \cdot n$	$\Delta x = x_i - \bar{x}$	$(\Delta x)^2$	$(\Delta x)^2 \cdot n$	$(\Delta x)^3$	$(\Delta x)^3 \cdot n$
0—16	8	21	168	—29	841	17 661	—24 389	—512 169
16—32	24	61	1464	—13	169	10 309	—2197	—134 017
32—48	40	29	1160	3	9	261	27	783
48—64	56	20	1120	19	361	7 220	6 859	137 180
64—80	72	10	720	35	1225	12 250	42 875	428 750
80—96	88	2	176	51	2601	5 202	132 651	265 302
96—112	104	3	312	67	4489	13 467	300 763	902 289
120	120	3	360	93	8649	25 947	804 357	2 413 171
		$N = 149$	Σ 5480			Σ 92 317		Σ 3 501 289

Once the frequency distribution has been generated, the parameters of distribution of the element contents are calculated, that is, the mean content and the sample variance.

The mean content is based on the formula

$$\bar{x} = \frac{\Sigma x_i}{N}, \tag{5}$$

where x_i is the content of an element and N is the total number of samples.

The formula for grouped data is

$$\bar{x} = \frac{x_1 \cdot n_1 + x_2 \cdot n_2 + \cdots + x_n \cdot n_n}{N} = \frac{\Sigma x_i \cdot n_i}{N}, \tag{6}$$

where $x_1, x_2, \ldots, x_n$ are the midpoints of the intervals,

$n_1, n_2, \ldots n_n$ are the frequencies of the content values within each, and N is the total number of samples.

From Table 62 we obtain

$$\bar{x} = \frac{5480}{149} = 36.77 \approx 37 \text{ ppm}$$

(this can be compared with the calculation for the ungrouped data in Table 61 which produced $\bar{x} = 38$ ppm).

Next, the sample variance, designated as s^2, is calculated:

$$s^2 = \frac{1}{N-1} \cdot \sum_{i=1}^{N} (x_i - \bar{x})^2 \tag{7}$$

or for the grouped data

$$s^2 = \frac{1}{N-1} \cdot \sum_{i=1}^{n} (x_i - \bar{x})^2 \cdot n_i. \tag{8}$$

The sample standard deviation, s, is the square root of the sample variance. The sample variance (or the sample standard deviation) characterizes the measure of scatter of element contents around the sample mean.

For the example in Table 62

$$s^2 = \frac{92,317}{148} = 623.76; \qquad s = 24.97.$$

In order to standardize the scatter of contents, the coefficient of variation is often used. It reflects the degree of scatter relative to the mean. The coefficient of variation is calculated as

$$v = \frac{s}{\bar{x}} \cdot 100 \tag{9}$$

$$\text{which for our example, } v = \frac{24.97}{37} \cdot 100 = 67.5\%.$$

In practice, average contents of chemical elements are determined from a limited number of samples. It is necessary, therefore, to determine the accuracy of the average content which is calculated.

To determine the accuracy of the mean in mathematical statistics, the probability of the error is expressed as:

$$\pm\lambda = \frac{z \cdot s}{\sqrt{N}}, \tag{10}$$

where $\pm\lambda$ is the error involved in calculating the mean for a specified probability;
s is the sample standard deviation of the contents;
N is the number of samples in the set;
z is the argument of the normalized Laplace function, which varies depending on the given probability with which the error is being evaluated (Table 63). It is customary to use the 5% significance level which corresponds to the probability 0.95 (95 percent). In this case $z = 1.96$ (roughly $z = 2$) and

$$\pm\lambda_{5\%} = \frac{2s}{\sqrt{N}}. \tag{11}$$

In the given example

$$\pm\lambda_{5\%} = \frac{2 \cdot 23,4}{\sqrt{149}} = \frac{46,8}{12,2} = 3,88 \approx 4,$$

that is, $\bar{x} = 37 \pm 4$ ppm.

The error in the mean determined at the 5 percent significance level, defines the confidence interval between 33 and 41 ppm for the lithium distribution in granites.

Very frequently exploration geologists use mean estimates of element contents without knowing their accuracy. This may result in erroneous inferences and conclusions which, if not recognized, may have a negative impact on exploration.

The necessity to determine the accuracy of the derived characteristics (in particular, the accuracy of average contents of particular elements) for the analysis of the results of geochemical studies, can be illustrated by the following example.

Average contents of thorium ($\bar{x}_{Th} = 26$ ppm) and copper ($\bar{x}_{Cu} = 52$ ppm) were determined in metasomatically altered granites for 12 systematically selected samples.

Table 63. Normalized Laplace function $\Phi(z)=\frac{1}{\sqrt{2\pi}}\int_0^r e^{-\frac{z^2}{2}}dz$

z	Hundredth fractions for r									
	0	1	2	3	4	5	6	7	8	9
00	0,0000	040	080	120	160	199	239	279	319	359
0,1	398	438	478	517	557	596	636	675	714	753
0,2	793	832	871	910	948	987	026	064	103	141
0,3	0,1179	217	255	293	331	368	406	443	480	517
0,4	554	591	628	664	700	736	772	808	844	879
0,5	915	950	985	019	054	088	123	157	190	224
0,6	0,2257	291	324	357	389	422	454	486	517	549
0,7	580	611	645	673	703	734	764	794	823	852
0,8	881	910	939	967	995	023	051	078	106	133
0,9	0,3159	186	212	238	264	289	315	340	365	389
1,0	413	437	461	485	508	531	554	577	599	621
1,1	643	665	686	708	729	749	770	790	810	830
1,2	849	869	888	907	925	944	962	980	997	015
1,3	0,4032	049	066	082	099	115	131	147	162	177
1,4	192	207	222	236	251	265	279	292	306	319
1,5	332	345	357	370	382	394	406	418	429	441
1,6	452	463	474	484	495	505	515	525	535	545
1,7	554	564	573	582	591	599	608	616	625	633
1,8	641	649	656	664	671	678	686	693	699	706
1,9	713	719	726	732	738	744	750	756	761	767
2,0	772	778	783	788	793	798	803	808	812	817
2,1	821	826	830	834	838	842	846	850	854	857
2,2	860	864	867	871	874	877	880	883	886	889
2,3	892	895	898	900	903	906	908	911	913	915
2,4	918	920	922	924	926	928	930	932	934	936
2,5	937	939	941	942	944	946	947	949	950	952
2,6	953	954	956	957	958	959	960	962	963	964
2,7	265	966	967	968	969	970	971	971	972	973
2,8	974	975	975	976	977	978	978	979	980	980
2,9	0,4981	981	982	983	983	984	984	985	985	986
3,0	986	986	987	987	988	988	988	989	989	989
3,1	990	990	990	991	991	991	992	992	992	992
3,2	993	993	993	993	994	994	994	994	994	994
3,3	995	995	995	995	995	995	996	996	996	996
3,4	996	996	996	996	997	997	997	997	997	997
3,5	997	997	997	997	977	998	998	998	998	998
3,6	998	998	998	998	998	998	998	998	998	998
3,7	998	998	999	999	999	999	999	999	999	999
3,8	999	999	999	999	999	999	999	999	999	999
3,9	999	999	929	999	999	999	999	999	999	999
4,0	999	999	999	999	999	999	999	999	999	999
4,1	999	999	999	999	999	999	999	999	999	999
4,2	999	999	999	999	999	999	999	999	999	999
4,3	999	999	999	999	999	999	999	999	999	999
4,4	999	999	999	999	999	999	999	999	999	999
4,5	999	999	999	999	999	999	999	999	999	999
5,0	999	999	999	999	999	999	999	999	999	999

In a discussion of this work we can give the resulting average estimates, as is often done, and let the reader judge the representativeness of the data reported. However, it can be shown that in doing this the reader and the author may be misled as to the usefulness of the information published.

Below are given the values of the parameters of the distribution of thorium and copper in biotite-amazonite-albite aplogranite:

	$\overline{x}$, ppm	s, ppm	v, %
Th	26	4.05	16
Cu	52	72.6	140

Let us determine, with probability 0.95 (the 5 percent significance level), the error in the means calculated using (11):

$$\text{for thorium} \pm \lambda_{5\%} = \frac{2 \cdot 4,05}{\sqrt{12}} = 2,3;$$

$$\text{for copper} \pm \lambda_{5\%} = \frac{2 \cdot 72,5}{\sqrt{12}} = 42,0.$$

Thus, the relative error in the determination of the mean of thorium content ($\overline{x}_{Th} = 26 \pm 2.3$ ppm) is less than 10 percent, whereas for copper ($\overline{x}_{Cu} = 52 \pm 42$ ppm) it exceeds 80 percent. In 95 times out of 100 for similar data $\overline{x}_{Cu}$ may range from 10 to 92 ppm, which points to the low reliability of the derived mean content of copper in the rock. The low reliability of $\overline{x}_{Cu}$ in the example is the result of scatter (non-uniformity) in the distribution of copper contents in the aplogranite massif, and the relatively small number of samples collected. (The large scatter in the distribution of the contents may also be due to the insufficient accuracy (reproducibility) of the analytical method.)

If a more accurate estimate of the average content of copper in a rock is desired (at least with a relative error equal to ±20% or ±30%), an additional number of samples should be collected. Knowing the standard distribution of the element contents and using (11) it is possible to calculate the necessary number of samples in the set.

To avoid serious errors, it is essential to evaluate the accuracy of the geological parameters estimated in the determination of the quality of the ores detected as a result of exploration. Evaluation of ore potential in exploration is usually based on analytical results obtained from a limited number of samples. An evaluation of the probable error involved in the determination of the mean in such cases contributes considerably to the objectivity of any subsequent interpretation.

Determination of the Distribution Law

The two characteristics used to determine whether a statistical series comes from a normal or a lognormal population are the asymmetry and peakedness of the sample distribution. Asymmetry is a measure which expresses the degrees of skewness of a distribution, that is, whether the largest deviations are greater than or less than the mean. Its value is calculated as:

$$A = \frac{m_3}{s^3}$$

where m_3 is the third central moment defined by

$$m_3 = \frac{\sum_{i=1}^{N} (x_i - \bar{x})^3}{N},$$

$$\text{so that A } = \frac{\sum_{i=1}^{N} (x_i - \bar{x})^3}{s^3 \cdot N}. \quad (12)$$

For a normal distribution curve A is equal to zero.

To characterize the peakedness of an empirical distribution curve relative to the normal curve, an index called the *kurtosis* (E) is used:

$$m_4 = \frac{\sum_{i=1}^{N} (x_i - \bar{x})^4}{N}; \qquad E = \frac{\sum_{i=1}^{N} (x_i - \bar{x})^4}{s^4 \cdot N} - 3. \quad (13)$$

For a normal curve the kurtosis is equal to zero because the ratio is equal to 3.

To determine the relative conformity of a given distribution to the normal law, it is necessary to obtain estimates of the skewness and kurtosis. In view of the possible error involved in such estimates which depend primarily on the number of samples, it should be determined whether the deviations of A and E from zero are of a chance nature, or are significant. The standard deviations of A and E are calculated using:

$$\sigma_A = \sqrt{\frac{6}{N}}; \qquad \sigma_E = 2\sqrt{\frac{6}{N}}. \quad (14)$$

Let us assume that all factual values of skewness and kurtosis for a given population of sampling data are equal to A_1 and E_1, respectively. Then, according to the major rule of mathematical statistics (the three sigma rule), the deviations of these empirical values from zero in the case of normal distribution must not exceed $3\sigma_A$ and $3\sigma_E$ in absolute values (tripled standard deviations of these values).

Consequently, an applicability condition for a normal distribution is the conformity to the inequalities

$$\frac{A}{\sqrt{\frac{6}{N}}} \leqslant 3; \qquad \frac{E}{2\sqrt{\frac{6}{N}}} \leqslant 3. \quad (15)$$

If these conditions are satisfied, it can be stated that the skewness and kurtosis are insignificant and that the distribution obeys the normal law. Textbooks on mathematical statistics and on probability theory also consider other methods of estimating whether or not the distributions conform to the normal law. However, these methods are more involved.

In geochemical studies, verification of the conformity of distributions of contents to the normal law often reveals appreciable positive skewness, which usually can be noticed during visual inspection of tabular data and becomes quite conspicuous when the distribution curve is plotted. In this case, the distribution being studied does not conform to the normal law, and its parameters cannot be used in the standard statistical analysis (if there is a definite positive skewness in the distribution curve, there is no need to calculate the standard deviation or variance).

In this case, a mean of the logarithms of contents is calculated:

$$\overline{\lg x} = \frac{1}{N}\sum_{i=1}^{n} \lg x_i \cdot n_i, \tag{16}$$

as well as the sample variance and sample standard deviation of the logarithms of contents

$$s_{\lg}^2 = \frac{1}{N-1}\sum_{i=1}^{n} (\lg x_i - \overline{\lg x})^2 \cdot n_i; \tag{17}$$

$$s_{\lg} = \sqrt{\frac{1}{N-1}\sum_{i=1}^{n} (\lg x_i - \overline{\lg x})^2 \cdot n_i}\,. \tag{18}$$

The coefficient of variation of contents in the case of lognormal distribution is determined from the expression:

$$v = \left(e^{\mu^2\sigma_{\lg}^2} - 1\right), \tag{19}$$

for which tables are available, based on standard deviations of logarithms.

In order to check the conformity of a distribution to a lognormal law, the skewness and kurtosis are calculated using:

$$A_{\lg} = \frac{\sum_{i=1}^{n} (\lg x_i - \overline{\lg x})^3 \cdot n_i}{N \cdot s_{\lg}^3}; \qquad E_{\lg} = \frac{\sum_{i=1}^{n} (\lg x_i - \overline{\lg x})^4 \cdot n_i}{N \cdot s_{\lg}^4} - 3. \tag{20}$$

The standard deviations of $A_{\log}$ and $E_{\log}$ are calculated as in the processing of statistical series of contents.

The condition for a lognormal distribution is the conformity to the inequalities:

$$\frac{A_{\lg}}{\sqrt{\frac{6}{N}}} \leqslant 3; \qquad \frac{E_{\lg}}{2\sqrt{\frac{6}{N}}} \leqslant 3. \tag{21}$$

Consequently, the following are statistical estimates of lognormal distribution parameters:

$$\overline{\lg x};\ s_{\lg}^2;\ s_{\lg};\ A_{\lg};\ E_{\lg}.$$

In the above example, the skewness for the curve of the lithium content distribution is quite conspicuous even when the set is grouped according to the contents within intervals. In such cases, evaluations of skewness may be omitted, and the calculation of the mean content using logarithms of contents can be used. However, let us calculate the skewness, by way of example, using the data of Table 62:

$$s^3 = 12\,813; \qquad A = \frac{3\,501\,289}{12\,813 \times 149} = 1{,}8; \qquad \frac{1{,}8}{\sqrt{\frac{6}{149}}} = \frac{1{,}8}{\sqrt{0{,}04}} = 9.$$

As can be seen, the distribution of lithium contents in granites has a distinct positive skewness from which it is concluded that the distribution does not conform to the normal law.

In order to ascertain the conformity of distribution of lithium contents in granitoid rocks to the lognormal law and for the purpose of determining the distribution parameters, let us take decimal logarithms for the extreme members of the set represented in Table 61. The logarithms are 0.6990 and 2.0792, respectively. The difference between these two values is roughly 1.4. Consequently, if one wishes to obtain a grouped set of data containing at least 8 intervals, the size of the interval must be equal to 0.2 (Table 64).

Table 64. Calculation of parameters of lognormal distribution.

Interval $\log x$	Middle of the interval $\log x_i$	n_i	$n_i \lg x_i$	$\Delta \lg x = \lg x_i - \overline{\lg x}$	$(\Delta \lg x)^2$	$(\Delta \lg x)^2 \cdot n$	$(\Delta \lg x)^3$	$(\Delta \lg x)^3 \cdot n$	$(\Delta \lg x)^4$	$(\Delta \lg x)^4 \cdot n$
0,6—0,8	0,7	1	0,7	—0,79	0,62	0,62	—0,49	—0,49	0,39	0,39
0,8—1,0	0,9	7	6,3	—0,59	0,35	2,45	—0,21	—1,47	0,12	0,84
1,0—1,2	1,1	13	14,3	—0,39	0,15	1,95	—0,06	—0,78	0,02	0,26
1,2—1,4	1,3	35	45,5	—0,19	0,04	1,40	—0,01	—0,35	0,00	—
1,4—1,6	1,5	34	51,0	0,01	0,00	—	0,00	—	0,00	—
1,6—1,8	1,7	42	71,4	0,21	0,04	1,68	0,01	0,42	0,00	—
1.8—2,0	1,9	14	26,6	0,41	0,17	2,38	0,07	0,98	0,03	0,42
2,0—2,2	2,1	3	6,3	0,6	0,37	1,11	0,23	0,69	0,14	0,42
		N 149	Σ 222,1			Σ 11,59		Σ—1,00		Σ2,33

The sample mean, variance and standard deviation of the logarithms of contents is calculated as

$$\overline{\lg x} = \frac{222{,}1}{149} = 1{,}49; \qquad s^2_{\lg} = \frac{11{,}59}{149} = 0{,}0783; \qquad s_{\lg} = 0{,}280.$$

The values of the parameters of the lithium distribution obtained without grouping are equal to $\lg \overline{x} = 1.504$ and $s_{\lg} = 0.256$, that is, they are not greatly different from those obtained for the grouped set. Let us check the conformity of the distribution of logarithms of lithium contents to the normal law:

$$s_{lg}^3 = 0{,}0219; \qquad A_{lg} = \frac{-1{,}0}{0{,}0219,\ 149} = -0{,}306;$$

$$\frac{-0{,}306}{\sqrt{\frac{6}{149}}} = \frac{-0{,}306}{0{,}2} = -1{,}153;$$

$$s_{lg}^4 = 0{,}0061; \qquad E_{lg} = \frac{2{,}33}{0{,}0061(\cdot)149} - 3 = -0{,}443;$$

$$\frac{-0{,}443}{2\sqrt{\frac{6}{149}}} = -1{,}102.$$

As can be seen, the values for skewness and kurtosis for the distribution of logarithms of lithium contents in granitoid rocks are not significantly different from zero. Consequently, the distribution of lithium in granitoid rocks agrees with the lognormal law, which makes it possible to use the values of mean, variance, and standard deviation of the logarithms of lithium contents as distribution parameters in statistical calculations.

Calculation of Fluctuation Limits in Element Contents

In the practice of geochemical mapping and exploration one often has to deal with the problem of estimating the range of fluctuation in the contents of an indicator element in a given type of rock.

We considered above the major parameters which characterize the distribution of element contents in different geological formations (mean, variance, and standard deviation), as well as the methods of their calculation. In addition, it is recommended that fluctuation limits of element contents with the probability 0.01 for the limiting values, be calculated. This parameter, which is derived from the major distribution parameters, provides a graphic illustration of the variance for the contents of particular elements.

The following equations are used:

$$a_{\substack{max\\min}} = \bar{x} \pm ts \qquad \text{for the normal law;} \quad (22)$$

$$\lg a_{\substack{max\\min}} = \overline{\lg x} + ts_{lg} \qquad \text{for the lognormal law,} \quad (23)$$

where a_{max} and a_{min} are maximum and minimum limiting contents of the given element in a set (or logarithms of contents in the case of lognormal distribution). These values are determined with a given probability;

$\bar{x}$; $\lg \bar{x}$ are the means of the contents and of the logarithms of the contents, respectively;

s; s_{lg} are the standard deviations of the contents and of the logarithms of the contents, respectively.

The theoretical value of the coefficient t for a one-tailed probability of 0.01 is equal to 2.33. The values of t for finite sets consisting of less than 500 samples differ appreciably from the theoretical and are found from the tables of the Student t distribution available in most textbooks on mathematical statistics (see Table 65).

Let us determine the range of variation in the lithium content, with the probability 0.01 for the limiting values, in the above population which represents this element's distribution in granites from different continents.

The values of parameters of the lognormal distribution of lithium are as follows:

$\lg x = 1.49$ and $s_{\lg} = 0.28$.

$\lg a_{\max/\min} = 1.49 \pm t \cdot 0.280$; t (for a set of 149 samples) is equal to 2.35;

$$\lg a_{\max} = 1.49 + 2.35 \cdot 0.280 = 1.49 + 0.658 = 2.148.$$

From the table of antilogarithms we find $a_{\max} = 141$ ppm; $\lg a_{\min} = 1.49 - 0.658 = 0.832$, whence $a_{\min} = 6.79 \approx 7$ ppm; the range of variation is 7 to 141 ppm.

Thus, we calculated that in the population representing granites from different continents, only one or two samples out of a hundred may contain less than 7 ppm and more than 141 ppm of lithium.

Table 65. Values of q (in %) of the limits for the Student t-distribution depending on the number f of the degrees of freedom.

f \ q	10,0	5,0	2,5	2,0	1,0	0,5	0,3	0,2	0,1	Two-sided
	5,0	2,5	1,25	1,0	0,5	0,25	0,15	0,1	0,05	One-sided
1	2	3	4	5	6	7	8	9	10	11
1	6,314	12,706	25,452	31,821	63,657	127,3	212,2	318,3	636,6	
2	2,920	4,303	6,205	6,965	9,925	14,089	18,216	22,327	31,600	
3	2,353	3,182	4,177	4,541	5,841	7,453	8,891	10,214	12,922	
4	2,132	2,776	3,495	3,747	4,604	5,597	6,435	7,173	8,610	
5	2,015	2,571	3,163	3,365	4,032	4,773	5,376	5,893	6,869	
6	1,943	2,447	2,969	3,143	3,707	4,317	4,800	5,208	5,959	
7	1,895	2,365	2,841	2,998	3,499	4,029	4,442	4,785	5,408	
8	1,860	2,306	2,752	2,896	3,355	3,833	4,199	4,501	5,041	
9	1,833	2,262	2,685	2,821	3,250	3,690	4,024	4,297	4,781	
10	1,812	2,228	2,634	2,764	3,169	3,581	3,892	4,144	4,587	
12	1,782	2,179	2,560	2,681	3,055	3,428	3,706	3,930	4,318	
14	1,761	2,145	2,510	2,624	2,977	3,326	3,583	3,787	4,140	
16	1,746	2,120	2,473	2,583	2,921	3,252	3,494	3,686	4,015	
18	1,734	2,101	2,445	2,552	2,878	3,193	3,428	3,610	3,922	
20	1,725	2,086	2,423	2,528	2,845	3,153	3,376	3,552	3,849	
22	1,717	2,074	2,405	2,508	2,819	3,119	3,335	3,505	3,792	
24	1,711	2,064	2,391	2,492	2,797	3,092	3,302	3,467	3,745	
26	1,706	2,056	2,379	2,479	2,779	3,067	3,274	3,435	3,704	
28	1,701	2,048	2,369	2,467	2,763	3,047	3,250	3,408	3,674	
30	1,697	2,042	2,360	2,457	2,750	3,030	3,230	3,386	3,646	
~	1,645	1,960	2,241	2,326	2,576	2,807	2,968	3,090	3,291	

In practice, probability limits of fluctuations in element contents in geochemical populations are used to choose criteria for differentiating between geological complexes of different composition. Let us discuss this problem based on an example using lithium.

In one of the areas, fluctuations in the content of lithium were calculated with probability 0.01 for limiting valuès for two types of granitoid rocks occurring within the region. Granites of one of these types (fluctuations in the lithium content ranging between 20 and 500 ppm) are potentially promising for rare-element deposits; granites of the other type (fluctuations in the lithium content ranging from 5 to 50 ppm) are practically barren. In this case, the content of 50 ppm can be taken as a criterion; the probability of its appearance in barren granites is only one or two tests per hundred.

In order not to confuse promising granites with barren ones one should know the probability of encountering contents of lithium equal to 50 ppm or higher within potentially promising granitoid complexes.

The probability of contents higher (or lower) than the given one is calculated with the help of the Laplace integral function $\Phi(z)$, tables for which are also available in textbooks on mathematical statistics and the probability theory (see Table 63). The following formulas are used for the purpose:

$$z = \frac{a - \bar{x}}{s}; \text{ and } P_{x>a} = 0{,}5 - \Phi_{(z)}, \tag{24}$$

where a is the given content of interest for the investigator (or logarithms of the given content in the case of lognormal distribution);

$\bar{x}$ and s are the values of arithmetic mean and of the standard deviation of the contents (or logarithms of contents), respectively;

$\Phi(z)$ is the Laplace integral function;

$P_{x>a}$ is the probability of contents higher than the given a.

In our example, 50 ppm is the given content. In view of the fact that lithium is distributed in granites lognormally, the calculations must use the values of the parameters of distribution of logarithms of the lithium content in potentially promising granitoid rocks (lg $x = 2.005$ and $s_{\lg} = 0.296$):

$$\lg 50 = 1{,}699; \qquad z = \frac{1{,}699 - 2{,}005}{0{,}296} = -1{,}04.$$

From tables of the Laplace integral function we obtain:

$$\Phi(z) = -(0{,}351); \text{ whence } P_{x>50} = 0{,}5 - (-0{,}35); \quad P_{x>50} = 0{,}85.$$

Consequently, 85 samples out of a hundred in the geochemical population studied must contain at least 50 ppm of lithium, or each series of 10 samples collected from a granite massif in the process of mapping must comprise several lithium contents which are critical for the nonproductive granites. This criterion provides a sufficient contrast (1 sample per hundred and 85 samples per hundred) and can be recommended for constructing maps showing probable concentrations.

The problem of detecting anomalous contents of indicator elements, which were grouped in a particular geochemical set according to sampling results, is solved in a similar way. From the point of view of mathematical statistics any content of a chemical element, which is outside a given geochemical set, is considered anomalous. The presence of such contents in a geochemical set is of a chance nature, and this contingency can be substantiated with the aid of statistical analysis which, in the given case, reduces to an estimate of the probability of encountering an "anomalous" content in the given geochemical population.

In the process of statistical analysis of a geochemical set, the maximum contents encountered must be tested for anomalous features if they are separated from a neighboring lower content by a sufficiently large interval.

The probability of encountering a suspected "anomalous" content in a given population is evaluated using the procedure outlined above. Samples, for which the product of probability by the number of samples in the set ($P \cdot N$) is much less than unity, are considered anomalous.

Let us assume that the previously studied set contains an additional 150th value of lithium content equal to 500 ppm. There is an appreciable gap between this value and the next 120 ppm. This suggests that the content 500 ppm of lithium is anomalous.

If this anomalous content (500 ppm) is included, this will not cause any significant changes in the values of the parameters of the lithium distribution in the previous set ($\lg \overline{x} = 1.50$; $s_{\lg} = 0.296$), due to its considerable size (150 members).

However, an estimate of the probability of encountering a content of 500 ppm in the given population indicates its anomalous nature:

$$\lg 500 = 2{,}699; \quad z = \frac{2{,}699 - 1{,}50}{0{,}296} = 4{,}0; \quad \Phi(z) = 0{,}4999;$$

$$P_{x > a} = 0{,}5 - 0{,}04999 = 0{,}0001.$$

Consequently, the probability of a content of 500 ppm or higher in the given population is only one sample per ten thousand. The product $P \cdot N = 0.015$ is much less than unity, which indicates an anomalous character for the content of 500 ppm of lithium, which is not typical of the given geochemical set.

Let us check now whether the content of 150 ppm of lithium is anomalous for the geochemical set being studied:

$$\lg 150 = 2{,}176; \quad z = \frac{2{,}176 - 1{,}49}{0{,}280} = 2{,}45; \quad \Phi(z) = 0{,}4928;$$

$$P_{x > a} = 0{,}5 - 0{,}4928 = 0{,}0072.$$

The probability of encountering samples with 150 ppm or more of lithium in the given case amounts to about seven tests per thousand. The product $P(\cdot)N = 1.06$. Therefore, the lithium content equal to 150 ppm is not anomalous for the given set consisting of 149 samples (theoretically there may be one such sample). The fact that this content was not encountered in the given particular case is due to the chance nature of the sampling.

Comparison Between Distribution Series

In the practice of geochemical mapping and exploration which attempts to reveal common features or differences between certain geological formations, one often has to compare sets of samples collected from lithologically similar types of rocks. Significance or insignificance of the differences in the means and variances of the sets being compared should be established. Unless this comparison is made, sets which differ only by chance may erroneously be considered dissimilar or, conversely, sets which exhibit small but significant differences may be mistaken for being identical.

In order to compare the means of two sets, the equality of the variances s_1^2 and s_2^2 (or $s_{1\,\lg}^2$ and $s_{2\,\lg}^2$ are first ascertained. The Fisher F test is used for this purpose. The tables of critical values of F for the 5 percent and 1 percent level of significance are available in textbooks on mathematical statistics. The F statistic is defined as:

$$F = \frac{s_1^2}{s_2^2}, \tag{25}$$

with the larger variance in the nominator. The number of degrees of freedom is equal to the number of samples minus one ($k = N - 1$) determined for each set. If the resulting value of F is less than that found in the table for the given significance level, then the difference between the variances is considered to be insignificant.

In the latter case, comparison is performed between the means $\bar{x}_1$ and $\bar{x}_2$ (or between lg $\bar{x}_1$ and lg $\bar{x}_2$). To this end, the value of the Student test is calculated using

$$t = \frac{\bar{x}_1 - \bar{x}_2}{\sqrt{\frac{s_1^2}{N_1} + \frac{s_2^2}{N_2}}}. \tag{26}$$

If the derived value of t exceeds 1.96, the difference between the means is considered significant (which corresponds to the 5 percent significance level). Otherwise, the hypothesis of the difference between the means is rejected.

The tested equality between variances and between means is usually regarded as an indication of similarity between the sets being compared.

There have been cases in the practice of geological exploration, as well as in the geological literature, when major theoretical or practical conclusions were based on a limited amount of analytical data or even on isolated determinations only. The investigator, who does not take into account the chance nature of the analytical data, may often arrive at erroneous or improbable conclusions without even suspecting any error. We can illustrate this by examples.

In a study of the tantalum distribution in micas from pegmatites for the purpose of substantiating geochemical criteria of exploration for tantalum-bearing pegmatites from a pegmatite deposit, the parameters of the distribution of this element in muscovite and biotite were determined:

	n	$\bar{x}$, ppm	lg $\bar{x}$	$s^2_{\lg}$	$s_{\lg}$	v, %
Biotite	10	55	1600	0.205	0.453	59
Muscovite	12	37	1.416	0.190	0.436	55

With the knowledge that in the later stages of pegmatite formation, biotite is replaced by muscovite, it is possible to make the inference (based on the mean contents of tantalum in micas) that tantalum is removed in the process of muscovitization of biotite. Moreover, it is reasonable to claim that this removal is accompanied by the formation of columbite-tantalite which is characteristic of muscovite-containing pegmatites. The question that arises is whether this interpretation, based on the impressive differences in the average contents of tantalum in biotite and muscovite, are sufficiently well substantiated by the data. Is the difference of 18 ppm of tantalum significant in this case, and if so, is it due to the differences in the distribution of this element in biotite and muscovite?

The significance of the difference in the tantalum content in muscovite and biotite can be evaluated with the aid of Fisher's F test and Student's t test using (25) and (26).

	n	F	t
Biotite	10	1.08 (2.90)*	1.0 (1.96)
Muscovite	12		

*Given in parentheses are tabular values for the 5 percent significance level.

The calculated values of F and t (for a 5 percent significance level) are much less than theoretically predicted, which means it is not possible to regard the observed differences between the mean values of tantalum in biotites and muscovites as significant. Therefore, the present data are not sufficient to prove the removal of tantalum during muscovitization of biotite.

Similar errors frequently occur in sampling from mineral deposits which are characterized by non-uniform and very non-uniform distribution of the useful component. When such objects are being compared, the small differences in the means, due to the natural fluctuations of sampling, may be mistaken for real differences with a resulting false evaluation of a particular prospect. We will now consider three mercury occurrences, one of which was evaluated incorrectly in the exploration stage (Table 66).

Table 66. Evaluation of the parameters of mercury distribution in three mineral deposits located in limestones (lognormal distribution).

Ore Occurrence	Number of samples n	Arithmetic mean		$s^2_{\lg}$	$s_{\lg}$
		$\bar{x}$, %	$\lg \bar{x}$		
I	150	0,24	—0,830	0,210	0,46
II	20	0,09	—1,097	0,370	0,61
III	60	0,16	—0,950	0,230	0,48

Despite the apparent differences in the mean contents, there was no statistical justification for concluding that the mercury distributions in these ore occurrences are different. This conclusion can be made after the statistical comparison between the variances and between the means of the logarithms of contents.

Ore occurrences	F	t
I and II	1.77 (2.00)*	1.84 (1.96)
I to III	1.10 (1.60)	1.67 (1.96)

In both cases, the theoretical values of F as well as of t exceed empirical estimates derived from the sets. This suggests that the observed differences are insignificant. An additional 100 samples collected from the second occurrence made it possible to estimate more accurately the average content of mercury which increased to 0.19.

Graphic Procedure for Determining the Function and Parameters of Element Distribution

In addition to the above analytical method of determining the function and distribution parameters in geochemical studies, graphical methods also exist. The most convenient of these is the method involving the use of probability paper, which was first applied to geochemistry by N. K. Razumovskii.

This graph paper has the grid specially ruled in order to show element contents along the X-axis on a linear scale on the bottom and on a logarithmic scale on top. Plotted along the Y-axis are the accumulated relative frequencies of the contents. The line which connects the points on the probability grid represents the cumulative frequency curve (Fig. 93).

The straight line indicates a normal (if the contents are plotted on a linear scale) or lognormal (if the contents are plotted on a logarithmic scale) distribution of the observations.

Once the nature of the distribution function has been established, the basic parameters of the distribution, the mean ($\bar{x}$) and standard deviation (s), can be determined graphically.

To illustrate the method of graphic estimation, we will use as an example, the distribution of a readily extractable uranium, the so-called mobile uranium, in granites.

The uranium distribution in granites is characterized by the frequencies given in Table 67. The sum of the frequencies is nearly equal to 100 percent.

In this instance, the distribution function of uranium contents conforms to the lognormal law as indicated by the straight line of the plot on a logarithmic scale.

The cumulative curves may be used to determine the following parameters:

(a) the mean and standard deviation. In the example, the mean of the logarithms of contents is estimated by the value on the accumulated frequency curve which corresponds to the ordinate at 50 percent. The standard deviation is estimated by the difference of the values which correspond to the 84th percentile and the mean;

(b) the probability of encountering contents higher than those specified in the data; this is accomplished using $P = 100 - \alpha$, where α is the cumulated frequency corresponding to a given content. This is relevant to the procedure of determining anomalous concentrations with different significance levels. Determination of anomalous concentrations reduces to the solution of the inverse problem stated above. For example, in order to determine minimum anomalous contents with a 1 percent one-sided significance level with the help

*Given in parentheses are tabular values (F: with a 1% significance level; t: with a 5% significance level).

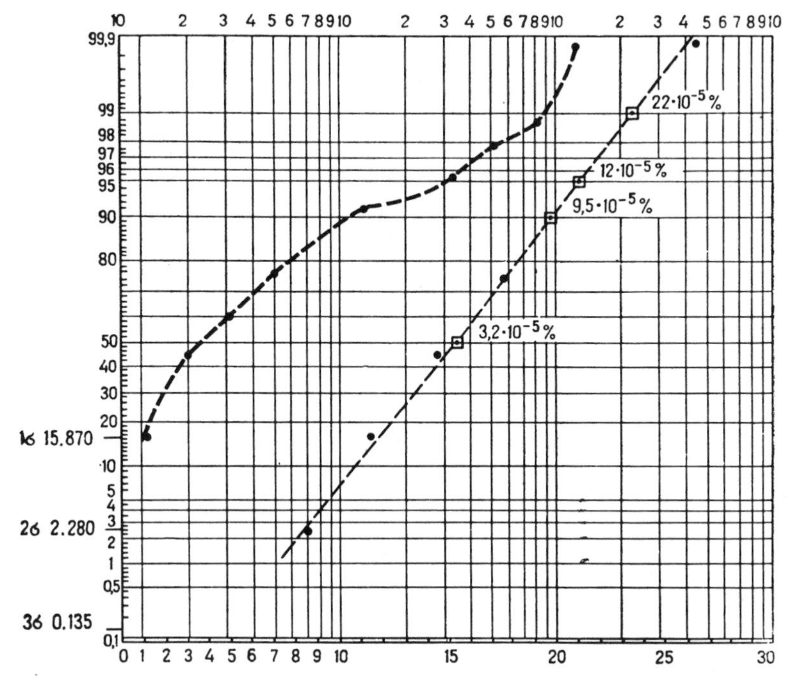

Fig. 93. The probability grid of Razumovskii which is used for a graphical determination of the law, and statistical estimates, of distribution parameters in a geochemical population.

Table 67. Uranium distribution in granites.

Interval of contents (in arithmetic progression) $n \cdot 10^{-5}$%	Frequency	Accumulated frequencies	Interval of contents (in geometrical progression*) $n \cdot 10^{-5}$%	Frequency	Accumulated frequencies
Up to 2	16,129	16,129	Up to 1	2,127	2,127
2—4	25,806	41,935	1—2	13,776	15,903
4—6	16,129	58,064	2—4	24,468	40,371
6—8	17,204	75,268	4—8	32,977	73,348
8—10	10,752	86,020	8—16	21,276	94,624
10—12	5,376	91,396	16—32	4,255	98,879
12—14	1,075	92,471	32—64	1,063	99,942
14—16	2,688	96,159			
16—18	2,688	97,847			
18—20	1,075	98,922			
20—22	1,075	99,997			

*Identification of intervals of contents in the geometrical progression is equivalent to the identification of intervals of contents in the arithmetic progression.

of the cumulative curve, it suffices to find the value of the abscissa, corresponding to the cumulative frequency of 99 percent; for a 5 percent significance level, the cumulative frequency would be 95 percent, and so forth. (The required cumulative frequency may be determined with the help of the formula $\alpha = 100 - P$, where P is the specified significance level, %.)

In the above example, the anomalous contents of uranium for the 1, 5, and 10 percent significance levels are equal to $22 \cdot 10^{-5}$, $12 \cdot 10^{-5}$ and $9.5 \cdot 10^{-5}$%, respectively (see Fig. 93). Thus, the graphical method for determining the function and the distribution parameters of the observed data is relatively simple. It is less accurate than analytical methods, however. Therefore, the latter must be used for control, as well as in cases where more accurate values of the parameters are preferred.

Determination of Correlation Between Variables

It often becomes necessary in geochemical studies to determine the correlation between variables, in particular between the contents of various elements in particular geologic formations. The correlation may be linear or curvilinear. In the first case, the variation of one variable is directly proportional to the variation of the other. In the case of curvilinear correlation, the condition of proportional dependence is violated, and a more complicated relationship results.

The correlation coefficient, ρ , is a measure of dependence between the random variables x and y. However, the use of this coefficient as a measure of dependence is justified only if the random variables x and y are distributed in conformity with one law: normal or lognormal. The correlation coefficient ranges from -1 to $+1$. The plus sign indicates positive correlation, which means that an increase in one of the variables being compared is accompanied by an increase in the other. In the case of negative correlation (minus sign), an increase (decrease) in one of the variables is accompanied by a decrease (increase) in the other. The closer the correlation coefficient is to unity, the stronger the correlation is between the variables.

$$\bar{x} = \frac{51,6}{12} = 4,3 \qquad \bar{y} = \frac{3600}{12} = 300;$$

$$\overline{x^2} = \frac{225,12}{12} = 18,76 \qquad \overline{y^2} = \frac{1\,290\,000}{12} = 107\,500;$$

$$\overline{xy} = \frac{16\,050}{12} = 1337,5 \qquad s_x = \sqrt{18,76-18,49} = \sqrt{0,27} = 0,52$$

$$s_y = \sqrt{107\,500-90\,000} = \sqrt{17\,500} = 132,3;$$

$$r = \frac{1337,5-1290}{0,52(.)132,3} = \frac{47,5}{68,796} = +0,69$$

In geochemical practice where the number of members in the set is limited, an approximate value of the correlation coefficient, r, is calculated, rather than the true value, ρ . The sample correlation coefficient r is defined as:

$$r = \frac{\overline{x \cdot y} - \bar{x} \cdot \bar{y}}{s_x \cdot s_y}, \tag{27}$$

where

$$s_x = \sqrt{\overline{x^2} - (\overline{x})^2}; \qquad s_y = \sqrt{\overline{y^2} - (\overline{y})^2}. \tag{28}$$

(27) is applicable in the case of a normal distribution of the data. In the case of a lognormal distribution, the above formulas incorporate lg x_i, lg x, lg y_i, and lg y instead of x_i, $\overline{x}$, y_i, and $\overline{y}$, respectively.

Let us consider, by way of example, a calculation of the correlation coefficient between the contents (in 12 samples) of potassium and rubidium in trachytes, in the case of a normal distribution (Table 68).

Table 68. Calculation of the correlation coefficient r.

Potassium content, %	Rubidium content, ppm	x^2	y^2	xy
3,3	100	10,89	10 000	330
3,6	250	12,96	62 500	900
3,9	150	15,21	22 500	585
4,1	150	16,81	22 500	615
4,2	300	17,64	90 000	1 260
4,2	350	17,64	122 500	1 470
4,4	450	19,36	202 500	1 980
4,6	150	21,16	22 500	600
4,6	450	21,16	202 500	2 070
4,5	400	20,25	160 000	1 800
5,0	350	25,00	122 500	1 750
5,2	500	27,04	250 000	2 800
51,6	3600	225,12	1 290 000	16 050

Once calculated, the sample correlation coefficient can be tested for significance. This is accomplished by using the tables containing critical values of the correlation coefficient for various degrees of freedom $f = n - 2$, where n is the number of samples. Such tables are available in supplements to textbooks and manuals on mathematical statistics. In our example $f = 12 - 2 = 10$ (Table 69, Fig. 94).

The tabular critical value of the correlation coefficient for the 5 percent level of significance at $f = 10$ is 0.576, which is less than the value for r calculated. We conclude, therefore, the correlation between the contents of potassium and rubidium to be real.

Table 69. Critical value of the correlation coefficient.

f	q				
	0,10	0,05	0,02	0,01	0,001
1	0,98769	0,99692	0,999507	0,999877	0,9999988
2	0,90000	0,95000	0,98000	0,990000	0,99900
3	0,8054	0,8783	0,93433	0,95873	0,99116
4	0,7293	0,8114	0,8822	0,91720	0,97406
5	0,6694	0,7545	0,8329	0,8745	0,95074
6	0,6215	0,7067	0,7887	0,8343	0,92493
7	0,5822	0,6664	0,7498	0,7977	0,8982
8	0,5494	0,6319	0,7155	0,7646	0,8721
9	0,5214	0,6021	0,6851	0,7348	0,8471
10	0,4973	0,5760	0,6581	0,7079	0,8233
11	0,4762	0,5529	0,6339	0,6835	0,8010
12	0,4575	0,5324	0,6120	0,6614	0,7800
13	0,4409	0,5139	0,5923	0,6411	0,7603
14	0,4259	0,4973	0,5742	0,6226	0,7420
15	0,4124	0,4821	0,5577	0,6055	0,7246
16	0,4000	0,4683	0,5425	0,5897	0,7084
17	0,3887	0,4555	0,5285	0,5751	0,6932
18	0,3783	0,4438	0,5155	0,5614	0,6787
19	0,3687	0,4329	0,5034	0,5487	0,6652
20	0,3598	0,4227	0,4921	0,5368	0,6524
25	0,3233	0,3809	0,4451	0,4869	0,5974
30	0,2960	0,3494	0,4093	0,4487	0,5541
35	0,2746	0,3246	0,3810	0,4182	0,5189
40	0,2573	0,3044	0,3578	0,3932	0,4896
45	0,2428	0,2875	0,3384	0,3721	0,4648
50	0,2306	0,2732	0,3218	0,3541	0,4433
60	0,2108	0,2500	0,2948	0,3248	0,4078
70	0,1954	0,2319	0,2737	0,3017	0,3799
80	0,1829	0,2172	0,2565	0,2830	0,3568
90	0,1726	0,2050	0,2422	0,2673	0,3375
100	0,1638	0,1946	0,2301	0,2540	0,3211

Rank Correlation

Semi-quantitative methods of analysis, in particular semi-quantitative emission spectrographic analysis, are widely used in geochemical prospecting for mineral deposits. The data from such analyses frequently do not obey the normal (lognormal) law. This is why the procedure described above is generally not applied to them.

A technique of calculating the coefficient of rank correlation can be used for determining the correlation between the contents of different elements, using semi-quantitative analytical data.

This technique is characterized by a simplicity in the calculations, and its use is not restricted by the distribution law of the observed data. Let us consider, by way of example, the procedure for calculating the rank correlation coefficient between the contents of lead and zinc in endogenic halos around polymetallic orebodies.

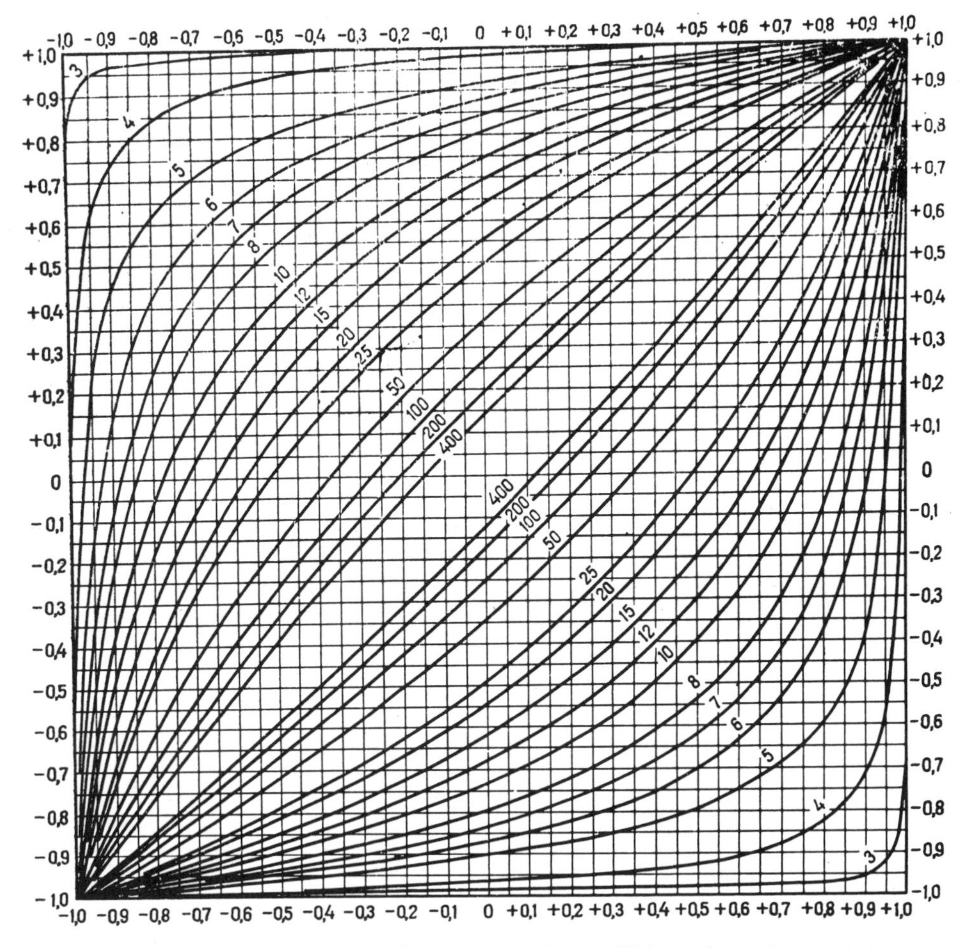

Fig. 94. Confidence intervals for the correlation coefficient of a main population (confidence probability is 0.95).

Column 1 in Table 70 contains the sample number. Column 2 lists the contents of lead. In column 3, the samples are ranked according to increasing contents. In cases where several samples have the same contents, which is typical of semi-quantitative emission spectrographic analysis, each of the samples is assigned the rank order equal to the mean of the successive ranks. For example, three samples, nos. 1, 6, and 8, with rank numbers 2, 3, and 4 in Table 70 have equal contents (0.01 to 0.03) of lead. Their mean rank is equal to 3 $\frac{2 + 3 + 4}{3}$ as indicated in column 4.

Table 70. Calculation of the rank coefficient.

No.	Lead content, %	Rank number		Zinc content, %	Rank number		Difference Δ between rank numbers	Squared difference Δ^2
		preliminary	adjusted		preliminary	adjusted		
1	2	3	4	5	6	7	8	9
1	0,01—0,03	4	3	0,03—1	8	8,5	5,5	30,25
2	0,03—0,1	9	9	0,03—1	9	8,5	0,5	0,25
3	0,03	5	6,5	0,3	11	11	4,5	20,25
4	1,0	12	12	1,0	12	12	0	0
5	0,03	6	6,5	0,03	7	6,5	0	0
6	0,01—0,03	3	3	0,01	3	2,5	0,5	0,25
7	0,1	10	10,5	0,01	4	4	6,5	42,25
8	0,01—0,03	2	3	Not detected	1	1	2,0	4,0
9	0,03	7	6,5	0,01—0,03	5	5	1,5	2,25
10	0,03	8	6,5	0,03	6	6,5	0	0
11	0,1	11	10,5	0,1	10	10	10,5	0,25
12	0,01	1	1	0,01	2	2,5	1,5	2,25
								$\Sigma(\Delta)^2 = 102$

Rank orders for zinc are determined in a similar way (columns 6 and 7). The differences (column 8) and the squared differences (column 9) for lead and zinc are then calculated. The rank correlation is determined using

$$r_n = 1 - \frac{6\Sigma\,(\Delta^2)}{n\,(n^2 - 1)}, \tag{29}$$

where $\Sigma\,(\Delta^2)$ is the sum of the squares of differences between ranks, and
n is the number of samples.

Because in the above example some of the ranks are tied, the equation for calculating rank correlation is given by

$$r_n = 1 - \frac{6\,(\Sigma\,(\Delta)^2 \pm T_x + T_y)}{n\,(n^2 - 1)}, \tag{30}$$

where T_x and T_y represent corrections for each of the tied ranks. These corrections are determined by

$$T_x = \sum_{1}^{t} \frac{t_i^3 - t_i}{12}. \tag{31}$$

In our example for lead, the tied ranks were: (i = 3); 0.01 to 0.03; 0.03 (t = 4); 0.01 (t = 2).

$$T_x = \frac{(3^3 - 3) + (4^3 - 4) + (2^3 - 2)}{12} = \frac{94}{12} \approx 8.$$

Similarly for zinc:

$$T_y = \frac{(2^3 - 2) + (2^3 - 2) + (2^3 - 2)}{12} = \frac{18}{12} = 1{,}5.$$

The calculated rank correlation therefore is

$$r_n = 1 - \frac{6\,(102 + 8 + 1{,}5)}{12 \cdot 143} = 1 - 0{,}39 = +0{,}61.$$

The significance of the correlation is evaluated using the sums of squares of the differences between the rank orders ($\Sigma(\Delta)^2$) and the corrections for the frequency of contents.

Table 71. Critical values of the sum of squares of the differences between serial numbers.

Number of samples	Significance level, %		Number of samples	Significance level, %	
	5	1		5	1
5	0—40		24	1370—3230	1115—3485
6	4—66	0—70	25	1570—3630	1287—3913
7	12—100	4—108	26	1789—4061	1475—4375
8	22—146	10—158	27	2028—4524	1681—4871
9	40—200	24—216	28	2287—5021	1906—5402
10	61—269	39—291	29	2569—5551	2149—5971
11	88352	58—382	30	2873—6117	2414—6576
12	121—451	84—488	31	3199—6721	2700—7220
13	163—565	115—613	32	3550—7362	3008—7904
14	213—697	154—756	33	3926—8042	3338—8630
15	272—848	201—919	34	4328—8762	3693—9387
16	342—1018	257—1103	35	4757—9523	4073—10207
17	423—1209	322—1310	36	5213—10327	4476—11064
18	515—1423	398—1540	37	5698—11174	4908—11964
19	621—1659	484—1796	38	6213—12065	5366—12912
20	740—1920	583—2077	39	6758—13002	5853—13907
21	873—2207	695—2385	40	7334—13086	6367—14953
22	1022—2520	820—2722			
23	1187—2861	960—3088			

The critical values depending on the number of samples, for the 5 percent and 1 percent levels of significance are available in the corresponding tables (with probabilities 95 percent and 99 percent). If the derived sum is less than the lower limit of the tabular values (Table 71), then the correlation coefficient is significant and represents a positive correlation. If it exceeds the upper limit, the correlation coefficient is significant but represents a negative correlation. If the sum lies within the interval listed in the table, the correlation is judged to be insignificant.

In our example, the sum of squared differences between the rank orders with corrections equals 111.5. From the tables of critical values, this sum indicates a significant positive correlation between the contents of lead and zinc. This is the case only at the 5 percent significance level. Experience has shown, however, that the 5 percent level of significance (probability of 95 percent) is acceptable in the interpretation of geochemical data.

The methods of correlation analysis makes it possible to solve the following major problems in geochemical exploration:

(a) to identify indicator elements by way of evaluation of the strength of correlation between the contents of economically important elements and their accessories in various natural materials, as well as for determining the correlation between the element contents in the series: plant-soil, soil-bedrock, plant-soil-bedrock;

(b) to estimate changes due to spatial location of the material being studied relative to an orebody according to the character of variation of the correlation between the indicator elements;

(c) to characterize the features of genesis of geological formations with the help of analysis of the correlation between the indicator elements.

The presence of correlation between an economic (ore) element and its accessories enables the latter to be used as indicators of orebodies in exploration. For example, a distinct correlation between lithium, which is a typomorphous element of the tantalum-bearing aplogranites, and fluorine, rubidium, beryllium, and tin within endogenic geochemical halos formed in the vicinity of the aplogranite massifs made up of schist (Table 72), is an important criterion in the search for tantalum mineralization.

Table 72. Correlation between indicator elements in endogenic geochemical halos of tantalum-bearing aplogranite massifs (after L. A. Leonteva).

		Li	Rb	Be	Sn	F
Schist	Li		0,91	0,90	0,57	0,72
	Rb	0,94		0,88	0,72	0,79
	Be	0,00	0,00		0,68	0,84
	Sn	HC	HC	HC		0,75
	F	0,80	0,77	HC	HC	

Sandstone

Note: HC means "weak correlation".

Of considerable interest is the utilization of variations in the character of the correlation between the elements within endogenic halos for an evaluation of the depth of a buried eroded orebody. Application of these geochemical criteria in the future will make it possible to estimate the potential reserves of newly discovered mineral deposits and to assist in directing geological exploration to the most promising prospects.

This chapter describes only the elementary procedures of mathematical statistics methods used in the processing of geochemical information obtained in geochemical exploration. When more sophisticated mathematical analysis of geochemical data is required, more advanced texts should be used. Computers are recommended for speeding up the calculations.

It should be mentioned, in conclusion, that the use of statistical methods offers a substantial aid in the processing of data in geochemical studies. Experience has shown that the neglect of the use of statistical methods and the use of computers can lead to faulty interpretations. However, the other extreme should also be avoided, that is, one should not be carried away by formal application of statistical methods of data processing without an appropriate thorough analysis of the geological and geochemical premises which determine the rational use of a particular method in the practice of geochemical studies.

Appendix

Table used for determining the size of a geochemical set with a guaranteed probability 0.95 that the specified indicator value appears in no less than a specified number of times, depending on the theoretical probability of encountering this indicator value in the geochemical population (for $m \geqslant 1, 3, 5, 8, 10$ times).*

$$N = f(p;\ P_{m \geqslant 1;\ 3;\ 5;\ 8;\ 10}),$$

where, N is the size of a geochemical set containing, with the probability 0.95, at least m samples with a specified indicator or a higher value;

p is the probability of encountering one value equal to or higher than the indicator value in the geochemical population;

$Pm \geqslant 1;\ 3;\ 5;\ 8;\ 10$ is the probability of encountering at least m samples with the specified indicator or higher contents in the geochemical set consisting of N samples. This probability is equal to 0.95.

An Example Of How The Table Is Used

Given the probability of encountering the values of the Mg/Li ratio less than or equal to 30 in parent granites of the lithium, beryllium, tungsten, and tantalum deposits is 0.27 or higher. It is to be determined how many samples must be collected from the productive granites of this group in order to ensure a guaranteed probability 0.95 of encountering at least three samples with the Mg/Li ratio equal to or less than 30 in the geochemical set.

Solution. According to the condition, $p = 0.27$; $m \geqslant 3$; $P_{m \geqslant 3} = 0.95$.

Using the table we find the horizontal line which corresponds to $p = 0.27$, and from the vertical column $m \geqslant 3$ at its intersection with the horizontal line $p = 0.27$ obtain the size of the set $N = 21$.

Consequently, 21 samples need to be collected in order to obtain a geochemical set containing at least three samples with critical values of the Mg/Li ratio.

*The Table is based on the tables of the number of tests depending on the guaranteed probability of the given event occurring at least a specified number of times and the probability of the given event per one test (Chernitskii, 1957).

p \ m	$\geqslant 1$	$\geqslant 3$	$\geqslant 5$	$\geqslant 8$	$\geqslant 10$	p \ m	$\geqslant 1$	$\geqslant 3$	$\geqslant 5$	$\geqslant 8$	$\geqslant 10$
0,004	750	1550	2280	3300	3900	0,32	8	17	26	38	46
0,01	300	620	910	1320	1560	0,34	7	16	24	36	43
0,02	147	310	455	660	780	0,36	7	15	23	34	40
0,03	98	206	304	440	520	0,38	6	14	22	32	38
0,04	76	154	228	330	390	0,40	6	14	20	30	36
0,05	58	123	182	260	310	0,42	5	13	19	28	34
0,06	48	100	150	216	256	0,44	5	12	18	27	32
0,07	41	87	128	185	220	0,46	5	12	17	26	31
0,08	36	76	112	162	193	0,48	5	11	17	25	30
0,09	32	68	99	144	171	0,50	4	10	16	23	28
0,10	28	61	89	130	154	0,52	4	10	15	22	27
0,11	25	54	81	116	140	0,54	4	9	14	21	25
0,12	23	50	73	106	128	0,56	4	9	13	20	24
0,13	22	46	68	97	118	0,58	3	9	13	19	23
0,14	20	42	63	90	109	0,60	3	8	12	18	22
0,15	18	40	59	83	102	0,62	3	7	12	18	21
0,16	17	37	55	78	95	0,64	3	7	11	17	20
0,17	16	35	62	74	90	0,66	3	7	11	16	20
0,18	15	33	49	70	84	0,68	3	7	11	16	19
0,19	14	31	46	66	80	0,70	2	6	10	16	18
0,20	13	29	43	62	76	0,72	2	6	8	14	17
0,21	13	27	40	59	71	0,74	2	6	8	14	17
0,22	12	26	38	56	68	0,76	2	6	8	14	16
0,23	11	25	37	54	64	0,78	2	5	8	13	16
0,24	11	24	35	52	62	0,80	2	5	8	13	15
0,25	10	23	34	49	59	0,82	2	5	8	12	15
0,26	10	22	33	47	57	0,84	2	5	7	11	14
0,27	10	21	31	46	55	0,86	2	5	7	11	14
0,28	9	20	30	44	53	0,88	1	4	7	11	13
0,29	9	20	29	42	51	0,90	1	4	7	10	13
0,30	8	19	28	41	49						

Bibliography

Publications In Russian

ABRAMSON, G. YA. and S. V. GRIGORIAN (1972) Multiformational halos at the Tyrnyauz mineral deposit. In, *Lithogeochemical Exploration Methods For Hidden Mineralizations,* pp. 20-21. Moscow.

BARNES, H. L. and G. K. CZAMANSKE (1970) Solubility and transport of ore minerals. In, *Geochemistry Of Hydrothermal Ore Deposits,* pp. 286-325. Moscow.

BARSUKOV, V. L. (1964) Metallogenetic specialization of granitoid intrusions. In, *Chemistry Of The Earth's Crust,* Vol. II, pp. 196-214. Moscow.

BARSUKOV, V. L. and L. I. PAVLENKO (1956) Distribution of tin in granitoid intrusions. *Doklady AN SSSR* 109 (No. 3), 417-419.

BARTH, T. F. W. (1956) *Theoretical Petrology,* 414 pages. Foreign Literature Publishing House. Moscow.

BEKZHANOV, G. R., A. E. ERMEKBAEV, and M. K. SERDYUKOV (1972) The status and results of lithochemical exploration in Kazakhstan. In, *Lithochemical Exploration For Mineral Deposits,* pp. 8-15. Alma-Ata.

BERENGILOVA, V. U. (1968) Secondary dispersion halos of tantalum deposits. In, *Geochemical Exploration For Endogene Deposits of Rare Elements.* Nedra. Moscow.

BEZVERKHNII, M. P. (1968) Effectiveness of exploration using dispersion trains and the evaluation of prospective metal reserves in ore showings. In, *Lithochemical Exploration For Mineral Deposits Based On Their Supergene Halos And Dispersion Trains,* pp. 130-131. Alma-Ata.

BEUS, A. A. (1958) The role of complex compounds in the transport and concentration of rare elements in endogenic processes. *Geokhimiya* No. 4, 307-313.

BEUS, A. A. (1960) *Geochemistry Of Beryllium And Genetic Types Of Beryllium Deposits,* 330 pages. Publishing House of the AN SSSR. Moscow.

BEUS, A. A. (1963) Evolution in the chemical composition of high-temperature postmagmatic solutions. In, *Problems Of Postmagmatic Ore Formation,* Vol. I, pp. 276-281. Prague.

BEUS, A. A. (1966) Principles of geochemical prediction of ore potential in geological complexes. In, *Problems In Applied Geochemistry.* Moscow.

BEUS, A. A. (1968) Geochemical criteria. In, *Theoretical Principles Of Exploration For Mineral Deposits,* pp. 127-145. Moscow.

BEUS, A. A. (1972) Geochemistry Of The Lithosphere, 296 pages. Nedra. Moscow.

BEUS, A. A. and YU. P. DIKOV (1967) *Geochemistry Of Beryllium In Processes Of Endogenetic Mineral Formation,* 160 pages. Moscow.

BEUS, A. A. and A. A. SITNIN (1968) Geochemical methods in the evaluation of potential tantalum reserves in granitoid rocks during exploration at scales of 1:50,000 and 1:200,000. In, *Geochemical Exploration For Endogenetic Deposits Of Rare Elements,* pp. 239-244. Moscow.

BEUS, A. A. and C. LEPELTIER (1971) *Mineral Exploration Within The Framework Of The UN Development Program,* 12 pages. VIEMS Publishing House. Moscow.

BOGOLYUBOV, A. N. (1968) Quantitative relation between halos and mechanical dispersion trains in mountainous regions. In, *Lithochemical Exploration For Mineral Deposits Based On Their Supergene Halos And Dispersion Trains,* pp. 128-130. Alma-Ata.

BORUTSKII, J. (1969) United Nations Mineral Survey In Madagascar. (Unpublished U.N. report).

CHERNITSKII, P. N. (1957) *Probability Tables.* Voenizdat. Moscow.

DEMENITSKAYA, R. M. (1967) *The Earth's Crust And Mantle,* 280 pages. Nedra. Moscow.

EREMEEV, A. N. (1963) Principles of the exploration techniques used for deep hidden mineral deposits. In, *Deep Mineral Exploration,* pp. 16-32. Moscow.

FERSMAN, A. E. (1933-1934) *Geochemistry,* Vol. I and II. ONTI. Moscow.

FERSMAN, A. E. (1955) *Selected Works,* Vol. I. Publishing House of the AN SSSR. Moscow.

FURSOV, V. Z. (1970) Mercury in the atmosphere over mercury deposits. *Dokl. AN SSSR* 194 (No. 6), 209-211.

FURSOV, V. Z., N. B. VOL'FSON, and A. G. KHVALOVSKII (1968) Results from studying mercury vapors in the Tashkent earthquake zone. *Dokl. AN SSSR* 179 (No. 5), 208-210.

GOLEVA, G. A. (1965) *Hydrogeochemical Methods In Mineral Exploration*, 285 pages. Nedra. Moscow.

GOLEVA, G. A. (1971) Geochemistry of aqueous dispersion halos from mercury deposits and the forms of mercury migration in subsurface waters. In, *Problems In Applied Geochemistry*, Vol. 2, pp. 113-126. Moscow.

GRABOVSKAYA, L. I. (1965) *Biogeochemical Methods Of Exploration*. VGF Publishing House. Moscow.

GRIGORIAN, S. V. (1963) Endogenic dispersion halos formed by chemical elements in the vicinity of lead-zinc ore bodies. In, *Endogenic Dispersion Halos From Some Hydrothermal Mineral Deposits*, pp. 16-42, Gosgeoltekhizdat. Moscow.

GRIGORIAN, S. V. and E. M. YANISHEVSKII (1968) *Endogenic Geochemical Halos In The Vicinity Of Ore Deposits*, 197 pages. Nedra. Moscow.

GRIGORIAN, S. V. and N. I. BESPALOV (1970) Exploration for endogenic mineralization using primary halos in Rudny Karamazar. In, *Scientific Principles of Geochemical Methods Of Exploration For Deep Ore Deposits*, pp. 198-252. Irkutsk.

GRIGORIAN, S. V., G. E. FEDOTOVA, and V. S. DEGTYAREV (1969) *Geochemical Exploration For Blind Multielement Mineralization In The Kurusai Ore Field*, Vol. 1, pp. 27-59. Moscow.

GRIGORIAN, S. V. and L. N. OVCHINNIKOV (1972) On the interpretation of geochemical anomalies. In, *Lithochemical Exploration For Mineral Deposits*, pp. 68-81. Alma-Ata.

GRIGORIAN, S. V., R. G. OGANESYAN, and N. V. LIKARCHUK (1973) On geochemical criteria for rejecting zones of dispersed ore mineralization. In, *Criteria For The Interpretation Of Geochemical Anomalies*, pp. 54-63. Moscow.

HELGESON, H. C. (1967) *Complexing In Hydrothermal Solutions*, 183 pages. Moscow.

Instructions On Geochemical Exploration Techniques. (1965). 227 pages. Nedra. Moscow.

KABLUKOV, A. D., et al. (1964) *Dispersion Halos Of Uranium And Its Associates In Exploration For Hydrothermal Uranium Deposits*. 234 pages. Nedra. Moscow.

KARASIK, M. A. and A. P. BOL'SHAKOV (1965) Mercury vapors in the Nikitovskoye ore field. *Dokl. AN SSSR*, 161 (No. 5), 204-206.

KHAIRETDINOV, I. A. (1971) Gaseous halos of mercury. *Geokhimiya* No. 6, 668-683.

KORZHINSKII, D. S. (1953) An outline of metasomatic processes. In, *Fundamental Problems In The Theory Of Ore Deposits*, pp. 332-452. Moscow.

KORZHINSKII, D. S. (1964) Acidity regime in polymagmatic processes. In, *Problems In Ore Genesis*, pp. 9-18. Moscow.

KRASNKOV, V. I. (1959) *Basic Principles Of Procedures Used In Exploration For Ore Deposits*, (Second edition, 1965). Gostoptekhizat. Moscow.

KVYATKOVSKII, E. M., N. F. MAIOROV, and T. I. NYUPPENEN (1972) Geochemical methods of exploration for copper-nickel deposits of the Kola Peninsula. In, *Scientific Principles Of The Methods Used For Mineral Exploration And The Evaluation Of Ore Potential In Precambrian Igneous And Metamorphic Complexes. Apatites*, pp. 119-125.

LEBEDEV, V. I. (1957) *Principles Of Energy Analysis In Geochemical Processes*, 342 pages. LGU Publishing House. Leningrad.

LEONTEVA, L., BEUS, A. A. and LEONTEVA, L. A. (1968) Endogene geochemical halos of apogranite deposits of tantalum. In, *Geochemical Exploration For Endogene Deposits of Rare Elements*, pp. 75-98. Nedra. Moscow.

LUGOV, S. F. (1964) On metallogenetic specialization of Mesozoic granitoids in the Chukchi Peninsula. In, *Metallogenetic Specialization Of Igneous Complexes*, pp. 187-197. Moscow.

MALYUGA, D. P. (1963) *Biogeochemical Methods Of Prospecting*, 415 pages. Publishing House of AN SSSR. Moscow.

NETREBA, A. V., A. I. FRIDMAN, and I. A. PLOTNIKOV (1971) On large-scale mapping of closed ore-bearing areas in the Northern Caucasus using gas surveys as a geochemical method. *Geokhimiya* No. 8.

NOCKOLDS, S. R. and R. ALLEN (1958) *Geochemical observations*. 176 pages. Foreign Literature Publishing House. Moscow.

OVCHINNIKOV, L. N. and S. V. GRIGORIAN (1970) Patterns in the composition and structure of primary geochemical halos formed in the vicinity of sulfide deposits. In, *Scientific Principles Of Geochemical Exploration Methods For Deep Ore Deposits*, Part I, pp. 3-36. Irkutsk.

O'HARA, M. J. (1973) Discussion of the paper by D. H. Greene. In, *Petrology Of Igneous And Metamorphic Rocks On The Ocean Floor,* pp. 258-259. Moscow.

PEREL'MAN, A. I. (1968) *Geochemistry Of Epigenetic Processes,* 272 pages. Moscow.

PEREL'MAN, A. I. and YU. V. SHARKOV (1957) Identifying provinces and regions with different environments for mineral exploration in the USSR. In, *Geochemical Mineral Exploration,* pp. 104-106. Moscow.

POLFEROV, D. V. (1962) *Geochemical Exploration For Copper-Nickel Sulfide Ores,* 35 pages. (Bull. nauchn.-tekhn. inform., No. 1 (35), ONTI MGiON). Moscow.

POLFEROV, D. V., S. I. SUSLOVA, and S. A. SHVARTSMAN (1968) *Geochemical Criteria For The Presence Of Ore In Basic To Ultrabasic Massifs,* 69 pages. ONTI VITR. Leningrad.

POLIKARPOCHKIN, V. V. (1968) Evaluation of ore showings based on their lithochemical dispersion trains. In, *Lithochemical Exploration For Mineral Deposits Based On Their Supergene Halos And Dispersion Trains,* pp. 127-128. Alma-Ata.

RAZUMOVSKII, N. K. (1941) On the significance of the lognormal law of frequency distribution in petrology and geochemistry. *Doklady Akad. Nauk. SSSR,* 33, 48-49.

RAZUMOVSKII, N. K. (1948) The lognormal distribution law and its properties. *Zapiski Leningradskogo Gornogo Instituta* (Leningrad Mining Institute) 20, 101-121.

RAZUMOVSKII, N. K. (1962) On the identification of anomalies during exploration over the background element distribution in rocks. In, *Problems of Exploration Geophysics,* Vol. 1, Gostoptekhizat. Moscow.

RINGWOOD, A. E. (1972) Composition and evolution of the upper mantle. In, *The Earth's Crust And Upper Mantle,* pp. 7-26. Moscow.

RONOV, A. B. and A. A. YAROSHEVSKII (1967) Chemical composition of the Earth's crust. *Geokhimiya* No. 11, 1285-1309.

SAUKOV, A. A. (1963) *Geochemical Methods Of Mineral Exploration,* 48 pages. MGU Publishing House. Moscow.

SMIRNOV, V. I. (1954) *Geological Basis for Exploration and Prospecting for Ore Deposits.* (Second edition, 1957). MGU (Moscow State University). Moscow.

SOLOVOV, A. P. (1959) In, *Theory And Practice Of Metallometric Surveys,* pp. 116-134. Alma-Ata.

SOLOVOV, A. P. (1963) Classification of dispersion halos from ore deposits. In, *Exploration For Deep Mineral Deposits,* pp. 11-15. Gosgeoltekhnizdat. Moscow.

SOLOVOV, A. P. (1966) Parameters of primary halos from endogentic deposits. *Geology Of Ore Deposits,* 3, 72-83.

SOLOVOV, A. P. (1972) Evaluation of supergene geochemical anomalies during exploration at a scale of 1:5000. In, *Lithochemical Exploration For Mineral Deposits,* pp. 53-68. Alma-Ata.

SOLOVOV, A. P. and A. V. GARANIN (1968) Geochemical spectra of anomalies and discriminant analysis. In, *Lithochemical Exploration For Ore Deposits Based On Their Supergene Halos And Dispersion Trains,* pp. 84-86. Alma-Ata.

SOLOVOV, A. P., A. V. GARANIN, and V. S. GOLUBEV (1971) Theoretical principles of geochemical exploration methods for blind ore bodies. In, *Scientific Principles Of Geochemical Methods Of Exploration For Deep Ore Deposits,* Part II, pp. 245-298. Irkutsk.

SOLOVOV, A. P. and A. V. GARANIN (1972) Geochemical spectra of anomalies and differences between similar objects. In, *Lithochemical Exploration For Mineral Deposits,* pp. 148-165. Alma-Ata.

TAUSON, L. V. (1961) *Geochemistry Of Rare Elements In Granitoid Rocks,* 231 pages. Publishing House of the AN SSSR. Moscow.

VERTEPOV, G. I. and S. V. GRIGORIAN (1963) Endogene halos of chemical elements around uranium ore bodies. In, *Endogene Halos of Some Hydrothermal Deposits,* pp. 48-51. Gosgeoltekhizdat. Moscow.

VINOGRADOV, A. P. (1950) *Geochemistry Of Rare And Dispersed Elements In Soils,* 254 pages. Publishing House of the AN SSSR. Moscow.

VINOGRADOV, A. P. (1959) *Chemical Evolution Of The Earth.* Publishing House of the AN SSSR. Moscow.

VINOGRADOV, A. P. (1962) Average contents of the chemical elements in the major types of crustal igneous rocks. *Geokhimiya* No. 7, 641-664.

Publications In English

ANDERSON, D. L., C. SAMMIS, and T. JORDAN (1972) Composition of the mantle and core. In, *The Nature Of The Solid Earth,* pp. 41-66. New York.

ANDREWS-JONES, D. A. (1968) The application of geochemical techniques to mineral exploration. Colorado School Mines, *Mining Industr. Bull.* 11, No. 6.

BUGROV, V. (1974) Geochemical sampling techniques in the Eastern Desert of Egypt. *J. Geochem. Explor.* 3, 67-76.

CANNON, H. L. (1960) Botanical prospecting for ore deposits. *Science* 132, 591-598.

FREDERIKSON, A. F., C. A. LEHNERTS, and H. E. KELOY (1971) Mobility, flexibility highlight a mass-spectrometer-computer technique for regional exploration. *Engng. and Mining. J.* 172, No. 6, 116-118.

GOTT, G. B., J. H. MCCARTHY, G. H. VAN SICLE, and J. B. MCHUGH (1969) Distribution of gold and other metals in the Cripple Creek district, Colorado. *U.S. Geol. Surv. Prof. Paper* 625-A.

GOVETT, G. J. S., V. AUSTRIA, W. W. BROWN, and W. E. HALE (1966) Geochemical prospecting by determination of cold-extractable copper in stream-silt and soil. *Information Circ.*, Vol. 21, p. 24. Manila.

GREEN, J. (1969) Abundances and distribution of elements. *Douglas Adv. Res. Lab., Res. Com.*, No. 91, p. 80. California.

HAWKES, H. E. and J. S. WEBB (1962) *Geochemistry In Mineral Exploration.* Harper and Row.

JEDWAB, J. (1955) Caracterisation spectrochimique des granites. *Bull. Soc. Belge. Geol.* 64, 78-91.

MCCARTHY, J. H. (1972) Mercury vapor and other volatile components in the air as guides to ore deposits. *J. Geochem. Explor.* 1, 143-162.

MORRIS, H. T. and T. S. LOVERING (1952) Primary patterns of heavy metals in carbonate and quartz monzonite wall rocks. *Econ. Geol.* 47, 698-716.

ONISHI, H., and E. B. SANDELL (1957) Meteoritic and terrestrial abundance of tin. *Geochim. Cosmochim. Acta.* 12, 213-221.

SLAWSON, W. F. and M. P. NACKOWSKI (1958) Lead in potassium feldspars from Basin and Range quartz monzonites (abstract). *Bull. Geol. Soc. Amer.* 69, 1644.

TOOMS, J. S. and J. S. WEBB (1961) Geochemical prospecting investigations in the Northern Rhodesian Copperbelt. *Econ. Geol.*, 16, 815-846.

WARREN, H. V. and R. E. DELAVAULT (1949) Further studies in biogeochemistry. *Bull. Geol. Soc. Amer.* 60, 531-560.

Glossary

In view of differences in the usage of certain terms between Russian and English, and because some terms routinely used in Russian are essentially unknown in English, this short glossary has been prepared to assist those reading this book. Editor.

anomaly ratio. The ratio of any single (or average) anomalous value to the average background value; it is equivalent to "contrast".

boundary value. extreme values calculated with a certain probability.

clarke. The Russian equivalent of "the average content (crustal abundance) of an element". This term was introduced by Fersman.

decade. An "historical" term, first used by Vernadskii, in which the elements are grouped into categories based on a 10-fold (hence "decade") difference in their crustal abundances. The 12 decades into which all elements may be grouped are listed in Table 4.

dispersed mineralization. Non-economic concentrations of elements and minerals. The term is not equivalent to disseminated because some disseminated mineralization, such as porphyry copper deposits, can be economic.

distribution parameters. The main statistical parameters of mean, variance, and standard deviation (under certain conditions, they may include other statistical parameters).

endogenic. Deposits, halos, etc. formed within the Earth's crust under conditions characterized by high temperatures and pressure.

exogenic. Deposits, halos, etc. formed in the surficial and other supergene environments which are characterized by low temperatures and pressures.

hydrographic network (or system). stream system.

hydrographic survey. drainage basin survey.

large-scale (mapping). detailed (mapping).

linear productivity. The average content of an element or groups of elements (in percent), times a specific linear distance (in meters), over a halo, ore zone, etc. It is expressed in meter percent (m%).

lithogeochemical survey. In the English literature, a lithogeochemical survey is equivalent to a "rock" survey. However, in the Russian literature, lithogeochemical is used in a much broader sense and may also include stream sediment and soil surveys, in addition to rock survey.

minimum anomalous content. threshold.

polymetallic deposit. Deposits composed essentially of lead, zinc and copper sulphides of various origins; although the term "multi-element" (which indicates elements other than Pb, Zn and Cu) may be used in some places in this book, it generally means polymetallic.

small-scale (mapping). regional (mapping).

standard cross-section. Traverses run over an ore body or mineral occurrence whose geochemical and other characteristics are known; orientation surveys are conducted over a "standard cross section" for comparison with geochemical anomalies in other areas.

supergene zone. This term, as used in the Russian literature, includes the upper part of the Earth's crust in which weathering takes place and fluid transport of all types (surface waters, groundwaters, brines) occurs; it also includes the biosphere, and the lower parts of the atmosphere, certainly at least to those elevations where vapor surveys and some airborne geophysical surveys (e.g., gamma ray) are conducted. The term "supergene" is far broader in the Russian literature than it is in the English.

typomorphic (ous) minerals. Minerals whose chemical composition (or some physical properties such as habit) are characteristic for certain mineral-forming processes. For example, the trace element content and crystal shape of cassiterites from pegmatites are different from cassiterites found in other types of deposits.

zonality. The concept of zonality in the Russian literature means the breakdown into natural-occurring zones by whatever criteria are adopted. The concept is not exclusive to geochemistry, but is also used in geophysics, mineralogy, etc. In exploration geochemistry it is applied to natural variations in elemental contents, which are broken down into zones of any shape (e.g., vertical, horizontal).

Equivalents

weight %		*ppm*		*ppb*		
1.0%	=	10,000 ppm			=	10^{-0}%
0.1%	=	1,000 ppm			=	10^{-1}%
0.01%	=	100 ppm			=	10^{-2}%
0.001%	=	10 ppm			=	10^{-3}%
0.0001%	=	1 ppm	=	1,000 ppb	=	10^{-4}%
0.00001%	=	.1 ppm	=	100 ppb	=	10^{-5}%
0.000001%	=	.01 ppm	=	10 ppb	=	10^{-6}%
0.0000001%	=	.001 ppm	=	1 ppb	=	10^{-7}%
0.00000001%	=	.0001 ppm	=	.1 ppb	=	10^{-8}%

For waters (dilute solutions)

1 mg/l	=	10^{-3} g/l	=	1000 ppb or 1 ppm
		10^{-4} g/l	=	100 ppb
		10^{-5} g/l	=	10 ppb
1 μg/l	=	10^{-6} g/l	=	1 ppb
		10^{-7} g/l	=	0.1 ppb

Index